Acid Mine Drainage

Acid Mine Drainage

Acid Mine Drainage

From Waste to Resources

Edited by

GEOFFREY S. SIMATE AND SEHLISELO NDLOVU

CRC Press
Taylor & Francis Group
Boca Raton London New York

CRC Press is an imprint of the
Taylor & Francis Group, an **informa** business

First edition published 2021
by CRC Press
6000 Broken Sound Parkway NW, Suite 300, Boca Raton, FL 33487-2742

and by CRC Press
2 Park Square, Milton Park, Abingdon, Oxon, OX14 4RN

Library of Congress Cataloging-in-Publication Data

Names: Simate, Geoffrey S., editor. | Ndlovu, Sehliselo, editor.
Title: Acid mine drainage: from waste to resources / edited by Geoffrey S.
  Simate and Sehliselo Ndlovu.
Description: First edition. | Boca Raton, FL: CRC Press/Taylor & Francis Group, LLC,
  2021. | Includes bibliographical references and index. | Summary: "Acid Mine
  Drainage (AMD) is basically the flow of water polluted with metals and other
  substances from existing/old mining areas and is considered as one of the sources
  of pollution. A wide range of technologies are available for preventing AMD
  generation and/or treating the AMD before discharge, but there is a shift towards
  recovery of industrially useful materials and products from AMD. Therefore,
  this book explores novel methods developed for the reuse and/or recovery of
  industrially useful materials from AMD including discussing generation, prediction,
  prevention, and remediation processes. It includes legislation and policy framework
  governing AMD and its environmental/health impacts"—Provided by publisher.
Identifiers: LCCN 2020043319 (print) | LCCN 2020043320 (ebook) |
  ISBN 9781138392915 (hardback) | ISBN 9780429401985 (ebook)
Subjects: LCSH: Acid mine drainage—Purification. | Extraction (Chemistry) |
  Water reuse. | In situ processing (Mining) | Minerals.
Classification: LCC TD899.M5 A335 2021 (print) | LCC TD899.M5 (ebook) |
  DDC 628.1/6832—dc23
LC record available at https://lccn.loc.gov/2020043319
LC ebook record available at https://lccn.loc.gov/2020043320

ISBN: 978-1-138-39291-5 (hbk)
ISBN: 978-0-429-40198-5 (ebk)

Typeset in Palatino
by KnowledgeWorks Global Ltd.

Contents

Part I Overview of Acid Mine Drainage

Part II Prevention and Remediation Processes of Acid Mine Drainage

Part III   Reuse, Recycle and Recovery Processes of Valuable Materials from Acid Mine Drainage

Preface

The mining industry has significantly contributed to the economic growth and prosperity of many nations. The industry provides enormous amounts of mineral products for both industrial and household consumption. Despite the criticality of the mining industry to national and global economic growth, the industry is associated with environmental challenges. One such challenge, which is one of the well-known harmful legacies of mining operations in many parts of the globe, is acid mine drainage (AMD). AMD is an environmental problem that has been studied extensively, and continues to receive considerable coverage in the media. AMD is basically the flow of acidic water polluted with metals and other substances from existing and old mining areas. It is considered as one of the main pollutants of surface and ground water in many countries that have historic or current mining activities and its potential impact on natural resources and human health has become increasingly evident. Indeed, AMD is one of the most serious and pervasive challenges facing the mining and minerals industry.

Whilst a wide range of technologies are available for preventing AMD generation and/or treating AMD before discharge into the environment, most of these technologies consider AMD as a nuisance that needs to be quickly disposed of after minimum required treatment. However, in the recent past, there has been an emerging paradigm shift towards environmental responsibility and sustainable development. In fact, the recovery of industrially useful materials and products from AMD is one of the emerging pragmatic approaches to mitigating the challenges associated with AMD. Therefore, the main focus of this book is to bring together a number of studies in which novel methods have been developed for the recovery and utilisation of industrially useful materials from AMD. The book also discusses the contribution of mining activities and hydrological processes to AMD formation, and covers aspects of prediction, prevention and remediation processes. Furthermore, the book gives an overview of the legislation and policy framework governing AMD and its environmental and health impacts. This book also contains the basics of life-cycle assessments including case studies of life-cycle assessments on AMD remediation technologies.

The book is divided into three parts. Part I is an introduction which covers mining and hydrological perspectives to AMD formation and dissemination, AMD prediction, the genesis and chemistry of AMD, legislation and policy frameworks governing the management of AMD, and its environmental and health impacts. Part II discusses AMD prevention and remediation processes. Some aspects of life-cycle assessments on AMD remediation technologies are also included. The reuse, recycle and recovery of industrially useful materials from AMD, which is the main focus of this book, are addressed in Part III.

This book is a result of a highly engaged set of distinguished researchers from the academia and industry. Therefore, it is an invaluable reference for engineers and researchers in industry and academics on the current status and future trends of the contribution of mining activities and hydrological processes to AMD formation, prediction, prevention and remediation of AMD, as well as the recovery and utilisation of industrially useful materials from AMD.

To write a book with varsity knowledge as that contained in this book requires enormous contribution from various sources either directly or indirectly. As editors and authors we are indebted to many sources from a number of researchers that are listed in the references at the end of each chapter. The editors would like to thank the University of the Witwatersrand, Johannesburg, South Africa, for the conducive environment rendered during the compilation of this book. Finally, we are grateful to CRC Press/Taylor & Francis Group who provided valuable guidance from the publisher's perspective.

Editors

Geoffrey S. Simate, PhD, is an academic/researcher in the School of Chemical and Metallurgical Engineering. Professor Simate is also an assistant dean for research in the Faculty of Engineering and the Built Environment at the University of the Witwatersrand, Johannesburg, South Africa. He has a PhD degree and an MSc (Eng) degree in chemical engineering from the University of the Witwatersrand and a BEng honours degree in chemical engineering from the University of Birmingham, Birmingham, United Kingdom. He also possesses a master of management degree in innovation studies from Wits Business School and a postgraduate diploma in higher education from University of the Witwatersrand. Professor Simate has over 20 years of both lecturing/research and industrial experience combined. His industrial experience revolves around hydrometallurgical industries where he has held various senior metallurgical engineering positions. He has been a National Research Foundation (NRF) rated scientist since 2014 and is an editorial board member of the journal *Metals*, and more recently was appointed as an editorial and advisory board member of the *African Journal of Engineering and Environmental Research*. He has published several technical journal articles, eight book chapters previously and one edited book including this one where he has single-handedly written five chapters and co-authored one. Professor Simate is also a co-author of a specialist textbook on waste production and utilisation in the metal extraction industry published by CRC Press in 2017.

Sehliselo Ndlovu, PhD, is a professor in the School of Chemical and Metallurgical Engineering at the University of the Witwatersrand in Johannesburg, South Africa. She holds a Diploma of Imperial College (DIC) in hydrometallurgy and a PhD in minerals engineering from Imperial College, London, United Kingdom. She has extensive research experience in the field of metallurgical Engineering. Her specialization is in extractive metallurgy, and in particular, mineral processing, hydrometallurgy, biohydrometallurgy and the treatment of industrial and mining effluents, and she has published journal papers and book chapters extensively in these areas. She is also a co-author of a specialist textbook on waste production and utilisation in the metal extraction industry published by CRC Press in 2017. Professor Ndlovu is a rated researcher with the National Research Foundation (NRF), South Africa, and currently holds the Department of Science and Innovation (DSI) and the National Research Foundation (NRF) funded South African Research Chairs Initiative (SARChI research chair) in Hydrometallurgy and Sustainable Development at the University of the Witwatersrand. Professor Ndlovu is a former president and also a fellow of the Southern African Institute of Mining and Metallurgy (SAIMM), an organization that looks after the interests of mining sector professionals in Southern Africa.

Contributors

Willis Gwenzi is a professor of Biosystems and Environmental Engineering at the University of Zimbabwe. Gwenzi earned the following qualifications: (1) PhD (Biosystems and Environmental Engineering, University of Western Australia, Perth, Australia), (2) MSc (Water Resources Engineering and Management, University of Zimbabwe/UNESCO-IHE, The Netherlands), (3) BSc Honours (Soil Science, University of Zimbabwe, Harare, Zimbabwe), and (4) Postgraduate Certificate (Applied Groundwater Modelling, UNESCO-IHE, The Netherlands). His research focuses on environmental remediation, hazardous waste disposal, environmental hydrology, low-cost water and wastewater treatment systems, emerging contaminants, waste management, and waste valorization including the development of novel (bio) materials from industrial wastes. He has published extensively in his areas of expertise as evidenced by over 60 research articles in international journals, five book chapters, and numerous conference papers.

Kevin Harding is an associate professor in the School of Chemical and Metallurgical Engineering at the University of the Witwatersrand in Johannesburg, South Africa. His research focuses on life-cycle assessment (LCA) and water footprinting, particularly in the metals and mining, and bioprocess industries. Professor Harding holds a BSc (Chemical Engineering) and PhD from the University of Cape Town and a Postgraduate Diploma in Higher Education from the University of the Witwatersrand. He has over ten years post-academic teaching and research experience, as well as corporate consulting experience. He is a founding member of the Industrial and Mining Water Research Unit (IMWaRU) and is currently acting on the board of directors for the Forum for Sustainability through Life Cycle Innovation (FSLCI).

James Manchisi is a lecturer in the School of Mines at the University of Zambia, Lusaka, Zambia, where he has been a faculty member since 2008. He also held the position of assistant dean for undergraduate students between 2008 and 2011 and 2016 and 2018. Manchisi holds a PhD in Chemical Engineering from the University of Birmingham, Birmingham, United Kingdom. In 2018, he was appointed as a Postdoctoral Research Fellow in the School of Chemical and Metallurgical Engineering at the University of Witwatersrand for two years. He currently teaches hydrometallurgical process engineering, chemical thermodynamics, and environmental stewardship courses at the University of Zambia. His research interests include the development of sustainable chemical processes for the extraction and refining of metallic products and the socio-economic considerations and environmental impacts of mining projects. Manchisi has collaborated actively with academics in other disciplines of sustainability sciences particularly the Education for Sustainable Development in Africa (ESDA) and the Next Generation of Researchers (NGR).

Part I

Overview of Acid Mine Drainage

Acid mine drainage (AMD) is formed by a series of complex geo-chemical and microbial reactions that occur when water comes into contact with sulphur-containing minerals. This part of the book addresses, amongst others, the following aspects: mining and excavation operations including hydrological processes and their effects on the occurrence of AMD; the prediction and occurrence of AMD; the chemistry of AMD generation; and the associated policies and legislation that have been created to regulate and manage AMD. The environmental and health impacts of AMD are also covered.

Part I

Overview of Acid Mine Drainage

1

Acid Mine Drainage Formation, Dissemination and Control: Mining and Hydrological Perspectives

Willis Gwenzi

CONTENTS

1.1 Introduction

The mining industry plays a key role in the global economy through the provision of industrial raw materials, contribution to gross domestic product, and employment creation (Golev et al., 2016). Moreover, according to a World Bank report, minerals such as gold are estimated to have a multiplier factor of about 1.7 to 1.8, indicating that for each mining job, the sector creates an additional 1.7 to 1.8 jobs through expenditure effects and backward linkages (Chuhan-Pole et al., 2017). It must be noted, however, that the mine project cycle consists of several steps and activities, including: (1) prefeasibility studies, (2) exploration, (3) design and engineering, (4) construction, (5) extraction and processing, and (6) mine decommissioning and closure (Durucan et al., 2006). Without proper planning, these mining activities can cause significant adverse impacts on the biophysical environment. Specifically, mining activities may cause land degradation via: (1) disruption of ecosystems and loss of biodiversity, (2) changes in surface and groundwater hydrology, and (3) environmental pollution, including acid mine drainage (AMD) (Ochieng et al., 2010).

AMD is formed through the oxidation of sulphidic rock materials including waste rock, mine tailings, and even in situ rocks exposed to water and oxygen during excavation and drilling operations (Pope et al., 2018; Wright et al., 2018). Two pre-conditions are necessary for the formation of AMD: (1) sulphidic rock materials with acid-generating capacity exceeding acid-neutralizing capacity, and (2) a combination of water and oxygen, which promotes the oxidation process (Skousen et al., 2019; Campbell et al., 2020).

Unfortunately, AMD is a global environmental problem, which has been widely reported in several countries in nearly all continents. To date, AMD has been reported in Africa (Fosso-Kankeu et al., 2017; Gwenzi et al., 2017; Ochieng et al., 2017; Mungazi and Gwenzi, 2019), Europe (Grande et al., 2018), North America (Campbell et al., 2020), South America (Galhardi and Bonotto, 2016), Asia (Hao et al., 2017), and the Pacific/Oceania (Wright et al., 2018). At the moment, several cases of AMD problems exist in South Africa (Ochieng et al., 2010), Zimbabwe (Gwenzi et al., 2017; Mungazi and Gwenzi, 2019), Australia (Wright et al., 2018), China (Hao et al., 2017), the United States (Campbell et al., 2020), and Canada (Genty et al., 2016). In these case studies, AMD has been reported in both underground and surface mining operations, waste rock dumps, and mine tailings (Gwenzi et al., 2017; Ochieng et al., 2010; Mungazi and Gwenzi, 2019). Moreover, AMD has been detected in a wide range of mining operations, including coal, gold, and uranium mining, among others (Choudhury et al., 2017; Humphries et al., 2017; Casagrande et al., 2020).

The environmental, human and ecological health risks of AMD are well-documented (Liao et al., 2016;) and discussed in detail in Chapter 5. Some of the impacts of AMD include: (1) soil pollution, including that of agricultural soils used for food production (Liao et al., 2016; Fernández-Caliani et al., 2019), (2) surface and groundwater pollution (Ochieng et al., 2010; Wright et al., 2018), and (3) disruption of aquatic ecosystems and functions (Leppänen et al., 2017). For example, significant pollution of soils, surface water, and groundwater by AMD has been reported at Iron Duke Mine in Zimbabwe (Gwenzi et al., 2017; Mungazi and Gwenzi, 2019). In Zimbabwe, anecdotal evidence also indicates significant human health effects due to downstream contamination of drinking water sources by AMD from underground coal mine workings dating back to the 1960s (Gwenzi et al., 2018b). This and several other studies show that AMD may have a long latent or lag time, between exposure to oxidative conditions and the manifestation of AMD (Kanda et al., 2017; Wright et al., 2018). This is because AMD only manifests when the acid-generating capacity exceeds the acid-neutralizing capacity derived from calcium and magnesium oxides and carbonates (Kanda et al., 2017).

The significant impacts of AMD and the challenges associated with its remediation once it occurs justify the need for AMD prevention during the whole mining project cycle. However, existing literature, including reviews on AMD, is dominated by studies on the formation, hydrochemistry, environmental health impacts, and remediation and control using either passive or active methods (Buxton, 2018; Park et al., 2018; Skousen et al., 2019; Tabelin et al., 2019). By comparison, limited attention has been paid to the role of mining activities including excavations, drilling, blasting, and metallurgical processes (e.g., comminution, pyrometallurgy, and hydrometallurgy) on the AMD generation. Similarly, a clear hydrological perspective on the generation, dissemination and control of AMD is largely missing.

Therefore, the main objective of this chapter is to discuss how mining activities and hydrology contribute to AMD formation, dissemination and prevention. The specific objectives are: (1) to discuss the role of mining activities, including excavations, drilling and metallurgical processes on hydrology, and subsequent AMD formation, (2) to discuss how hydrology controls AMD formation, mobilisation and dissemination, (3) to summarize how a fundamental understanding of hydrology is used as a basis to design engineered cover systems for preventing AMD formation, and (4) to highlight constraints and knowledge gaps on AMD.

1.2 Mining Operations and Acid Mine Drainage

1.2.1 The Mining Project Cycle

Figure 1.1 depicts the mine project cycle consisting of key sequential phases, each entailing specific activities. A detailed discussion of the key phases of a mining project cycle is presented in literature (Robertson et al., 2017; Gorman and Dzombak, 2018). In summary, a typical mining project cycle consists of the following steps.

1.2.1.1 Prefeasibility Stage

This step entails project ideation and scoping, including identification of prospective mining sites. Preliminary geological investigations during this stage are limited to analysis of data based on desktop studies. Preliminary environmental scoping may be conducted during this phase. However, compared to subsequent steps, activities in this stage have limited environmental impacts and contribution to the formation of AMD.

1.2.1.2 Detailed Feasibility/Exploration Phase

This stage involves conducting environmental and social impact assessments (ESIAs) covering exploration, mining, metallurgical processing, and mine closure (Joyce et al., 2018; European Community, 2019). However, by nature, such ESIAs are overly broad and limited to scoping on potential social and environmental health impacts, hence have limited capacity to adequately address the potential generation and impacts of AMD. Field geological exploration and sampling involving excavation of test pits, drill holes, and subsequent sampling and assaying are also conducted during this phase. Activities conducted in this stage and, subsequent ones, have potential significant effects on the biophysical environment and AMD formation (Figure 1.1).

1.2.1.3 Design and Engineering

This stage involves the design of the mine layout and key infrastructure, including surveying of mining pits, roads, processing plants, and waste

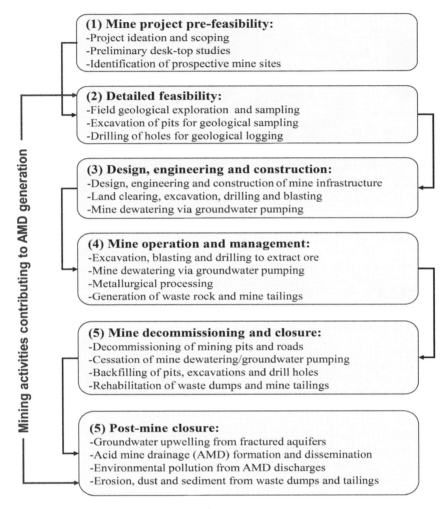

FIGURE 1.1
Mining activities contributing to the generation of acid mine drainage.

and wastewater disposal facilities including waste rock dumps and tailings dams (Gorman and Dzombak, 2018). This stage may include excavations and drilling during geotechnical site investigations, and land clearing to create access roads and set up site camps.

1.2.1.4 Construction

This stage entails significant land and vegetation clearing, excavation, drilling and blasting, and traffic movements during the construction of infrastructure. Dewatering via groundwater pumping may occur to facilitate

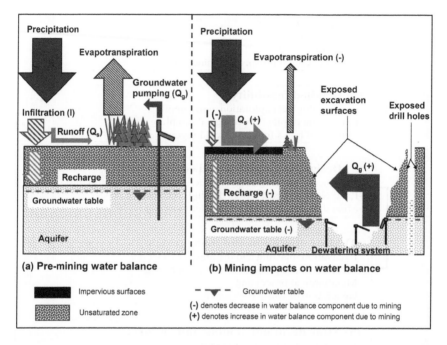

FIGURE 1.2
Conceptual depiction of the pre-mining water balance (a) compared to the hydrological impacts of mining activities (b). The size of the arrow is qualitatively indicative of the magnitude of the water balance components relative to the pre-mining state.

construction of infrastructure. Typical infrastructure constructed during this stage includes groundwater dewatering systems, mine access points, roads, metallurgical processing plants, and tailings dams (Gorman and Dzombak, 2018). Compared to the pre-mining state, this phase is also characterized by an increase in build-up areas and impervious surfaces, which in turn influences hydrological response (Figure 1.2).

1.2.1.5 Mine Operation and Management

This is the main phase in a mining project cycle, during which significant changes in the biophysical environment occur. Key activities during this stage include large-scale excavations, drilling and blasting during the extraction of ore (Hartlieb et al., 2017; Winn, 2020). In this phase of the mining project cycle, the removal of overburden rock generates large quantities of waste rock and run-of-mine. In cases where the ore body occurs below the groundwater table, mine dewatering via pumping may also occur. Frequent movement of vehicular mining equipment such as excavators, bulldozers and loaders occurs, causing soil compaction, soil detachment and dust generation. It must also be noted that subsequent mineral processing may require large quantities of water from surface or groundwater sources. In addition, mineral processing itself generates large quantities of solid waste including waste rock, mine tailings, and wastewater (Giblett and Morrell, 2016; Habib et al., 2020).

1.2.1.6 Mine Decommissioning and Closure

This is the final stage of a mining project entailing the decommissioning of mining pits and infrastructure. During this stage, the mine dewatering/groundwater pumping ceases, and groundwater upwelling/rising may occur. In some cases, rehabilitation of waste dumps and mine tailings are undertaken during this stage (Sanders et al., 2019).

In summary, mining activities entailing land/vegetation clearing, excavations, drilling and blasting have significant effects on the surface and groundwater hydrological systems. Mineral processing and waste disposal may release sulphidic wastes and wastewaters which may undergo oxidation to release AMD and associated contaminants. In subsequent sections, the impacts of the mining activities on hydrology and AMD formation are discussed.

1.2.2 Hydrological Impacts of Mining

Mining operations have significant impacts on local hydrological processes and water balances. Figure 1.2 depicts the hydrological impacts of mining operations relative to the pre-mining state. Specifically, mining activities reduce water infiltration, while increasing surface water runoff and soil erosion. Mining operations also remove vegetation cover, thereby reducing evapotranspiration. Mine dewatering increases groundwater pumping/abstraction, resulting in significant declines in groundwater levels and potential ground subsidence. Impervious surfaces associated with build-up areas reduce groundwater recharge. The mechanisms and processes responsible for the changes in surface water and groundwater hydrology are summarized in the subsequent sections.

1.2.2.1 Land Clearing and Impervious Surfaces

Vegetation cover protects soils against wind and water erosion, while plant roots increase soil aggregation by binding soil particles together. Vegetation cover also attenuates raindrop impacts and surface water runoff velocity, thereby increasing infiltration during rainfall events. Hence, land clearing, including the removal of vegetation, exposes soil to wind and water, leading to soil detachment and subsequent erosion (Asabonga et al., 2017; Jarsjö et al., 2017). The replacement of natural surfaces and vegetation cover by build-up areas consisting of mine buildings and impervious surfaces such as roads and roofs alters the water balance and affect hydrological processes (Figure 1.2). First, the removal of vegetation reduces interception of rainfall by vegetation covers, root uptake of soil moisture, and transpiration (Awotwi et al., 2019). Impervious surfaces associated with roads and build-up areas reduce infiltration and promote rapid generation of runoff water, resulting in peak water flows or flash floods (Figure 1.2). The removal of vegetation reduces root uptake of soil moisture, hence reduces evapotranspiration. In turn, reduced root uptake of soil moisture may potentially increase soil moisture storage and increase the risk of drainage (Ketcheson et al., 2016).

1.2.2.2 Vehicular Traffic Movements

Frequent movement of both heavy and light vehicular equipment including excavators, bulldozers, loaders and passenger vehicles has potential impacts on soil physical properties and hydrological behaviour (Weyer et al., 2019). The effects of traffic movement on soils and hydrology depend on soil type and moisture conditions. For example, movement of vehicular traffic on moist soils results in soil compaction via increased soil bulk density and reduced porosity (Strahm et al., 2017). Soil compaction is more pronounced on fine-textured soils such as clays and loams compared to sandy soils. Reduced porosity increases surface water runoff volumes and velocity during rainfall events. On dry soils, movement of vehicular traffic detaches soil particles and sediments. This increases soil erodibility or susceptibility to wind erosion and dust generation (Patra et al., 2016).

1.2.2.3 Excavation, Drilling and Blasting

Excavation, drilling and blasting alter the natural landforms and surface topography, which may in turn change the surface and groundwater hydrology (Hartliebet al., 2017; Winn, 2020). Furthermore, excavations conducted to remove overburden materials to expose the ore body may create surface and underground cavities and pits, which act as artificial depressions. These cavities and pits alter the flow directions of surface water runoff and may even change river flow regimes and flow directions (Newman et al., 2017; Zhang et al., 2017).

Drilling and blasting cause fractures or cracks in geological rock formations (Zhang et al., 2018). Recent concerns in the mining industry have also focused on the impacts of hydraulic fracturing/fracking (Taherdangkoo et al., 2017; Wu et al., 2019). Hydraulic fracturing or fracking involves injecting pressurized water, sand and chemicals into a rock formation through a well, in order to induce rock fragmentation and increase permeability of rock formations (Bao and Eaton, 2016; He et al., 2016). In cases where the ore body is submerged under groundwater, drilling and blasting may cause fractures or cracks in groundwater-bearing rock formations or aquifers. Fractures and cracks alter groundwater flow directions and result in groundwater upwelling, where groundwater rises to the surface under hydraulic pressure (Toner et al., 2017; Morrison et al., 2019). Groundwater upwelling is particularly common in confined or semi-confined aquifers, where groundwater occurs under hydraulic pressure.

1.2.2.4 Mine Dewatering via Groundwater Pumping

Mine dewatering entails the pumping of water from mine working areas such as mine pits to facilitate mining operations (Szczepiński, 2019). Such water may originate from two sources: (1) surface water runoff from surrounding areas, and/or (2) groundwater in cases where mining occurs below the groundwater table. Mine dewatering systems include wells, well galleries, boreholes and pumping units. Subsequently, groundwater pumping lowers the groundwater table and alters groundwater flow directions. Excessive dewatering or groundwater pumping may significantly alter the groundwater balance, leading to aquifer drying or groundwater recession. Excessive groundwater pumping also creates a large void space in the subsurface. Therefore, as the geological material settles to fill up the void space, this causes ground subsidence (Morrison et al., 2019).

1.2.2.5 Mineral Processing and Waste/Wastewater Generation

A wide range of mineral processing technologies exist, including pyrometallurgy, and hydrometallurgy (Giblett and Morrell, 2016; Habib et al., 2020). Mineral processing requires large quantities of water (e.g., washing, cooling) and may increase surface water and groundwater abstractions. In turn, mineral processing, particularly hydrometallurgy, generates large quantities of wastewaters, which may contribute to surface water and groundwater pollution.

1.2.2.6 Mine Decommissioning and Closure

The termination of mine dewatering/groundwater pumping may be accompanied by groundwater upwelling and flooding of mine pits. Excessive surface water runoff, and water and wind erosion may also occur from waste rock dumps, and mine tailings, thereby generating dust and sediments.

1.2.3 Mining Operations and Acid Mine Drainage Formation

Mining operations may contribute to AMD through the following: (1) directly by altering and exposing the previously buried/submerged sulphidic geological materials to weathering agents, and (2) indirectly by changing surface and groundwater hydrological flows and regimes. Table 1.1 presents a summary of mechanisms and processes related to mining operations.

TABLE 1.1

Contribution of Various Mining Activities to AMD Formation and Dissemination

Mining Activity	Impact	Contribution to AMD
Vegetation clearing during construction	Increases soil erosion and runoff	Disseminates sulphidic materials and AMD
Traffic movement during haulage	Sediment detachment and compaction	Increases erosion and transport of sulphidic sediments
Mine dewatering via groundwater pumping	Lowers groundwater table and alters flow directions	Promotes AMD formation by exposing previously submerged geological materials to weathering
Drilling and blasting of rock masses	Causes hydraulic fracturing or fracking of aquifers and changes water flow directions	Fractures promote AMD transport and expose previously submerged geological materials, while increasing surface area
Excavation of mining pits	Breaks down and exposes previously buried geological materials	Promotes AMD formation by breaking down rock materials and exposing them to oxidizing agents
Milling/crushing of mineral ores	Reduces particle size and increases surface area	Promotes AMD formation due to increased surface area
Metallurgical processing, including hydrometallurgy	Size reduction and increases in surface area, and use of strong extracting solutions	Promotes AMD formation in mine tailings and colloids in mine effluents. Strong acids/oxidizing agents accelerate weathering
Mine tailings / wastewater disposal	Environmental pollution	Promotes AMD formation from mine tailings and colloids in wastewater
Post-mine closure	Environmental pollution	AMD formation and dissemination via groundwater upwelling, erosion and runoff from waste dumps and mine tailings

1.2.3.1 Vegetation Clearing and Impervious Surfaces

Increased surface water runoff and soil erosion caused by land clearing and impervious (Asabonga et al., 2017; Jarsjö et al., 2017; Awotwi et al., 2019) promote the mobilisation and transport of sulphidic sediments and fine-textured mine wastes (Table 1.1). Traffic movements may also detach sulphidic mine wastes and sediments, thereby increasing their susceptibility to water and wind erosion (Patra et al., 2016). These processes mobilise and disseminate sulphidic geological materials and sediments, resulting in off-site impacts. Reduced evapotranspiration caused by vegetation clearing also increases the risk of deep drainage into buried sulphidic mine wastes and thus the formation of AMD.

1.2.3.2 Excavations, Drilling and Blasting

Excavations, drilling, and blasting expose previously buried/submerged sulphidic rock materials to weathering agents (Scheiber et al., 2018). Fractures, cracks and drill holes promote oxygen ingress into the sub-surface, causing in situ oxidation and AMD formation (Figure 1.2). Fractures and cracks also increase the permeability of rock formations to both water and oxygen (Bao and Eaton, 2016; He et al., 2016). This in turn increases connectivity and preferential flow pathways for the transport of AMD and associated contaminants.

Large quantities of waste rock or run-of-mine materials are generated during excavation, blasting and drilling to remove overburden geological material (Ruiseco et al., 2016; Pearce et al., 2019). Excavation, drilling, and blasting also cause rock fragmentation into particles relatively smaller than the original geological materials (Singh et al., 2016; Iravani et al., 2018). The reduced particle size and resulting increase in surface area create ideal conditions for subsequent weathering processes and AMD formation. Excavation, drilling and blasting, transport, and stockpiling of ores and waste rock also disperse sulphidic materials over large areas.

1.2.3.3 Mine Dewatering

The decline in groundwater levels caused by mine dewatering may expose previously submerged sulphidic rock formations to weathering conditions including oxidation and AMD formation. Groundwater upwelling may mobilise and transport AMD and contaminants from oxidized underground sources to the surface and aquatic systems as reported in a number of studies (Gomo, 2018; Migaszewskiet al., 2019). The impact on groundwater upwelling on AMD is most pronounced in water-limited environments such as the tropics. This is because, cyclic phases of groundwater upwelling in the wet summer season, followed by a decline in groundwater levels in the dry winter season, may subject sulphidic geological materials to wetting and drying cycles. In fact, the repeated cycles of drying and wetting of excavation pits and drill holes followed by groundwater rise or upwelling promote AMD generation. This phenomenon is well-pronounced in old coal mine workings in Hwange, Zimbabwe, for example, where coal mining dates back to the 1960s (Gwenzi et al., 2018b).

1.2.3.4 Mineral Processing and Waste/Wastewater Generation

Mineral processing entails the use of comminution circuits for mechanical size reduction using grinding and milling to increase surface area for subsequent mineral extraction in some cases (e.g., gold extraction) and/or to meet specifications for certain applications in other cases (e.g., coal) (Giblett and Morrell, 2016; Habib et al., 2020; Martínez et al., 2020). Three processes are critical in the context of AMD: (1) size reduction may generate sulphidic solid wastes in the form of waste rock or dust, (2) the use of strong extracting solutions in metallurgical processing generates sulphidic mine tailings and wastewaters laden with fine particles and colloids, and (3) strong extracting solutions such as peroxides and acids may promote weathering and AMD formation. Fine particles and colloids are also prone to both wind and water erosion.

1.2.3.5 Mine Decommissioning and Closure

In most developed countries using the mine bonding system such as Australia and Canada, among others, the environmental regulatory authority in the ministry responsible for environment often takes over the responsibility for post-mine closure monitoring and management (Cheng and Skousen, 2017; Sanders et al., 2019). In contrast, for most developing countries (e.g., Zimbabwe, Zambia, etc.), this phase marks the period of the most significant environmental impacts. This is partly because there is lack of clarity on the responsible authority between the ministry of mines on one hand, and the ministry of environment on the other hand. Moreover, due to the lack of a mine bond system, resources for such monitoring and management of mines in the post-closure phase are often unavailable after mine closure. Environmental regulatory agencies in most developing countries also often lack expertise and resources to mitigate AMD and its associated impacts, because no such bond systems exist. The mine or environmental bond system consists of an upfront, gradual set-aside or the allocation of financial resources to cater for mine closure and associated environmental health risks, rehabilitation and clean-up (Pepper et al., 2014; Cheng and Skousen, 2017). At the end of the mining operation, such bonds are either: (1) relinquished or forfeited to the state or its delegated regulatory authority that then takes over the responsibility of environmental management, or (2) repaid to the mining company in cases where mine rehabilitation and closure are done to the satisfaction of the regulatory authority. Due to the challenges associated with restoration of mined sites, and the potential for long-term adverse impacts (e.g., AMD) in the post-mine closure period, the former option is more common than the latter.

In summary, compared to the operational phase, the post-mine closure stage is characterized by relatively minimal environmental monitoring and control, thus, leading to significant AMD-related environmental impacts. Specifically, mine excavation pits, drill holes and cavities, which are typical relicts of mining operations, act as potential hotspots for the formation and transport of AMD and the associated contaminants. In addition, the termination of mine dewatering accompanied by frequent groundwater upwelling and flooding of mine pits promote AMD formation and dissemination. Indeed, several legacy cases of post-mine closure AMD occur globally, pointing to the lack of adequate control systems to manage AMD in several countries (Favas et al., 2016; Gwenzi et al., 2017, 2018b; Kim and Choi, 2018; Tabelin et al., 2019).

1.3 Hydrological Controls on Acid Mine Drainage

1.3.1 Acid Mine Drainage Generation

AMD is generated via the oxidation of sulphidic geological minerals containing pyrite, pyrrhotite and chalcopyrite (Kocaman et al., 2016). The oxidation process is catalysed by bacteria that oxidize metals and sulphur, including: (1) *Desulfovibrio desulfuricans*, (2) *Thiobacillus* spp., and (3) *Sulfolobus acidocalderius* (Chen et al., 2016). Equations 1.1 to 1.4 depict a series of processes involved in the formation of AMD (Wolkersdofer, 2006). In Equation 1.1, the oxidation of sulphide minerals occurs in the presence of water and oxygen to release ferrous iron (Fe^{2+}), sulphate and acidity or protons (H^+). The Fe^{2+} generated in Equation 1.1 is then rapidly oxidised to Fe^{3+} in the presence of oxygen and protons according to Equation 1.2. Subsequently, the Fe^{3+} formed reacts with water to form ferric hydroxide ($Fe(OH)_3$) and more acidity (Equation 1.3). Equation 1.4 shows the overall governing equation obtained by summing up Equations 1.1 to 1.3.

$$FeS_2 + \frac{7}{2}O_2 + H_2O => Fe^{2+} + 2SO_4^{2-} + 2H^+ \tag{1.1}$$

$$Fe^{2+} + \frac{1}{4}O_2 + H^+ => Fe^{3+} + \frac{1}{2}H_2O \tag{1.2}$$

$$Fe^{3+} + 3H_2O => Fe(OH)_{3(s)} + 3H^+ \tag{1.3}$$

$$FeS_2 + \frac{15}{4}O_2 + \frac{7}{2}H_2O => Fe(OH)_{3(s)} + 2SO_4^{2-} + 4H^+ \tag{1.4}$$

The protons generated during the formation of AMD account for the acidic pH of AMD. High Fe^{3+} and its oxyhydroxides ($Fe(OH)_3$) generate the reddish brown colour of AMD. Equations 1.1 and 1.4 explain the origins of high concentrations of sulphates detected in AMD (Mungazi and Gwenzi, 2019). Highly acid pH conditions are also responsible for the release of other contaminants such as metals, metalloids, radionuclides, and rare earth elements (REEs). These other contaminants are released via two processes: (1) simultaneously during the oxidation of sulphide-bearing geological materials, and/or (2) dissolution induced by highly acidic conditions (Consani et al., 2017; Balci and Demirel, 2018; Skousen et al., 2019).

Overall, Equations 1.1 to 1.4 clearly demonstrate that hydrology plays a critical role in AMD generation in two ways: (1) on the one hand, the oxidation process is mediated by water, and (2) on the other hand, permanently saturated and submerged conditions restrict the oxidation process, because oxygen has a very low solubility and diffusivity in water. Permanently saturated or submerged conditions may occur in natural systems in cases where sulphidic geological materials occur below the groundwater table. Permanently saturated/submerged conditions can also be created artificially, and this underpins the application of wet/water covers for the prevention of AMD (Aubertin et al., 2016; Karna and Hettiarachchi, 2018) as discussed in Chapter 6. In most natural geochemical settings, the availability of oxygen, rather than water, is often considered the rate-limiting condition in AMD generation. Therefore, strong wetting and drying cycles associated with distinct wet and dry seasons experienced in water-limited environments such as the tropics are ideal for AMD generation. Such alternation of wet and dry or freeze/thaw conditions will promote oxidation and formation of AMD in pits, drill holes, old mine workings, waste dumps, and mine tailings (Jouini et al., 2020). The hydrochemistry of AMD formation depicted in Equations 1.1 to 1.4 shows that restricting water and oxygen is critical in the prevention of AMD generation (refer to Chapter 6 for more details).

1.3.2 Dissemination of Acid Mine Drainage

Hydrological processes control the mobilisation and dissemination of both AMD and the associated contaminants into other environmental compartments (Fosso-Kankeu et al., 2017; Santisteban et al., 2016; Oldham et al., 2019). The environmental compartments include soils, surface aquatic systems such as rivers, streams and reservoirs, and groundwater systems. A number of hydrological processes such as surface water runoff, sub-surface water flow, and groundwater flow mobilise and disseminate AMD in the environment (Chaubey and Arora, 2017). Surface water runoff and erosion mobilise and transfer AMD and associated colloids from various sources into surface aquatic systems such as rivers, streams and reservoirs (Ravengai et al., 2004; Masocha et al., 2020).

Groundwater flow, which is governed by Darcy's law, also transport AMD and contaminants from regions of high total hydraulic heads to those with low heads, in response to a total head gradient (Francisca et al., 2012). Surface water-groundwater interactions or exchanges provide the hydrological connectivity between surface aquatic systems and the groundwater systems (Chaubey and Arora, 2017). Therefore, the surface water-groundwater interactions facilitate the transfer of AMD and contaminants between surface aquatic systems and groundwater. Surface and underground cavities, holes, and cracks created during excavations, drilling, and blasting act as connecting and preferential pathways for the flow of AMD and contaminants.

Equations 1.5 to 1.9 show that hydrological fluxes/flows in both surface and groundwater systems are strongly coupled to contaminant transport (Francisca et al., 2012; Chaubey and Arora, 2017). Three key processes are responsible for contaminants transport by water (Francisca et al., 2012; Chaubey and Arora, 2017): (1) advection/convention/mass flow, (2) diffusion, and (3) dispersion. Advection or mass flow is driven by flowing surface water or groundwater flow, and flow velocity (Equations 1.6–1.7). Diffusion is governed by Fick's law, and transfers contaminants from regions of high to low concentrations, in response to a concentration gradient (Franscisca et al., 2012). Dispersion is caused by turbulences and flow heterogeneities induced by spatial variability in porosity and hydraulic conductivity or permeability of groundwater-bearing rock formations. Diffusion and dispersion occur concurrently, hence, are often collectively termed hydrodynamic dispersion. The governing equations for mass flow and hydrodynamic dispersion have a flow velocity as a variable, indicating the importance of flow velocity in contaminant transport (Equations 1.5–1.7):

$$\frac{dc}{dt} = -V\frac{dc}{dx} + D_L\frac{d^2c}{dx^2} \tag{1.5}$$

$$\frac{dc}{dt} = -\frac{V}{R_d}\frac{dc}{dx} + \frac{D_L}{R_d}\frac{d^2c}{dx^2} \tag{1.6}$$

$$\frac{dc}{dt} = -V\frac{dc}{dx} + D_L\frac{d^2c}{dx^2} - \lambda c \tag{1.7}$$

In Equations 1.5 to 1.7, the first and second terms on the right-hand side represent advection/mass flow and hydrodynamic dispersion, respectively. In these equations, V = water flow velocity (L/T), c = contaminant concentration (mg/L^3), x = distance (L), and D_L = longitudinal dispersion coefficient, calculated as a product of longitudinal dispersivity (a_L) and velocity (V) (i.e., $D_L = a_L \times V$), R_d = retardation factor, and λ = first-order degradation or decay constant.

Depending on the nature of contaminants and biogeochemical conditions, two transport phenomena may occur: (1) reactive transport, and (2) non-reactive transport. Equation 1.5 depicts the governing equation for non-reactive transport of conservative (i.e., non-reactive) contaminants such as chloride. Non-reactive transport occurs when conservative contaminants such as chlorides only undergo transport without any reactions. Thus, barring the effects of dilution, the plume of a non-reactive contaminant is often characterized by a relatively fixed concentration along the transport pathway.

Reactive transport occurs when non-conservative (i.e., reactive) contaminants in AMD such as metals and nutrient ions (e.g., nitrates, phosphates, sulphates) simultaneously undergo transport and reactions. Typical reactive processes include retardation via adsorption onto solid matrices, and biochemical decay or degradation. Equations 1.6 and 1.7 show reactive transport, where the contaminant undergoes retardation (Equation 1.6) and decay (Equation 1.7), respectively (Francisca et al., 2012; Sadrnejad and Memarianfard, 2017; Sethi and Di Molfetta, 2019). An example of contaminants that undergo reactive transport is metals, which undergo adsorption on the solid matrix such as soils and sediments, a process that slows the transport processes. Other reactive contaminants include nutrients such as nitrates, phosphates and sulphates, which are taken up by microbes and aquatic plants during the transport process. Hence, depending on the nature of the reactions, the concentrations of reactive or non-conservative contaminants may undergo attenuation or increase along the transport pathway.

In summary, regardless of the nature of the transport phenomena (reactive or non-reactive), surface water and groundwater flows control the transport and dissemination of AMD and contaminants. The dissemination via hydrological processes explains the potential off-site environmental, human and ecological health impacts of AMD. Collectively, these processes demonstrate how hydrology controls the formation, mobilisation and subsequent dissemination of AMD and contaminants.

1.3.3 Acid Mine Drainage Impacts on Hydrochemistry

Besides altering surface and groundwater flow directions and water balances, AMD also has adverse impacts on surface and groundwater hydrochemistry. Several studies drawn from various countries have reported the impacts of AMD on hydrochemistry, including extremely acidic conditions and high concentrations of dissolved contaminants (Grande et al., 2018; Gwenzi et al., 2017; Xia et al., 2017; Wen et al., 2018; Mungazi and Gwenzi, 2019). For example, extremely high concentrations of metals, sulphates and acidity were reported at Iron Duke Mine in Mazowe, Zimbabwe (Gwenzi et al., 2017; Mungazi and Gwenzi, 2019). According to Williams and Smith (2000), AMD at Iron Duke Mine has some of the highest values ever reported in the world as evidenced by extremely low pH (0.52–0.82), and very high concentrations of both sulphates (SO_4^{2-}) (220 853–355 425 mg/L) and iron (Fe) (85 672–132 929 mg/L. High concentrations of other anions such as chloride and fluoride have also been reported (Alsaiari and Tang, 2018; Li et al., 2019), while other studies detected high concentrations of toxic metalloids (e.g., arsenic) and radionuclides (Galhardi and Bonotto, 2017; Ilay et al., 2019; Manjón et al., 2019). Recent studies also reported the enrichment of rare earth elements (REEs) in AMD and receiving waters (Migaszewski et al., 2019; Soyol-Erdene et al., 2018). For example, a study conducted in Poland detected extreme enrichment of arsenic and REEs in AMD with concentrations of up to 1548 mg L^{-1} for arsenic and up to 24.84 mg L^{-1} for REEs (Migaszewski et al., 2019).

The changes in hydrochemistry induced by AMD have potential ecological and human health risks (Ochieng et al., 2017). For example, metals, metalloids, and radionuclides are well-known toxins that have adverse impacts on human and ecological health (Martínez-Alcalá and Bernal, 2020; Iryna, 2017). The REEs pose several human health risks, including genotoxicity, teratogenicity, and carcinogenicity (Gwenzi et al., 2018a; Squadrone et al., 2019). Gwenzi et al. (2018b) discussed the contribution of mining and mineral processing including AMD to REEs detected in aquatic systems and the associated health risks.

1.3.4 Applications of Hydrology Principles in Acid Mine Drainage Prevention

The fact that hydrology controls the formation, mobilisation and dissemination of AMD forms the underlying principle for the design and operation of wet and dry engineered covers for AMD prevention. A detailed discussion of wet and dry covers is presented in Chapter 6 of this book. Therefore, only an overview of the application of hydrology in the design and operation principles of wet and dry covers is highlighted here.

1.3.4.1 Wet or Water Covers

Wet or water covers prevent AMD generation by restricting oxygen ingress by maintaining permanent saturation (Aubertin et al., 2016; Karna and Hettiarachchi, 2018). The use of water to restrict oxygen ingress is based on the fact that oxygen has low solubility and diffusivity in water. According to Equations 1.1–1.4, the lack of oxygen prevents oxidation of sulphidic materials and formation of AMD. Wet covers are commonly used in temperate and permafrost environments, where precipitation far exceeds potential evapotranspiration (Aubertin et al., 2016). Such environments lack strong seasonality patterns, hence it is possible to maintain permanent saturated conditions over the sulphidic wastes. However, wet covers are not ideal in water-limited environments such as the tropics, due to strong seasonality associated with distinct wet and dry seasons, which induce strong wetting and drying cycles.

1.3.4.2 Evapotranspirative 'Store-Release' Covers

Evapotranspirative covers are also known as dry, water balance or store-release covers (Gwenzi, 2010). The design objective of the covers is to reduce drainage, and to some extent oxygen ingress into buried sulphidic wastes (Knidiri et al., 2017). This is achieved by: (1) storing water in a storage layer consisting of soil or benign geological material during or soon after rainfall events, and (2) then releasing it back into the atmosphere as water vapour via evapotranspiration (Gwenzi, 2010). Accordingly, vegetation plays a key role in the hydrology and performance of store-release covers. Uptake of soil water by plant roots and the subsequent loss of such water via transpiration account for the soil dewatering process and reduction of drainage. In principle, evapotranspirative or store release covers are designed to enhance soil moisture storage and the evapotranspiration components of the water balance. Evapotranspirative covers are ideal in water-limited environments such as the tropics, where potential evapotranspiration far exceeds precipitation. These climatic conditions are ideal for maximum soil water loss via evapotranspiration. A detailed review of global literature on the design and performance of store-release covers is presented in earlier studies such as by Gwenzi (2010).

1.4 Future Perspectives on Acid Mine Drainage

The current chapter highlighted how several mining activities along the mine project cycle alter hydrology and promote the generation and dissemination of AMD. However, several constraints and knowledge gaps warrant further public and research attention.

1.4.1 Effectiveness of Environmental and Social Impact Assessments

The hydrogeochemistry of AMD is quite complex, hence AMD often has a long latent or lag period. Therefore, the occurrence and impacts of AMD often manifest long after the mine closure, when no resources are available for implementing mitigation and control measures. Moreover, most ESIAs lack details to address potential long-term AMD risks, yet such ESIAs and their associated environmental management plans (EMPs) are used as a basis for project approval and post-closure mining. Therefore, in most countries, current ESIAs and EMPs have limited potential to address AMD risks.

1.4.2 Rapid Acid Mine Drainage Prediction

Most methods for assessing AMD potential such as pH measurements and short-term leaching tests may underestimate the AMD generating potential of sulphidic geological materials (Kanda et al., 2017). Yet, alternative methods for evaluating AMD are often expensive, time-consuming, and require well-equipped geochemical laboratories. Therefore, such expensive and robust methods are rarely used in feasibility studies such as ESIAs and routine environmental surveillance systems. Hence, the development and validation of rapid and universal methods for assessing AMD potential warrant further research. However, some of the advanced methods for predicting AMD formation are covered in Chapter 2.

1.4.3 Acid Mine Prainage Prevention during the Mine Project Cycle

The bulk of AMD prevention and control methods seem to adopt an 'end-of-pipe' approach (Gwenzi et al., 2017; Mungazi and Gwenzi, 2019; Skousen et al., 2019). Specifically, particular attention is paid to: (1) treatment of AMD using active and passive methods, and (2) use of wet and dry covers to prevent AMD generation from waste rock and mine tailings. By comparison, AMD prevention receives limited attention during the various phases in the mining project cycle highlighted in the current chapter. Given the high cost of AMD control and its health risks, AMD control and prevention may need to be integrated in the whole mine project cycle as part of safety, health and environment (SHE) programme. Some of the techniques for preventing AMD are covered in Chapter 6.

1.4.4 Translating Existing Research into Appropriate Solutions

A significant body of literature exists on the formation, hydrochemistry and control of AMD (Skousen et al., 2019). However, the bulk of literature applies to laboratory and pilot studies, while data on large-scale field applications are still limited, partly due to high costs associated with current technologies. Given the global scale of AMD, and the existence of several legacies associated with it, the next decade should focus on developing appropriate and scalable solutions for AMD prevention and control.

1.4.5 Acid Mine Drainage Control in Underground Workings

Compared to waste dumps and mine tailings, the prevention and control of AMD in current and old mine workings appear to pose serious challenges to both researchers and regulators. First, the spatial extent of AMD in underground mine workings is not known with certainty in most countries. Second, even in most countries, particularly developing nations, where underground AMD and its impacts on human and ecological heath have been detected, feasible control measures appear to be lacking. Hence, further research is required to address AMD in current and old mine workings.

1.4.6 Mapping of Acid Mine Drainage at Country and Regional Levels

Limited data exist on the spatial distribution of AMD hotspots at country and regional levels. Yet such information is critical for devising control measures to safeguard human and ecological health. For example, in sub-Saharan Africa, several surface water and aquifer systems are shared among several countries, hence water pollution risks posed by AMD constitute a shared trans-boundary problem. Thus, there is a need for coordinated efforts to map AMD problems close to major national and regional surface water and aquifer systems. The availability of remote sensing tools, coupled with GIS spatial tools can enable such mapping at various spatial and temporal scales (Masocha et al., 2020).

1.4.7 Hydrological Impacts of Mining

The hydrological impacts of mining activities such as land clearing, excavation, drilling and blasting on water flows and balances are largely based on inferential evidence. Hence, systematic studies are required to understand the hydrological impacts of mining activities on surface and groundwater hydrological processes and water balances at mine, local and catchment levels. Such studies should also investigate impacts on hydrochemistry, including the occurrence of residual chemicals used for drilling, blasting and hydraulic fracturing/fracking.

1.5 Concluding Remarks

The current chapter discussed the impacts of mining activities on hydrology and AMD generation. Specifically, the role of mining activities including excavation, blasting and drilling on the hydrological systems and AMD formation were discussed. The impacts include the alteration of surface and groundwater flow directions, as well as the lowering of groundwater levels during dewatering or groundwater pumping to expose submerged ores. Furthermore, excavations and drill holes provide access points for the ingress of oxygen, thereby promoting in situ AMD generation in old underground mine workings. In situ AMD generation from old mine workings is particularly pronounced in water-limited environments characterized by strong wetting and drying cycles. Ex situ AMD generation occurs in waste rock dumps and mine tailings, which constitute a significant portion of mine processing wastes. Metallurgical processes involve comminution and use of strong extracting solutions which are likely to increase surface area for oxidation and formation of AMD.

The role of hydrology in AMD formation, mobilisation and dissemination was discussed. For example, on the one hand, hydrology plays a critical role in the mobilisation and dissemination of AMD and contaminants via reactive and non-reactive contaminant transport. Yet, on the other hand, the manipulation of hydrological processes and the water balance is used as a basis for the prevention of AMD using wet and evapotranspirative or store release covers. The lack of data on sub-surface AMD remediation noted in this chapter highlighted the need for further research. Moreover, most available data are limited to laboratory or pilot scale studies, while case studies of large-scale applications and their evaluations are largely lacking. Therefore, despite several research efforts on AMD, the legacy of AMD problems still exists globally.

References

Alsaiari, A. and Tang, H.L. (2018). Field investigations of passive and active processes for acid mine drainage treatment: are anions a concern? Ecological Engineering 122: 100–106.

Asabonga, M., Cecilia, B., Mpundu, M.C. and Vincent, N.M.D. (2017). The physical and environmental impacts of sand mining. Transactions of the Royal Society of South Africa 72(1): 1–5.

Aubertin, M., Bussière, B., Pabst, T., James, M. and Mbonimpa, M. (2016). Review of the reclamation techniques for acid-generating mine wastes upon closure of disposal sites. Geo-Chicago 2016: 343–358.

Awotwi, A., Anornu, G.K., Quaye-Ballard, J.A., Annor, T., Forkuo, E.K., Harris, E. and Terlabie, J.L. (2019). Water balance responses to land-use/land-cover changes in the Pra River Basin of Ghana, 1986–2025. Catena 182: 104129.

Balci, N. and Demirel, C. (2018). Prediction of acid mine drainage (AMD) and metal release sources at the Küre Copper Mine Site, Kastamonu, NW Turkey. Mine Water and the Environment 37(1): 56–74.

Bao, X. and Eaton, D.W. (2016). Fault activation by hydraulic fracturing in western Canada. Science, 354(6318):1406–1409.

Beckett, C. (2017). Rethinking remediation: mine closure and community engagement at the Giant Mine, Yellowknife, Northwest Territories, Canada. PhD dissertation, Memorial University of Newfoundland.

Buxton, G.A. (2018). Modeling the effects of vegetation on fluid flow through an acid mine drainage passive remediation system. Ecological Engineering 110: 27–37.

Campbell, K.M., Alpers, C.N. and Nordstrom, D.K. (2020). Formation and prevention of pipe scale from acid mine drainage at iron Mountain and Leviathan Mines, California, USA. Applied Geochemistry 104521.

Casagrande, M.F.S., Moreira, C.A. and Targa, D.A. (2020). Study of generation and underground flow of acid mine drainage in waste rock pile in an uranium mine using electrical resistivity tomography. Pure and Applied Geophysics 177(2): 703–721.

Chaubey, J. and Arora, H. (2017). Transport of contaminants during groundwater surface water interaction. In: Development of Water Resources in India. Springer, Cham, pp. 153–165.

Chen, L.X., Huang, L.N., Méndez-García, C., Kuang, J.L., Hua, Z.S., Liu, J. and Shu, W.S. (2016). Microbial communities, processes and functions in acid mine drainage ecosystems. Current Opinion in Biotechnology 38: 150–158.

Cheng, L. and Skousen, J.G. (2017). Comparison of international mine reclamation bonding systems with recommendations for China. International Journal of Coal Science & Technology 4(2): 67–79.

Choudhury, B.U., Malang, A., Webster, R., Mohapatra, K.P., Verma, B.C., Kumar, M. and Hazarika, S. (2017). Acid drainage from coal mining: effect on paddy soil and productivity of rice. Science of the Total Environment 583: 344–351.

Chuhan-Pole, P., Dabalen, A.L. and Land, B.C. (2017). Mining in Africa: Are Local Communities Better Off? World Bank, Washington, DC.

Consani, S., Carbone, C., Dinelli, E., Balić-Žunić, T., Cutroneo, L., Capello, M. and Lucchetti, G. (2017). Metal transport and remobilisation in a basin affected by acid mine drainage: the role of ochreous amorphous precipitates. Environmental Science and Pollution Research 24(18): 15735–15747.

Durucan, S., Korre, A. and Munoz-Melendez, G. (2006). Mining life cycle modelling: a cradle-to-gate approach to environmental management in the minerals industry. Journal of Cleaner Production 14(12-13): 1057–1070.

European Community. (2019). Development of a Guidance Document on Best Practices in the Extractive Waste Management Plans Circular Economy Action. European Community, Brussels.

Favas, P.J.C., Sarkar, S.K., Rakshit, D., Venkatachalam, P. and Prasad, M.N.V. (2016). Acid mine drainages from abandoned mines: hydrochemistry, environmental impact, resource recovery, and prevention of pollution. In: Environmental Materials and Waste. Academic Press, Amsterdam, pp. 413–462.

Fernández-Caliani, J.C., Giráldez, M.I. and Barba-Brioso, C. (2019). Oral bioaccessibility and human health risk assessment of trace elements in agricultural soils impacted by acid mine drainage. Chemosphere 237: 124441.

Fosso-Kankeu, E., Manyatshe, A. and Waanders, F. (2017). Mobility potential of metals in acid mine drainage occurring in the Highveld area of Mpumalanga Province in South Africa: implication of sediments and efflorescent crusts. International Biodeterioration & Biodegradation 119: 661–670.

Francisca, F.M., Carro Pérez, M.E., Glatstein, D.A. and Montoro, M.A. (2012). Contaminant transport and fluid flow in soils. Horizons in Earth Research 6: 97–131.

Galhardi, J.A. and Bonotto, D.M. (2016). Hydrogeochemical features of surface water and groundwater contaminated with acid mine drainage (AMD) in coal mining areas: a case study in southern Brazil. Environmental Science and Pollution Research 23(18): 18911–18927.

Galhardi, J.A. and Bonotto, D.M. (2017). Radionuclides (222 Rn, 226 Ra, 234 U, and 238 U) release in natural waters affected by coal mining activities in southern Brazil. Water, Air, & Soil Pollution 228(6): 1–19.

Genty, T., Bussière, B., Paradie, M. and Neculita, C.M. (2016). Passive biochemical treatment of ferriferous mine drainage: Lorraine mine site, Northern Quebec, Canada. In Proceedings of the International Mine Water Association (IMWA) Conference, July, pp. 11–15.

Giblett, A. and Morrell, S. (2016). Process development testing for comminution circuit design. Minerals & Metallurgical Processing 33(4): 172–177.

Golev, A., Lebre, E. and Corder, G. (2016). The contribution of mining to the emerging circular economy. AusIMM Bulletin, p. 30.

Gomo, M. (2018). Conceptual hydrogeochemical characteristics of a calcite and dolomite acid mine drainage neutralised circumneutral groundwater system. Water Science 32(2): 355–361.

Gorman, M.R., and Dzombak, D.A. (2018). A review of sustainable mining and resource management: transitioning from the life cycle of the mine to the life cycle of the mineral. Resources, Conservation and Recycling 137: 281–291.

Grande, J.A., Santisteban, M., de la Torre, M.L., Dávila, J.M. and Pérez-Ostalé, E. (2018). Map of impact by acid mine drainage in the river network of The Iberian Pyrite Belt (Sw Spain). Chemosphere 199: 269–277.

Gwenzi, W. (2010). Vegetation and soil controls on water redistribution on recently constructed ecosystems in water-limited environments. PhD Dissertation, University of Western Australia. Perth.

Gwenzi, W., Mangori, L., Danha, C., Chaukura, N., Dunjana, N. and Sanganyado, E. (2018a). Sources, behaviour, and environmental and human health risks of high-technology rare earth elements as emerging contaminants. Science of the Total Environment 636: 299–313.

Gwenzi, W., Mapanda, F. and Mabidi, A. (2018b). Environmental feasibility of the implementation of Liberation Coal Mining in the Gwayi-Shangani Floodplain. A technical report prepared for the Environmental Management Agency (EMA), Harare.

Gwenzi, W., Mushaike, C. C., Chaukura, N. and Bunhu, T. (2017). Removal of trace metals from acid mine drainage using a sequential combination of coal ash-based adsorbents and phytoremediation by bunchgrass (Vetiver [*Vetiveria zizanioides* L]). Mine Water and the Environment 36(4): 520–531.

Habib, A., Bhatti, H. N. and Iqbal, M. (2020). Metallurgical processing strategies for metals recovery from industrial slags. Zeitschrift für Physikalische Chemie 234(2): 201–231.

Hao, C., Wei, P., Pei, L., Du, Z., Zhang, Y., Lu, Y. and Dong, H. (2017). Significant seasonal variations of microbial community in an acid mine drainage lake in Anhui Province, China. Environmental Pollution 223: 507–516.

Hartlieb, P., Grafe, B., Shepel, T., Malovyk, A. and Akbari, B. (2017). Experimental study on artificially induced crack patterns and their consequences on mechanical excavation processes. International Journal of Rock Mechanics and Mining Sciences 100: 160–169.

He, Q., Suorineni, F.T. and Oh, J. (2016). Review of hydraulic fracturing for preconditioning in cave mining. Rock Mechanics and Rock Engineering 49(12): 4893–4910.

Humphries, M.S., McCarthy, T.S. and Pillay, L. (2017). Attenuation of pollution arising from acid mine drainage by a natural wetland on the Witwatersrand. South African Journal of Science 113(1-2): 1–9.

İlay, R., Baba, A. and Kavdır, Y. (2019). Removal of metals and metalloids from acidic mining lake (AML) using olive oil solid waste (OSW). International Journal of Environmental Science and Technology 16(8): 4047–4058.

Iravani, A., Åström, J.A. and Ouchterlony, F. (2018). Physical origin of the fine-particle problem in blasting fragmentation. Physical Review Applied 10(3): 034001.

Iryna, P. (2017). Toxicity of radionuclides in determining harmful effects on humans and environment. Journal of Environ Science 1(2): 115–119.

Jarsjö, J., Chalov, S.R., Pietroń, J., Alekseenko, A.V. and Thorslund, J. (2017). Patterns of soil contamination, erosion and river loading of metals in a gold mining region of northern Mongolia. Regional Environmental Change 17(7): 1991–2005.

Jouini, M., Neculita, C.M., Genty, T. and Benzaazoua, M. (2020). Freezing/thawing effects on geochemical behavior of residues from acid mine drainage passive treatment systems. Journal of Water Process Engineering 33: 101087. Available at https://doi.org/10.1016/j.jwpe.2019.101087.

Joyce, S., Sairinen, R. and Vanclay, F. (2018). Using social impact assessment to achieve better outcomes for communities and mining companies. In Mining and Sustainable Development. Routledge, pp. 65–86.

Kanda, A., Nyamadzawo, G., Gotosa, J., Nyamutora, N. and Gwenzi, W. (2017). Predicting acid rock drainage from nickel mine waste pile and metals levels in surrounding soils. Environmental Engineering and Management Journal 16(9): 2089–2096.

Karna, R.R. and Hettiarachchi, G.M. (2018). Subsurface submergence of mine waste materials as a remediation strategy to reduce metal mobility: an overview. Current Pollution Reports 4(1): 35–48.

Ketcheson, S.J., Price, J.S., Carey, S.K., Petrone, R.M., Mendoza, C.A. and Devito, K.J. (2016). Constructing fen peatlands in post-mining oil sands landscapes: challenges and opportunities from a hydrological perspective. Earth-Science Reviews 161: 130–139.

Kim, S. M. and Choi, Y. (2018). SIMPL: a simplified model-based program for the analysis and visualization of groundwater rebound in abandoned mines to prevent contamination of water and soils by acid mine drainage. International Journal of Environmental Research and Public Health 15(5): 951.

Knidiri, J., Bussière, B., Hakkou, R., Bossé, B., Maqsoud, A. and Benzaazoua, M. (2017). Hydrogeological behaviour of an inclined store-and-release cover experimental cell made with phosphate mine wastes. Canadian Geotechnical Journal 54(1): 102–116.

Kocaman, A.T., Cemek, M. and Edwards, K.J. (2016). Kinetics of pyrite, pyrrhotite, and chalcopyrite dissolution by Acidithiobacillus ferrooxidans. Canadian Journal of Microbiology 62(8): 629–642.

Leppänen, J.J., Weckström, J. and Korhola, A. (2017). Paleolimnological fingerprinting of the impact of acid mine drainage after 50 years of chronic pollution in a southern Finnish lake. Water, Air, & Soil Pollution 228(6): 1–13.

Li, Y., Wang, S., Sun, H., Huang, W., Nan, Z., Zang, F. and Li, Y. (2019). Immobilization of fluoride in the sediment of mine drainage stream using loess, Northwest China. Environmental Science and Pollution Research 27(7): 6950–6959.

Liao, J., Wen, Z., Ru, X., Chen, J., Wu, H. and Wei, C. (2016). Distribution and migration of heavy metals in soil and crops affected by acid mine drainage: public health implications in Guangdong Province, China. Ecotoxicology and Environmental Safety 124: 460–469.

Manjón, G., Mantero, J., Vioque, I., Galván, J., Díaz-Francés, I. and García-Tenorio, R. (2019). Some naturally occurring radionuclides (NORM) in a river affected by acid mining drainages. Chemosphere 223: 536–543.

Martínez, R., Bednarek, M. and Zulawska, U. (2020). Validation of a sustainable model for the mining-metallurgical industry in Mexico. In: Multidisciplinary Digital Publishing Institute Proceedings 38(1): 12. Available at Doi:10.3390/proceedings2019038012.

Martínez-Alcalá, I. and Bernal, M.P. (2020). Environmental impact of metals, metalloids, and their toxicity. Metalloids in Plants: Advances and Future Prospects, pp. 451–488.

Masocha, M., Dube, T., Mambwe, M. and Mushore, T.D. (2019). Predicting pollutant concentrations in rivers exposed to alluvial gold mining in Mazowe Catchment, Zimbabwe. Physics and Chemistry of the Earth, Parts A/B/C, 112, 210–215.

Migaszewski, Z.M., Gałuszka, A. and Dołęgowska, S. (2019). Extreme enrichment of arsenic and rare earth elements in acid mine drainage: case study of Wiśniówka mining area (south-central Poland). Environmental Pollution 244: 898–906.

Morrison, K.G., Reynolds, J.K. and Wright, I.A. (2019). Subsidence fracturing of stream channel from longwall coal mining causing upwelling saline groundwater and metal-enriched contamination of surface waterway. Water, Air and Soil Pollution 230(2): 37. Available at https://doi.org/10.1016/j.jwpe.2019.101087.

Mungazi, A.A. and Gwenzi, W. (2019). Cross-layer leaching of coal fly ash and mine tailings to control acid generation from mine wastes. Mine Water and the Environment 38(3): 602–616.

Newman, C., Agioutantis, Z. and Leon, G.B.J. (2017). Assessment of potential impacts to surface and subsurface water bodies due to longwall mining. International Journal of Mining Science and Technology 27(1): 57–64.

Ochieng, G.M., Seanego, E.S. and Nkwonta, O.I. (2010). Impacts of mining on water resources in South Africa: a review. Scientific Research and Essays 5(22): 3351–3357.

Oldham, C., Beer, J., Blodau, C., Fleckenstein, J., Jones, L., Neumann, C. and Peiffer, S. (2019). Controls on iron (II) fluxes into waterways impacted by acid mine drainage: a Damköhler analysis of groundwater seepage and iron kinetics. Water Research 153: 11–20.

Park, I., Tabelin, C. B., Seno, K., Jeon, S., Ito, M. and Hiroyoshi, N. (2018). Simultaneous suppression of acid mine drainage formation and arsenic release by carrier-microencapsulation using aluminum-catecholate complexes. Chemosphere 205: 414–425.

Patra, A. K., Gautam, S. and Kumar, P. (2016). Emissions and human health impact of particulate matter from surface mining operation – a review. Environmental Technology and Innovation 5: 233–249.

Pearce, S., Brookshaw, D., Mueller, S. and Barnes, A. (2019, September). Optimising waste management assessment using fragmentation analysis technology. In: Proceedings of the 13th International Conference on Mine Closure. Australian Centre for Geomechanics, Crawley, pp. 883–896.

Pepper, M., Roche, C.P. and Mudd, G.M. (2014). Mining legacies – understanding life-of-mine across time and space. In: Proceedings of the Life-of-Mine Conference. Brisbane, Australia, 16–18 July. Australasian Institute of Mining and Metallurgy, Carlton, pp. 449–465.

Pope, J., Christenson, H., Gordon, K., Newman, N. and Trumm, D. (2018). Decrease in acid mine drainage release rate from mine pit walls in Brunner Coal Measures. New Zealand Journal of Geology and Geophysics 61(2): 195–206.

Ravengai, S., Owen, R. and Love D (2004) Evaluation of seepage and acid generation potential from evaporation ponds, Iron Duke Pyrite Mine, Mazowe Valley, Zimbabwe. Physics and Chemistry of Earth 29: 1129–1134.

Robertson, S. A., Blackwell, B., Haslam McKenzie, F. and Argent, N. (2017). Mine life-cycle planning and enduring value for remote communities. University of New England. Available at https://rune.une.edu.au/web/handle/1959.11/21387 [Accessed 15 May 2020].

Ruiseco, J. R., Williams, J. and Kumral, M. (2016). Optimizing ore–waste dig-limits as part of operational mine planning through genetic algorithms. Natural Resources Research 25(4): 473–485.

Sadrnejad, S.A. and Memarianfard, M. (2017). Contamination transport into saturated land upon advection-diffusionsorption including decay. Journal of Numerical Methods in Civil Engineering 1(3): 67–75.

Sanders, J., McLeod, H., Small, A. and Strachotta, C. (2019, September). Mine closure residual risk management: identifying and managing credible failure modes for tailings and mine waste. In: Proceedings of the 13th International Conference on Mine Closure. Australian Centre for Geomechanics, Crawley, pp. 535–552.

Santisteban, M., Grande, J.A., de La Torre, M.L., Valente, T., Perez-Ostalé, E. and Garcia-Pérez, M. (2016). Study of the transit and attenuation of pollutants in a water reservoir receiving acid mine drainage in the Iberian Pyrite Belt (SW Spain). Water Science and Technology: Water Supply 16(1): 128–134.

Scheiber, L., Ayora, C., Vázquez-Suñé, E. and Soler, A. (2018). Groundwater-Gossan interaction and the genesis of the secondary siderite rock at Las Cruces ore deposit (SW Spain). Ore Geology Reviews 102: 967–980.

Sethi, R. and Di Molfetta, A. (2019). Mechanisms of contaminant transport in aquifers. In: Groundwater Engineering. Springer, Cham, pp. 193–217.

Singh, P.K., Roy, M.P., Paswan, R.K., Sarim, M.D., Kumar, S. and Jha, R.R. (2016). Rock fragmentation control in opencast blasting. Journal of Rock Mechanics and Geotechnical Engineering 8(2): 225–237.

Skousen, J.G., Ziemkiewicz, P.F. and McDonald, L.M. (2019). Acid mine drainage formation, control and treatment: approaches and strategies. Extractive Industries and Society 6(1): 241–249.

Soyol-Erdene, T.O., Valente, T., Grande, J.A. and de la Torre, M.L. (2018). Mineralogical controls on mobility of rare earth elements in acid mine drainage environments. Chemosphere 205: 317–327.

Squadrone, S., Brizio, P., Stella, C., Mantia, M., Battuello, M., Nurra, N. and Mogliotti, P. (2019). Rare earth elements in marine and terrestrial matrices of Northwestern Italy: implications for food safety and human health. Science of the Total Environment 660: 1383–1391.

Strahm, B., Sweigard, R., Burger, J., Graves, D., Zipper, C., Barton, C. and Angel, P. (2017). Loosening compacted soils on mined lands. In: Adams, Mary Beth (Editor) The Forestry Reclamation Approach: Guide to Successful Reforestation of Mined Lands. Gen. Tech. Rep. NRS-169. Newtown Square, PA: US Department of Agriculture, Forest Service, Northern Research Station: 5-1–5-6: 1–6. Available at https://www.fs.usda.gov/treesearch/pubs/54354 [Accessed 18 May 2020].

Szczepiński, J. (2019). The significance of groundwater flow modeling study for simulation of opencast mine dewatering, flooding, and the environmental impact. Water 11(4): 848. Available at Doi:10.3390/w11040848.

Tabelin, C., Sasaki, A., Igarashi, T., Tomiyama, S., Villacorte-Tabelin, M., Ito, M. and Hiroyoshi, N. (2019). Prediction of acid mine drainage formation and zinc migration in the tailings dam of a closed mine, and possible countermeasures. In: MATEC Web of Conferences. EDP Sciences. 268: 06003. Available at https://www.matecconferences.org/articles/matecconf/abs/2019/17/matecconf_rsce18_06003/matecconf_rsce18_06003.html [Accessed 5 May 2020].

Taherdangkoo, R., Tatomir, A., Taylor, R. and Sauter, M. (2017). Numerical investigations of upward migration of fracking fluid along a fault zone during and after stimulation. Energy Procedia 125: 126–135.

Toner, J.D., Catling, D.C. and Sletten, R.S. (2017). The geochemistry of Don Juan Pond: evidence for a deep groundwater flow system in Wright Valley, Antarctica. Earth and Planetary Science Letters 474: 190–197.

Wen, J., Tang, C., Cao, Y., Li, X. and Chen, Q. (2018). Hydrochemical evolution of groundwater in a riparian zone affected by acid mine drainage (AMD), South China: the role of river–groundwater interactions and groundwater residence time. Environmental Earth Sciences 77(24): 794. Available at Doi: 10.1007/s12665-018-7977-2.

Weyer, V.D., De Waal, A., Lechner, A.M., Unger, C. J., O'Connor, T.G., Baumgartl, T. and Truter, W.F. (2019). Quantifying rehabilitation risks for surface-strip coal mines using a soil compaction Bayesian network in South Africa and Australia: to demonstrate the R2AIN Framework. Integrated Environmental Assessment and Management 15(2): 190–208.

Williams, T.M. and Smith, B. (2000). Hydrochemical characterization of acute acid mine drainage at Iron Duke Mine, Mazowe, Zimbabwe. Environmental Geology 39: 272–278.

Winn, K. (2020). Engineering geology and hydrogeology aspects of sedimentary Jurong formation in Singapore: implication on safe excavation of underground storage caverns. Geotechnical and Geological Engineering 38: 3535–3558.

Wolkersdorfer, C. (2006). Acid mine drainage tracer tests. 6th Proceedings of ICARD, 26–30. Available at http://mwen.info/docs/imwa_2006/2490-Wolkersdorfer-DE.pdf [Accessed 6 May 2020].

Wright, I.A., Paciuszkiewicz, K. and Belmer, N. (2018). Increased water pollution after closure of Australia's longest operating underground coal mine: a 13-month study of mine drainage, water chemistry and river ecology. Water, Air, & Soil Pollution 229(3): 55. Available at https://doi.org/10.1007/s11270-018-3718-0.

Wu, F., Xu, E., Wei, X., Liu, H. and Ding, Q. (2019). Laws of multi-fracture coupling initiation during blasting induced hydraulic fracturing. Natural Gas Industry B 6(3): 293–301.

Xia, D., Ye, H., Xie, Y., Yang, C., Chen, M., Dang, Z. and Lu, G. (2017). Isotope geochemistry, hydrochemistry, and mineralogy of a river affected by acid mine drainage in a mining area, South China. RSC Advances 7(68): 43310–43318.

Zhang, Y., Cao, S., Lan, L., Gao, R. and Yan, H. (2017). Analysis of development pattern of a water-flowing fissure zone in shortwall block mining. Energies 10(5): 734. Available at Doi:10.3390/en10050734.

Zhang, X.L., Jia, R. S., Lu, X.M., Peng, Y.J. and Zhao, W.D. (2018). Identification of blasting vibration and coal-rock fracturing microseismic signals. Applied Geophysics 15(2): 280–289.

2

Prediction of Acid Mine Drainage Formation

James Manchisi and Sehliselo Ndlovu

CONTENTS

2.1 Introduction

Many countries have now enacted national legislation, signed international conventions and regional agreements and protocols that recognise the use of environmental impact assessment (EIA) tool as a key legal instrument to manage environmental impacts of development projects and policies (Maest et al., 2005; Morgan, 2012). The International Association for Impact assessment (IAIA, 1999) defines EIA as "the process of identifying, predicting, evaluating, and mitigating the biophysical, social, and other relevant effects of development proposals prior to major decisions being taken and commitments made". The specific forms of impact assessments may include environmental, social, health, sustainability, regulatory, human rights, cultural and climate change (Morgan, 2012). Thus, EIA process is crucial for identifying and predicting the potential impacts of projects such as mining on the biophysical and social environments. In addition, the EIA is used as an environmental management tool to develop environmental management plans (EMPs) as measures to mitigate impacts. The basic EIA steps include screening, scoping, impact prediction and evaluation, mitigation and follow-up studies for implemented projects to provide feedback (Noble, 2011; Castilla-Gomez and Herrera-Herbert, 2015). The potential for acid mine drainage (AMD) to form at mine sites is one of the key questions to be answered in an EIA process. A detailed discussion of the activities for each step in the EIA process is given by Noble (2015).

Despite many socio-economic benefits, the legacy of mining is mostly associated with many environmental impacts such as land degradation, solid waste disposal challenges, biodiversity loss, AMD, air and water pollution. This chapter focuses on the problem of AMD at mine sites and gives a critical review of its management through prediction. It is evident that mining, quarrying, excavation and mineral beneficiation activities to recover mineral-based products such as base metals, uranium, precious metals, coal and industrial minerals often expose sulphidic materials to the outside environment with the potential to form AMD (Rae et al., 2007). As part of EIA process, it is now a regulatory requirement for mine owners in most countries to predict the AMD potential for all types of mineral wastes and prepare management plans for future mitigation and to protect the environment at all stages of the mine life cycle (Maest et al., 2005; Price, 2009; Elaw, 2010).

The issue of AMD at mine sites has been a challenging environmental problem for many years. It is known to have contaminated soils, polluted water, affected ecosystems and destroyed biological resources (Banks et al., 1997; Bell et al., 2001; Gordon, 1994; Gray, 1997). Rae et al. (2007) classify AMD discharge as either acidic, neutral or saline (alkaline). Although the environmental impacts of acidic drainage are well known, the neutral and alkaline drainages may also be harmful and difficult to manage if they are metalliferous in nature, that is, if they contain elevated levels of dissolved metal ions. However, alkaline drainage is rare relative to acidic or metalliferous drainage (Rae et al., 2007; Price, 2009).

In the past, mining firms managed AMD problems by developing remediation plans only after its occurrence and subsequent impacts. The consequences of this reactive approach were extensive damage to natural resources and huge remediation costs. It was later realised that the practice was uneconomic and environmentally unacceptable, and so, the mining sector changed and started to focus more on proactive mitigation measures based on accurate prediction of drainage chemistry to prevent AMD formation and its impact (USEPA, 1994; Price, 2009). Furthermore, the interests for early assessment and detection for potential of mineral waste materials to generate acid using various prediction methods arose out of concerns of lag times associated with the occurrence of AMD and the costly lessons learnt from its long-term (perpetual) care over the years (USEPA, 1994; Price, 2009). In addition, the prediction of drainage chemistry was driven by the legal requirements in many countries such as the United States, Australia, Canada, etc., which stipulated that mining permits would only be granted if applicants included plans with practical methods to avoid AMD (USEPA, 1994; Price, 2009). This requirement led to increased efforts to develop techniques to predict the drainage chemistry and understand methods to avoid pollution (Lawrence et al., 1989; Ridge and Seif, 1998).

Lawrence et al. (1989) noted that accurate prediction of AMD formation would provide the basis for the development of effective waste management plans to prevent acid generation and/or allow for use of innovative methods to contain and treat unavoidable AMD. Other key benefits of predicting AMD formation include a reduction in environmental damages and remediation costs, and timely planning for mitigation facilities. Prediction also ensures a sustainable development of mineral resources by preventing impacts of AMD on water resources, aquatic and terrestrial life, vegetation, human life and livelihoods. The proposed proactive approach based on prediction agrees with sustainability principles and solid waste management hierarchy, in which the sustainability preference is to prevent AMD formation rather than treatment (Rae et al., 2007; El-Hagger, 2007; Lottermoser, 2011).

The aim of this chapter is to review the current understanding and recent developments in the AMD prediction methods, analytical procedures and any limitations that may arise during the characterisation of drainage chemistry at mine sites. In addition, the chapter will assess the sustainable AMD management options that eliminate risks and/or prevent AMD formation through prediction tools.

2.2 Developments in Acid Mine Drainage Prediction

The earliest observations to understand AMD formation and its impacts date back to the 16th century when Diego Delgado recognised that the oxidation of pyrite at Rio Tinto mine generated acidic drainage that poisoned fish, and that the use of simple tests to indicate the formation of AMD provided evidence for the ecotoxicity of AMD (Lottermoser, 2015). Since then, the chemistry of acid generation (see Chapter 3) has been studied extensively and is fairly well understood. Following the introduction of laboratory tests to predict AMD, a lot of progress has been made in developing and improving the testing tools to identify and accurately predict AMD formation. Historically, the prediction approach has evolved from merely making qualitative observations and correlations between mineralogical characteristics and leachate composition, to the introduction of geochemical static and kinetic tests. These geochemical tests have formed the basis of the AMD prediction procedures still being utilised today (Lottermoser, 2015).

The Mine Environment Neutral Drainage (MEND) (2000) defined prediction as a set of integrated approaches that are followed in order to assess, in advance, the geochemical behaviour of mineral wastes at all phases of the mine life cycle. The geochemical reactions, described in detail in Chapter 3, that lead to AMD formation depend primarily on the mineralogical composition of mineral wastes and the presence of water and oxygen. Various prediction methods aim to determine whether mineral wastes at mine sites will react and oxidise to generate acid and to predict the drainage quality. However, due to high variations in the mineralogy of different waste materials from different mine sites, it may be a difficult and costly task to accurately predict the potential for AMD formation, and sometimes the reliability of data may be questioned (USEPA, 1994). Price, (2009) proposes a site-specific approach in predicting the drainage chemistry and suggests that all waste materials should be considered in the AMD prediction process.

The purpose of a prediction study is to evaluate the characteristics of the present and future drainage chemistry, and to determine the potential environmental impacts and the required remediation measures to address the problems (Price, 2009). Depending on the stage of a mining project in its life cycle, the specific prediction objectives may be to address the quality problems of existing drainage chemistry or to determine if any of the mineral waste materials will potentially form acidic drainage and estimate the timing of acid generation (MEND, 2000).

2.3 Overview of Acid Mine Drainage Prediction Methods

The potential of mineral wastes to form acid, mobilise pollutants and affect natural resources (land, soil, water, ecosystems, biodiversity and others) is often evaluated by field measurements, various laboratory tests and predictive modelling approaches. These techniques constitute what is sometimes known as the "Wheel Approach" for predicting the future drainage chemistry and potential environmental impacts as summarised in Figure 2.1 (Morin and Hutt, 1998; Maest et al., 2005; Elaw, 2010).

Price (2009) proposed three main steps to follow in order to identify and predict AMD formation, namely (1) determination of the environmental baseline conditions at the mine site (water quality objectives, environmental values, etc.), (2) measurement of the existing drainage chemistry and (3) determination of the potential future drainage chemistry. The types of analyses and tests that are conducted under each of these prediction steps have been

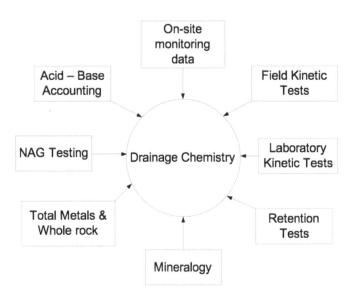

FIGURE 2.1
The Wheel approach to predict acid mine drainage formation. (From Morin and Hutt, 1998.)

reviewed in many publications (Sobek et al., 1978; Coastech Research, Inc., 1989; Skousen et al., 1990; USEPA, 1994; Lawrence and Wang, 1996; White et al., 1999; Morin and Hutt, 1997; 1998; 2001; Lapakko, 2002; Rae et al., 2007; Price, 2009; Lottermoser, 2015; Jones et al., 2016).

The Wheel approach given in Figure 2.1 is the accepted standard practice in AMD prediction and metal leaching studies whereby several geochemical static and kinetic tests are carried out and the results compared (Morin and Hutt, 1998). The most common static tests are acid-base accounting (ABA), mineralogy studies, total metal content and whole-rock analysis, retention (and reaction product) tests and net acid generation (NAG) tests. The main kinetic tests are laboratory-based kinetic tests (humidity cells, leach columns and the international kinetic database [IKD]) and field-based kinetic tests (large-scale bins or cribs, i.e., test pads or piles), mine wall stations and routine site monitoring (Morin and Hutt, 2001; 1997; 1998). The IKD contains pre-test characterisation and overall test results that help to compare kinetic data with that from other mine sites (Morin and Hutt, 2001). According to Morin and Hutt (2001) and Price (2009), there is no single test that can provide a basis for a reliable AMD prediction alone, but rather prediction data should be derived from various tests and sources.

2.3.1 Water Quality Survey

The best method to assess the present situation of the water quality and identify AMD presence is by measuring the chemistry of existing drainage through field observations, measurements and monitoring of surface seepage and groundwater quality (i.e., field water quality survey) discharged from all types of mineral wastes such as overburden material, ore stockpiles, heap and dump leach residue materials, tailings and waste rocks impoundments and bedrocks (walls of open pit mines, underground mine workings, etc.) (Rae et al., 2007; Price, 2009). Apart from water quality survey, the laboratory prediction tests on all mine waste materials are necessary to confirm AMD formation (ZCCM-IH, 2005). The poor effluent quality in the form of AMD may contain sulphuric acid, toxic metal ions and sulphates that present a risk to water resources, vegetation, aquatic life and human livelihoods.

One of the simplest and most useful field parameters to measure is effluent pH, which is an indicator of free acid. The pH value also gives an indication of whether (or not) sulphide oxidation has exhausted the acid neutralising capacity of the mineral wastes. Electrical conductivity (EC) is another useful parameter that indicates leachable soluble salts (salinity) (Jones et al., 2016). The typical chemical characteristics of AMD may be low pH (1.5–4.0), high concentration of metal ions ($Fe^{2+/3+}$, Cu^{2+}, Pb^{2+}, Zn^{2+}, Al^{3+}, Mn^{2+}, Cd^{2+}, $As^{2+/3+}$), high sulphate concentration (500–10,000 mg/L) and low turbidity (total suspended solids). Some of the key visual indicators of AMD presence are red coloured water, orange-brown precipitates (of iron oxides), death of fish and corrosion of steel/concrete structures (Price, 2009).

The pH, EC, oxidation/reduction (redox) potential and total dissolved solids (TDS) of AMD samples are easily measured by portable pH meters. The liquid samples are also commonly analysed for the concentration levels of metal ions and sulphates by techniques such as atomic absorption spectrophotometry (AAS) and inductively coupled plasma atomic emission spectroscopy (ICP-AES). The measured values may be compared with water and effluent discharge standards to determine whether the drainage meets the water quality objectives. All these measurements are carried out so as to characterise the status of any exposed sulphidic material.

2.3.2 Geochemical Static Tests

Many types of static tests have been developed to measure the geochemical properties of mineral samples at mine sites in order to predict the potential future drainage chemistry. They mainly aim to quantify the maximum capacities of mine wastes to either produce acid or consume acid by measuring the theoretical balance between acid producing and neutralising components of a waste material. The static tests are simple, low-cost and rapid laboratory test procedures that are conducted in a matter of hours or days, but at one point in time to evaluate the net acid generation potential (AP) of samples. These tests form the basis of predicting AMD formation (Morin and Hutt, 2001; Price, 2009; Lottermoser, 2015). According to Price (2009), static tests data is useful for identifying materials with little AMD potential and can also help to develop criteria to classify and segregate mineral wastes for separate disposal and mitigation procedures.

2.3.2.1 Acid-Base Accounting Method

The commonest static tests fall under the ABA methods and its various modifications. The characteristic tests in the ABA procedure include the determination of sulphur species (sulphide, sulphate and total sulphur), AP, acid neutralisation potential (NP) and the net acid NP. The other key analytical tests that must accompany ABA methods are the mineralogy, elemental composition, soluble components, paste pH and the related direct NAG test procedure (Sobek et al., 1978; USEPA, 1994; White et al., 1999; Morin and Hutt, 2001; Rae et al., 2007; Price, 2009; Lottermoser, 2015; Jones et al., 2016). A combined data interpretation from static tests, kinetic tests, drainage chemistry and baseline water quality is necessary to fully understand and predict the potential of AMD formation in the future. The objectives, procedures and limitations of various static tests are discussed in the sections that follow.

The earliest approaches in the development of the ABA method in the 1960s focussed on assessments of the potential of coal mine wastes (i.e., overburden and mine spoil) for revegetation based on rock types, acidity and alkalinity. This included an assessment of the need for lime and suitability for plant species. By the 1970s, the AP of rocks from coal seams was reported and a system to balance acid and alkaline producing potential of rocks was developed. Later, the important role of acid neutralising minerals was recognised and quantified, and this became known as the NP (West Virginia University, 1971; Grube et al., 1973; Smith et al., 1974; Smith et al., 1976; Skousen et al., 1990; Kania, 1998; Perry, 1998). Based on these early publications, Sobek et al. (1978) formally presented the detailed laboratory procedures for performing the ABA method, which is now termed the Sobek method and is frequently cited and considered as the source document and basis of AMD prediction methods.

The ABA method, also known as the EPA 600 ABA, is the best known and most widely used method for predicting the potential future drainage chemistry from mine wastes (Ferguson and Erickson, 1988; Morin and Hutt, 1997; 2001). The theoretical principle in the ABA method is that AMD formation in the future can be predicted by a quantitative determination of the total amount of acidity and alkalinity a particular mine waste material can potentially produce. The ABA method consists of geochemical analyses and calculations to evaluate the potential for mine wastes to produce net acidic drainage if exposed to air and water. It is, thus, a quantitative estimate of the balance between the acid generated from the oxidation of sulphide minerals and the acid consumption by the carbonate minerals. The capacity of a mineral sample to generate net acidity is based on measurement of sulphur species, the calculation of AP and the determination of the acid NP.

The difference between AP and NP values indicates the net acid production potential (NAPP) of the waste material (Morin and Hutt, 1997; 2001; Kania, 1998; Lottermoser, 2015). One of the inherent limitations of the ABA methods as discussed by Lottermoser (2015) is that they do not predict the time when acid generation would occur. In addition, the tests do not consider the role of microbial activity in catalysing acid formation. Furthermore, the ABA method provides no information about effluent quality and the individual mineral reaction rates for acid generation and neuralisation. The large differences between laboratory tests and actual mine site conditions have also been criticised (Morin and Hutt, 2001; Lottermoser, 2015; Parbhakar-Fox and Lottermoser, 2015). Thus, kinetic tests have been developed to compliment static tests data by calculating the rates of acid formation in order to fully assess the potential of future water quality. The details of the main laboratory procedures in the ABA method and its modifications are discussed in the sections that follow.

2.3.2.2 Sulphur Analysis and Acid Generation Potential Calculation

The acid generation potential or simply AP, also referred to as the maximum potential acidity (MPA), is the total acid that mine wastes can produce. Thus, AP is determined by calculating the theoretical amount of acid that can be produced if the total amount of sulphur in the mineral sample is oxidised to sulphuric acid (Lawrence and Wang, 1996). It is used to estimate the maximum acid production potential based on the concentration of sulphur and/or its forms (i.e., total sulphur (S), sulphide (S^{2-} / S_2^{2-}), sulphate (SO_4^{2-}) and organic sulphur). Thus, sulphur is the primary source of acid through oxidation (Perry, 1998; Price, 2009). In the original Sobek method, the AP is stoichiometrically calculated using the percent (%) total sulphur (S) content that occurs only as pyrite (FeS_2) as given in Equation 2.1. The AP is commonly expressed in units of kilogram $CaCO_3$ equivalent per tonne of the sample (kg/t) (or kg H_2SO_4 per tonne). Alternatively, a conversion factor of 30.6 is used in Equation 2.1 to estimate MPA values in kg H_2SO_4/t (Weber et al., 2005).

$$AP = \frac{\text{Sulphur content } (\%) \times 1000 \text{ kg}}{100} \times \frac{\text{Molecular weight of } CaCO_3}{\text{Atomic weight of sulphur}} \quad (2.1)$$

Thus, AP (kg $CaCO_3$/t) = % total sulphur × 31.25

The conversion factor (31.25) is derived from the stoichiometric consideration of the standard pyrite oxidation reaction (2.2) and the acid neutralisation reaction (2.3) for calcite (Morin and Hutt, 1997; 2001; Price, 2009). It is assumed, in Equation 2.2, that all sulphur occurs only as a sulphide (S_2^{2-}) and also that the sulphide occurs only in pyrite (FeS_2) with oxygen and water being the only oxidants. In addition, it is assumed that pyrite is completely oxidised to sulphate (SO_4^{2-}) and ferric hydroxide, and H^+ ions are neutralised by $CaCO_3$ (Morin and Hutt, 1997; 2001).

$$FeS_2 + (15/4)O_2 + (7/2)H_2O \rightarrow Fe(OH)_3(s) + 2SO_4^{2-} + 4H^+ \qquad (2.2)$$

$$2CaCO_3 + 4H^+ \rightarrow 2Ca^{2+} + 2H_2O + 2CO_2\left(pH < \sim 6.3\right) \qquad (2.3)$$

Overall reaction is given in Equation 2.4 as follows:

$$FeS_2 + (15/4)O_2 + (7/2)H_2O + 2CaCO_3 \rightarrow Fe(OH)_3(s) + 2SO_4^{2-}$$
$$+ 2Ca^{2+} + 2H_2O + 2CO_2 \qquad (2.4)$$

Therefore, from Equation 2.4, one mole of oxidised pyrite requires two moles of calcite to neutralise the acid formed. By mass basis, 1 g of sulphur present requires 3.125 g of calcite. Using the units of "parts per thousand" (ppt) of mine waste, for each 10 ppt sulphur (i.e., 1% sulphur) present, 31.25 ppt calcite is required to neutralise the acid (Morin and Hutt, 2001; Perry, 1998). Similarly, from Equation 2.2, for every one (1) mole of pyrite that is oxidised two moles of sulphuric acid are produced. Therefore, the MPA value of a mine waste material containing 1% S (as pyrite) is 30.6 kg of H_2SO_4 per tonne (Smart et al., 2002; Weber et al., 2005). Hence, the MPA value may be calculated from Equation 2.1 using a conversion factor of 30.6. Although different conversion factors for sulphur to AP exist if pyrite is oxidised by other oxidants, for example, 15.63 using Fe^{3+} and 125.0 using Mn, the standard and more accurate conversion factor is 31.25 (Morin and Hutt, 2001).

However, the use of total sulphur (mainly a sum of sulphate and sulphide species) to estimate AP in mine wastes is considered as a conservative approach because not all forms of sulphur produce acid (Morin and Hutt, 2001). For example, the sulphate form of sulphur or simply sulphate-sulphur in gypsum, anhydrite and barite minerals does not generate acid, but sulphate in jarosite, alunite and melanterite is acid generating. The sulphide form of sulphur or sulphide-sulphur that generate acid occurs mainly in iron sulphide minerals (pyrite, pyrrhotite, arsenopyrite) and chalcopyrite, but chalcocite and covellite yield less acid than pyrite while sphalerite and galena may not yield acid at all (refer to Chapter 3 as well). Furthermore, organic sulphur (in coal mines) does not produce net acidity (Price, 2009; Jones et al., 2016). Thus, the use of total sulphur may overestimate the AP in a sample. For these reasons, the modified ABA (or Sobek) method uses sulphide-sulphur rather than total sulphur in Equation 2.1 to estimate AP (Coastech Research, Inc., 1989).

The accurate determination of sulphur species is a crucial step in the prediction of AP. A variety of analytical methods exist to determine the concentration of sulphur species in mine wastes, and these include volatilisation (roasting/pyrolysis), wet chemical extraction, mineralogical analysis and solid phase elemental analysis (Price, 2009). The standard analytical procedure for total sulphur analysis involves volatilisation of the sample by roasting (or pyrolysis) at 1500–1700°C in a Leco high temperature induction furnace followed by sulphur dioxide analysis in the gas phase. In the American Society of Testing and Materials (ASTM) method (E-1915-97) as presented in ASTM (2000) and reviewed by Lapakko (2002), the mineral sample is ignited at 1500–1700°C to convert all sulphur species to sulphur dioxide gas which is then analysed by absorption spectrometric techniques. To measure the sulphide-sulphur content, the sample is heated at a relatively lower temperature of 550°C in a muffle furnace. Here, it is assumed that only sulphides are converted to sulphur dioxide and not sulphate-sulphur. The sulphate content is the difference between the total sulphur in the original sample material, measured at 1500–1700°C, and total sulphur (also measured at 1500–1700°C) in the residue material that is left after pyrolysis that is carried out at lower temperature of 550°C as given in the ASTM method E-1915-99 (ASTM, 2000; Lapakko, 2002; Price, 2009).

The response to pyrolysis treatment differs for different minerals. The sulphide in iron sulphides (pyrite (FeS_2), marcasite (FeS_2), arsenopyrite (FeAsS) and pyrrhotite ($Fe_{(1-x)}S$)) may be volatilised completely. However, the loss of sulphide from copper sulphides (bornite (Cu_5FeS_4), chalcopyrite ($CuFeS_2$)), pentlandite (($Fe,Ni)_9S_8$), galena (PbS) and sphalerite (ZnS) is reported to be minor, which then underestimates the total sulphide content. In addition, hydrated sulphate minerals (gypsum ($CaSO_4 \cdot 2H_2O$), jarosite ($KFe_3(SO_4)_2(OH)_6$)) may decompose partially, resulting in overestimation of the sulphide species (Bucknam, 1999; Li et al., 2007). Therefore, mineralogical analysis must always be taken into account when selecting a method for determining the sulphur species (Lapakko, 2002). The importance of mineralogy in AMD prediction is discussed further in Subsection 2.3.2.7.

The estimation of sulphur species can also be carried out by various wet chemical extraction procedures where mine wastes are digested either by water to measure soluble sulphates ($FeSO_4 \cdot 7H_2O$, $MgSO_4 \cdot 7H_2O$ and $CaSO_4 \cdot 2H_2O$) (Li et al., 2007), or by hydrochloric acid to remove sulphate-sulphur (Sobek et al., 1978; Tuttle et al., 2003; Ahern et al., 2004; Li et al., 2007), or by sodium carbonate to remove less soluble sulphate minerals (Bucknam, 1999) and by nitric acid to measure sulphide-sulphur (Sobek et al., 1978). The detailed laboratory procedures can be found in various relevant references. For example, in the EPA-600 method, a mine waste sample is milled and leached with 40% hydrochloric acid to determine sulphate-sulphur (acid soluble). The leach residue material is then analysed for residue total-sulphur by the high temperature pyrolysis technique in a Leco induction furnace. The sulphur lost or dissolved by leaching, that is, acid soluble sulphate-sulphur, is taken as the difference between total sulphur in the sample before and after extraction. The sulphate-sulphur is then subtracted from the total-sulphur to determine the sulphide-sulphur content, which is used to calculate the AP of the material. This method is not suitable if acid insoluble sulphate minerals are present since AP may be underestimated (Price, 2009).

2.3.2.3 Acid Neutralisation Potential

The acid NP or acid neutralisation capacity (ANC) method measures the capacity of mine waste materials (tailings, waste rocks) to neutralise the acid that is produced. The alternative to the ANC test is the costly acid buffering characteristic curve (ABCC) test that also measures the ANC of waste materials. The main sources of acid neutralisation are carbonate minerals (calcite ($CaCO_3$), dolomite ($Ca,Mg(CO_3)_2$), magnesite ($MgCO_3$)) and some reactive silicate minerals. The most common analytical methods used to measure NP are based on the Sobek method and its modified forms that use either bulk acid neutralisation or carbonate acid neutralisation techniques (Lawrence and Wang, 1996; Price, 2009). In the Sobek or US EPA-600 method (Sobek et al., 1978), NP is determined experimentally by adding hydrochloric acid to a finely ground sample and digesting under boiling conditions. The volume and strength of hydrochloric acid added is determined by the fizz test. The strength of the fizz resulting from the addition of drops of acid to the sample is an indicator of the amount of reactive carbonate minerals present. For example, for "no fizz", add 20 mL 0.1 N; "slight fizz", add 40 mL 0.1 N; "moderate fizz", add 40 mL 0.5 N and for a "strong fizz", 80 mL 0.5 N of hydrochloric acid is added (Price, 2009). After the sample has been cooled, the residue acid in the slurry is determined by titration with sodium hydroxide to pH 7.0 so that the amount of acid consumed by the waste material can be calculated (in kg $CaCO_3$/t). The NP value is then compared with AP value to calculate the NAPP of a sample. Although the basic laboratory procedure is the same, the NP method based on Sobek et al (1978) has been modified by Coastech Research, Inc. (1989) and Lawrence and Wang (1996) mainly to provide longer acid digestion time for samples. Furthermore, the pH end point during titration is different (pH 8.3 for the modified NP method). All these methods aim to ensure that sufficient acid is added to dissolve all carbonates and reactive silicates (Price, 2009).

According to Price (2009), the Sobek method is fast (3–4 h) and widely used to determine NP, while the modified NP method by Lawrence and Wang (1996) require over 25 hours which may be costly, time consuming and practically difficult to implement. The errors in these methods may come from misinterpretation of fizz test results, leading to incorrect determination of the amount of acid to be added, and any strong acid employed may dissolve minerals that do not contribute to acid neutralisation.

Other less common bulk NP methods include the British Columbia (BC) Research Initial test (Duncan and Bruynesteyn, 1979) and the Lapakko test (1994) in which NP is determined by the amount of acid required to attain pH values of 3.5 and 6.0, respectively. It must be noted that different methods can produce different NP values for the same sample due to differences in particle size, digestion conditions (acid used, pH, temperature and duration) and endpoint pH of titration (Lapakko, 1992; Lapakko, 1994).

Another technique that can be used to estimate the acid NP is called the carbonate neutralisation potential (CO_3-NP). This is the ANC that a waste material possesses if all the carbonates reacted like calcite. The carbonate concentration is calculated from total carbon (i.e., carbon present as carbonate, organic carbon and graphite), inorganic carbon or carbon dioxide as shown in Equation 2.5:

$$CO_3 - NP = \% \text{ C} \times 83.4 \left(Kg \text{ CaCO}_3 / t \right) \qquad (2.5)$$

where C is percentage of total carbon.

If all carbon occurs as carbonate, the total carbon is measured by Leco furnace facilities. For high concentration of non-carbonate carbon (coal, organic matter, graphite), the acid soluble analysis of carbonate-carbon is used whereby carbonate minerals are converted to carbon dioxide by hydrochloric acid (Price, 2009).

2.3.2.4 Net Acid Production Potential

The NAPP, as used in Australia and Asian countries, is calculated as the difference between the MPA and ANC values (Equation 2.6a) (Price, 2009; Jones et al., 2016). The NAPP value gives an acid-base account of the mine waste materials, which is then interpreted to predict AMD formation. When the NAPP value is negative (i.e., MPA < ANC), then the mine waste has sufficient neutralisation capacity to prevent acid formation. However, the material may generate acid if the NAPP value is positive (i.e., MPA > ANC). Thus, NAPP data may then be used to classify mine waste materials as either potentially acid forming (PAF) or non-acid forming (NAF). Figure 2.2 shows that the risk of acidic drainage is high if NAPP values are located in the risk domain. However, a related term that is commonly used in North America is the net neutralisation potential (NNP) which is the difference between NP and AP values (Equation 2.6b). Similar to the interpretation of NAPP, a mine waste may generate acid if NNP value is found to be negative (i.e., NP < AP) and may neutralise any acid formed if NNP value is positive (i.e., NP > AP) (Coastech Research, Inc., 1989; USEPA, 1994; White et al., 1999; Price, 2009).

$$NAPP = MPA - ANC \left(Kg \text{ H}_2SO_4 / t \right) \qquad (2.6a)$$

$$NNP = NP - AP \left(Kg \text{ CaCO}_3 / t \right) \qquad (2.6b)$$

Another ABA classification criterion is based on the net potential ratio (NPR), that is, ANC/MPA or NP/AP ratio that is used to assess the acid producing

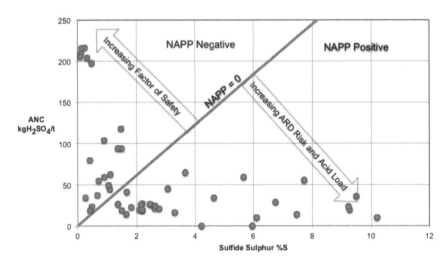

FIGURE 2.2
Acid-base accounting data interpretation plot. (From Jones et al., 2016.)

or acid consuming (ACM) potential of mine wastes (Morin and Hutt, 1994; Price, 2009; Jones et al., 2016). Lawrence and Wang (1996) stated that the use of NPR to interpret ABA test data is favoured because it clarifies the relative amounts of acid producing and ACM phases. A range of NPR values has been recommended in the literature to differentiate the potentially acid producing materials from ACM ones. Jones et al. (2016) cited NPR values of 1.5 to 3, but recommended safe values of ≥2 to ensure that the material will remain near-neutral in pH and that it will not result in AMD formation. According to the universal criteria given in Table 2.1 by Morin and Hutt (1997), an NPR < 1 or NNP < 0 t $CaCO_3$/1000 t indicates that the material has insufficient neutralisation capacity and, therefore, should eventually result in net acidic outcome. However, the material may remain near-neutral or alkaline indefinitely if NPR > 1 or NNP > 0 t $CaCO_3$/1000 t.

2.3.2.5 Paste pH

The analytical procedure for paste pH as presented by Sobek et al (1978) provides a simple method to assess whether a material will be acidic, neutral or alkaline at the time of analysis. It involves pulverising a sample (to d_p < 100 μm) and mixing it with distilled water at ~pH 5.3 to form a slurry from which paste pH is measured. As presented in Table 2.1, a paste pH of less than 5.0 indicates net acidity, pH values between 5.0 and 10.0 are regarded as near neutral and values above pH 10.0 indicate net alkalinity at the time of the test, but the future drainage pH is non-predictable in all cases.

TABLE 2.1

Universal ABA Criteria for Assessing and/or Predicting Drainage pH from Mine Wastes

	Criteria	Prediction / Current Condition
Paste pH		
I	Paste pH <5.0	Currently acidic; future non-predictable
II	5.0 ≤ /Paste pH ≤ 10.0	Currently near neutral; future non-predictable
III	Paste pH > 10.0	Currently alkaline; future non-predictable
NPR or NNP[a]		
A	NPR < 1.0 or NNP < 0.0 t CaCO$_3$/1000 t	Eventually acidic
B	1.0 ≤ NPR ≤ 2.0 or 0 ≤ NNP ≤ 20	Uncertain future
C	NPR > 2.0 or NNP > +20 t CaCO$_3$/1000 t	Indefinitely near neutral or alkaline

Source: Morin and Hutt, 2001.

[a] The static tests are based on sulphide-sulphur.

2.3.2.6 Net Acid Generation

According to Price (2009), the NAG test utilises hydrogen peroxide (H_2O_2) to rapidly oxidise sulphides in order to assess the capacity of mine waste materials to neutralise acid formed by sulphide oxidation, for example, pyrite (Equation 2.7). The NAG test is based on the original hydrogen peroxide method by Sobek et al. (1978). The acid formed may subsequently solubilise carbonate (or neutralising) minerals to give a net effect between acid forming and neutralising reactions which can then be measured directly in the NAG test. This implies that the test directly measures the net quantity of acid formed by a mine waste sample. The test is simple, rapid and relatively cheap, which helps to identify PAF waste materials at mine sites. The NAG test is often used in association with the NAPP values to classify the acid-generating potential of waste materials (Miller et al., 1994; Lapakko, 2002; Smart et al., 2002).

$$FeS_2 + 15/2H_2O_2 = Fe(OH)_3 + 4H_2O + 2SO_4^{2-} + 4H^+ \qquad (2.7)$$

Several types of the NAG test have been developed to accommodate the wide geochemical variability of mine waste materials. These include the single addition NAG test (low S_2^{2-}), sequential NAG test (high S_2^{2-}), partial ABA (for total S, NP and paste pH), kinetic NAG test and the ABCC (Smart et al., 2002; Price, 2009). The theory, detailed experimental procedures and limitations of these tests can be found in laboratory manuals by Smart et al. (2002)

and Price (2009). The NAG test procedures commonly used are the single addition NAG test and the sequential NAG test. The sequential NAG test is applicable if mine waste samples have high total sulphur or sulphide sulphur in order to measure the total acid-generating capacity, and also for samples with high ANC.

The NAG test is done by mixing hydrogen peroxide with finely milled mine waste samples and heating the mixture until effervescence stops followed by cooling and recording the pH (or NAG pH). The mixture is filtered and titrated with sodium hydroxide (NaOH) to pH 4.5 and 7.0. The first pH value of 4.5 accounts for acidity due to Fe, Al and most of the H^+ ion, and the pH between 4.5 and 7 indicates soluble metals (e.g., Cu and Zn). A pH after the sulphide oxidation reaction (NAG pH) of less than 4.5 indicates that the mine waste sample is PAF or potentially acid generating (PAG). However, if NAG pH is greater than 4.5, the sample may be considered as non-acid forming (NAF) or non-PAG. The amount of acid produced is determined by titration and expressed in the same units as NAPP (kg H_2SO_4/tonne) (Smart et al., 2002) as given by Equation 2.8. The classification criteria for mine waste material based on NAPP and NAG test data are shown in Table 2.2.

$$NAG = 49 \times V \times M / W \left(kg\ H_2SO_4 / tonne \right) \qquad (2.8)$$

where V = amount of NaOH (mL), M = concentration of NaOH (moles/L), W = weight of rock sample (g).

The data from the NAG test is useful to assess the risks of acid formation, sulphide reactivity and the quantity of acid that may form. This information helps to develop plans to implement mine waste management strategies to prevent AMD formation (Miller et al., 1994). However, one of the limitations of the NAG test is that hydrogen peroxide may decompose before it reacts with all sulphides and thus underestimate the total acid-generating

TABLE 2.2

Geochemical Classification Criteria

Primary Geochemical Material Type	NAPP (kg H_2SO_4/tonne)	NAG pH
Potentially acid forming (PAF)	>10	<4.5
Potentially acid forming – low capacity (PAF-LC)	0 to 10[a]	<4.5
Non-acid forming (NAF)	Negative	≥4.5
Acid consuming (ACM)	Less than –100	≥4.5
	Positive	≥4.5
Uncertain[b]	Negative	<4.5
	Positive	<4.5

Sources: Smart et al., 2002; Jones et al., 2016.

[a] Site-specific, 5–20 kg H_2SO_4/tonne.

[b] Further testing required.

potential. The use of pH 4.5 criteria also can imply that Non-PAG classification might include weakly acidic mine waste samples that can potentially solubilise toxic trace elements. In addition, the rapid oxidation of sulphides may not simulate expected field conditions (Price, 2009).

2.3.2.7 Mineralogical and Elemental Analyses

The prediction of AMD formation also depends on mineralogy as shown in the "Wheel approach" by Morin and Hutt (1998). Thus, mineralogical characterisation of mine waste materials is essential to understand the actual mineral phases that drive AMD formation. A mineralogical study should reveal mineral type, composition, abundance, particle sizes, liberation and grain size distribution of minerals. A combination of characterisation methods are often used such as optical mineralogy, petrographic, X-ray diffraction (XRD), scanning electron microscopes (SEM), micro-probes, quantitative evaluation of materials by scanning electron microscopy (QEMSCAN) and bulk elemental analyses (such as X-ray fluorescence (XRF), ICP-AES). The chemical analysis of elements is accomplished by whole rock or near-total solid phase analysis by hot acid digestion method and leachate analysis by XRF or ICP-AES techniques (Price, 2009). The aim of the characterisation is to identify the sulphide and carbonate minerals that are responsible for acid formation and neutralisation, respectively. However, silicates (e.g., anorthite, olivine and chlorite) also have long-term acid neutralising properties but may be ineffective due to slow reaction kinetics. The knowledge of mineralogy not only helps to indicate the mineral phases that may contribute to acid formation and NP in static tests, but also to establish the likely source of harmful elements in the leachates. The characterisation may also provide information about the concentration of elements of interest relative to background rocks/soils. The common acid-forming sulphides are pyrite (FeS_2), pyrrhotite (FeS), marcasite (FeS_2), chalcopyrite ($CuFeS_2$) and arsenopyrite ($FeAsS$). However, not all sulphides are acid-generating (see Chapter 3) during oxidation, for example, covellite, sphalerite, galena, chalcocite and bornite may not form acid when oxidised, but may release metals when exposed to acidic water (Morin and Hutt, 2001; Smart et al., 2002; Rae et al., 2007; Price, 2009).

2.3.3 Geochemical Kinetic Tests

According to Coastech Research, Inc. (1991), the kinetic prediction tests are follow-up geochemical test procedures that may be carried out on mine wastes if the results from static tests are either uncertain or predict potential for AMD formation. Therefore, the aim of kinetic tests is to understand whether, and/or when, AMD is likely to occur by predicting the long-term oxidation (or dissolution) rates of mine wastes (in months to years) (Coastech Research, Inc., 1991; Parbhakar-Fox and Lottermoser, 2015; Dold, 2017). The specific objectives (Coastech Research, Inc., 1991; Lapakko, 2003; Price, 2009; Lottermoser, 2015) may be to:

- Determine the rates of acid generation and neutralisation,
- Determine whether AMD is likely to occur and its time frame,
- Estimate the rate of metal leaching from mine wastes,
- Establish the chemical composition and leachate quality (drainage chemistry),
- Identify the main chemical weathering reactions, and
- Confirm results from static tests, hence, AMD generation potential.

A number of kinetic tests have been developed to complement static tests data in the prediction of AMD formation. These kinetic tests are categorised as either laboratory-based (e.g., humidity cells, leach columns, shake flasks, biokinetic tests, soxhlet extraction, BC research confirmation tests and the use of the international kinetic database) or field-based (e.g., large-scale field test pads (i.e., bins or cribs or piles), mine wall stations and routine site monitoring) (Coastech Research, Inc., 1991; Morin and Hutt, 2001). The basic concepts, principles and inherent limitations of the most popular kinetic tests with wide applications, i.e., humidity cell test, leach columns and biokinetic tests, are briefly discussed in the sections that follow.

2.3.3.1 Humidity Cell

Morin and Hutt (1997) reported that the humidity cell was the kinetic test of choice. This is now the recommended and most popular kinetic test. As discussed by Lapakko (2003), the original humidity cell testing method based on Sobek et al. (1978) has been revised by many researchers, with the notable method being the ASTM5744-96 (revised 2007–2018). The humidity cell test simulates the geochemical weathering of mine wastes (e.g., rocks or tailings) in order to estimate mainly the rate of acid generation and the quality of the leachate. The test results may indicate whether the sample will generate acidic, neutral or alkaline drainage and types of dissolved species (e.g., metals and sulphates) and rates of their release under controlled conditions. The test also helps to address any uncertainty in data from static tests.

The principle of the humidity cell test is that a suitable sample size of the waste material is crushed to a particular particle size and loaded into a column of suitable dimensions. The loaded column is then subjected to alternating cycles of dry air (3 days) and humid (or moist) air (3 days). Thereafter, water is percolated through the column to soak, rinse or leach the material (1 day) which is later discharged as a leachate. The quality or chemistry of the leachate is analysed for metal ions, sulphate, pH, conductivity, redox potential, acidity and alkalinity (Sobek et al., 1978; Lawrence et al., 1989; Morin and Hutt, 1997; 2001; Lapakko, 2003; Coastech Research, Inc., 2008, Lottermoser, 2010).

The ASTM D8187-18 method (ASTM, 2018) provides a detailed discussion on the interpretation of results from the humidity cell test. Basically, the results are analysed to assess the potential of the mineral wastes in forming AMD based on leachate composition, for example, a pH of 3–5 indicates that the sample is PAF. In addition, the acidity and alkalinity measurements refer to the balance between acid and alkaline generating minerals and a high redox potential (>500 mV) indicates a high sulphide oxidation rate by ferric ions. Plots of pH, acidity/alkalinity, metal ions with time help to establish the variation in leachate chemistry and deduce evidence for the potential to form AMD (Coastech Research, Inc., 2008).

The humidity cell test is simple and reliable. It predicts the leachate quality expected from waste materials, and thus, it is useful in helping to assess the initial plans that can prevent AMD formation. However, the humidity cell test is costly and requires a lot of time to be completed or to generate meaningful data (Coastech Research, Inc., 2008). Furthermore, the equilibrium weathering conditions that are expected in actual mine waste dumps might not be reproduced in the humidity cell test due to relatively limited reaction time that is employed under laboratory conditions (Parbhakar-Fox and Lottermoser, 2015).

2.3.3.2 Leach Columns

The leach column tests are designed to simulate leaching conditions in mine waste materials under partial or full saturation conditions and/or deprived of oxygen. These tests are often carried out on scale that is larger than that of humidity cell test (Lawrence and Day, 1997; MEND, 2000). The most commonly used column type is the free draining leach column test that utilises the Buchner funnel or drums depending on the size of the sample. The columns are loaded with crushed mine waste material and exposed to cycles of wetting and drying (with heat lamps) to oxidise the material and flush the reaction products. The loaded samples are wetted by addition of water (weekly) to the surface of the column and leachates collected at the column base (monthly). The testing period depends on the characteristics of the waste material and the objectives of the project. The leachate quality is analysed for metal solubility, oxidation kinetics, sulphide reactivity and the leaching characteristics of mine waste materials (Smart et al., 2002). However, columns can be operated such that they are sub-aerial type (well-drained waste materials) or sub-aqueous leach columns (waste materials are flooded) (Price, 2009).

Apart from determining the rates of acid formation, sulphide oxidation and depletion of NP, a column test may reliably predict the drainage quality and dissolved metals expected from mine wastes (Coastech Research, Inc., 2008). The column test set-up may be modified to accommodate different grain sizes and sample mass of waste materials and the frequency of leachate collection can also be varied (Parbhakar-Fox and Lottermoser, 2015). It is claimed by some studies that experimental conditions in the column test closely simulate actual field conditions than the humidity cell test (Bradham and Caruccio, 1990; Maest et al., 2005), but this view was refuted by Morin and Hutt (1997), who noted that laboratory conditions may not reproduce field conditions and that columns rarely attain steady-state conditions to give reliable estimates of oxidation rates. Furthermore, solution channelling in columns can be a challenge (MEND, 2000), and there is no provision to integrate mineralogical assessment during the test (Parbhakar-Fox and Lottermoser, 2015).

2.3.3.3 Biokinetic Tests

The effect of iron and sulphur oxidising bacteria (e.g., *Acidithiobacillus ferrooxidans*) on the rate of acid formation by mine waste materials has been studied by the shake flask test (USEPA, 1994), the biological acid producing potential (BAPP) test (Parbhakar-Fox and Lottermoser, 2015) which is based on the BC Research confirmation test (USEPA, 1994) and the relatively recent biokinetic test (Hesketh et al., 2010). In the BC Research confirmation test, a mine waste sample is milled to form an acidic slurry in shake flasks, which are then inoculated with *Acidithiobacillus ferrooxidans* and incubated for the bacterial adaptation followed by frequent monitoring of pH and other water quality parameters. A final pH value that is below 3.5 indicates microbial activity and the potential for the sample to form acid. For the pH above 3.5, the sample is considered NAF. One major criticism in this test is the use of a single bacterial species (Hesketh et al., 2010; Parbhakar-Fox and Lottermoser, 2015). The shake flask test is similar to the BC Research confirmation test except that consideration of the effect of bacteria on the kinetics of acid formation is optional.

The biokinetic test, as presented by the work of Hesketh et al (2010), is an improvement to the existing shake flask tests. The objective of this test is to determine the potential and likelihood of acid formation by mine wastes in the presence of bacteria and to determine the rates of acid formation and neutralisation reactions. In the test, the mine waste materials are milled to form an acidic slurry at pH 2 through sulphuric acid addition in flasks and inoculated with more than one bacterial species. The bacteria is a mixed culture of *Acidithiobacillus ferrooxidans*, *Leptospirillum ferriphilum*, *Acidithiobacillus caldus* and *Sulfobacillus benefaciens* that simulate typical microbial environments found at actual mine sites with AMD. The biokinetic test data (e.g., final pH, acid consumption) are able to validate the NAG pH, ANC and waste classification criteria under the static ABA methods. However, the timing is different between ACM and acid-generating reactions. The ANC of the waste material is depleted rapidly and controlled by the more reactive carbonate phases. The sulphides oxidise and form more acid over long term than that predicted by static tests. Thus, the biokinetic shake flask test is seen to be relatively simple, fast and of low-cost. However, there is still no practical use of this test to predict AMD formation (Parbhakar-Fox and Lottermoser, 2015).

2.4 Applications of Acid Mine Drainage Prediction Data

As discussed previously, all mine wastes should be characterised to predict the potential, rate and timing of AMD formation, and likely metals in leachates from sulphide oxidation. This forms the basis for developing waste management plans where AMD risk is evident. Therefore, the control and safe disposal of waste materials to minimise AMD formation may involve prediction for AMD potential, classification and confinement of any materials at risk of AMD. Consequently, any material classified as PAF from prediction tests have high AMD risk that must be properly identified. The material must be selectively disposed and encapsulated with benign materials, that is, those that are classified as NAF mine wastes with high ANC and low AMD risk (Jones et al., 2016). Thus, the integrated and sustainable AMD management through prediction dictates the use of source control technologies for AMD in order to prevent, minimise or reduce AMD formation through the following measures, some of which are discussed in Chapter 6:

- Identification of AMD risk and early integration of AMD issue in the life cycle of the mine,
- Blending of acid-generating wastes with acid-consuming materials,
- Encapsulation of PAF waste dumps by dry covers, caps, seals, etc.,
- Flooding/sealing underground mines (with water) after closure to stop oxygen ingress and thus, prevent further AMD generation,
- Underwater storage of tailings, and
- Application of cleaner production principles.

The awareness, adoption and use of AMD prediction tools by the mining industry is evident, however, post mine closure incidents of AMD continue to occur at many mine sites despite improved understanding of AMD prediction tests. Consequently, failure to predict AMD has resulted in environmental destruction, unplanned remediation costs and reputational damage to the industry. Lottermoser (2015) provides a comprehensive discussion on why some mine sites remain AMD liabilities after mine closure. For example, from a practical perspective, the mining industry puts too much emphasis on reactive or end-of-pipe approaches to manage environmental impacts through legal compliance, mine site rehabilitation and monitoring, etc., as opposed to the use of AMD prediction tests to prevent AMD formation at source which is regarded as a sustainable option and best practice. In addition, the industry often uses static tests to make major waste management plans rather than the more realistic, long-term, field-based kinetic tests that have not received much attention. Furthermore, mine waste dump facilities are susceptible to AMD partly because they tend to be constructed without strictly following the design and mine closure plans (e.g., capping, etc.).

Although AMD prediction tests are well established and have predicted AMD formation correctly in many case studies, they have also failed in many others due to their inherent limitations. For example, the simple waste classification criteria of PAF, NAF and uncertain (UC) are limited in that it does not account for parameters like the mobility of metal ions of interest at neutral pH, effects of microorganisms, reaction rates of individual minerals, particle size, texture, climate and many others (Lottermoser, 2015).

2.5 Concluding Remarks

The use of predictive tools to prevent AMD formation has historically been driven by the application of increasingly unacceptable end-of-pipe approaches to address the destructive nature of AMD and its huge reme-diation costs. It is now a regulatory requirement in most mining countries, as part of EIA process, to predict the potential of AMD formation for all solid waste materials at mine sites and to proactively develop plans for waste management and mitigation facilities for unavoidable AMD.

Amongst the many methods used, the ABA method is widely used to pre-dict AMD formation from mine wastes by measuring sulphur species, the cal-culation of AP and the determination of the acid NP. The difference between AP and NP values indicates the NAPP of the waste material. It is, thus, a quantitative estimate of the balance between the acid generated from mainly the oxidation of sulphide minerals and the acid consumption by carbonate minerals. The other tests that complement the ABA method are the mineral-ogy, elemental composition, paste pH and the NAG test. The results from all these tests are useful to identify waste materials with potential to form AMD and to classify and segregate the mine wastes for separate disposal and miti-gation. However, ABA methods do not predict the time when acid generation would occur and provide no information about effluent quality and rates for acid generation and neuralisation. Thus, several kinetic tests (e.g., humidity cells, leach columns, etc.) have been developed to complement ABA static tests in the prediction of AMD formation. They are aimed at determining the sulphide oxidation kinetics, rates of acid formation and neutralisation, assess water quality and predict when AMD is likely to occur.

Some of the recent developments in AMD prediction include the acid rock drainage index (ARDI) tool that was proposed by Parbhakar-Fox et al (2011). This tool addresses some limitations in the waste characterisation criteria using an integrated geochemistry-mineralogy-texture (GMT) approach to generate detailed and accurate prediction data at a relatively low cost. Other new tests have been developed such as the computed acid rock drainage (CARD) risk protocol that uses the automated mineralogy data to calculate surface area of minerals that form or neutralise acid. In addition, the existing testing tools are being validated, for example, the paste pH method by ASTM has been found to be the best paste pH procedure on drill cores (Lottermoser, 2015).

It is recommended that a site-specific approach in predicting the potential for AMD formation is developed and that all mine waste materials are considered. To obtain a reliable prediction, a combined data interpretation from the mineralogy, static tests, kinetic tests, drainage chemistry and water quality is necessary to fully understand and predict the potential of AMD formation in the future. Waste materials that are PAF may be blended with acid-consuming materials or encapsulated with benign materials to minimise AMD formation at source.

References

Ahern, C.R., McElnea, A.E. and Sullivan, L.A. (2004). Acid sulfate soils laboratory methods guidelines. Queensland Department of Natural Resources, Mines and Energy, Indooroopilly, Queensland, Australia. Available at https://www.environment.nsw.gov.au/resources/soils/acid-sulfate-soils-laboratory-methods-guidelines.pdf [Accessed 16th July 2018].

ASTM. (2018). ASTM D8187-18, Standard guide for interpretation of standard humidity cell Test results. ASTM International, West Conshohocken, PA. Available at www.astm.org.

ASTM. (2000). ASTM D5744-96, Standard test method for accelerated weathering of solid materials using a modified humidity cell. ASTM International, West Conshohocken, PA. Available at www.astm.org.

Banks, D., Younger, P.L., Arnesen, R.T., Iversen, E.R. and Banks, S.B. (1997). Minewater chemistry: the good, the bad and the ugly. Environmental Geology 32(3): 157–174.

Bell, F.G., Bullock, S.E.T., Halbich, T.F.J. and Lindsay, P. (2001). Environmental impacts associated with an abandoned mine in the Witbank Coalfield, South Africa. International Journal of Coal Geology 45: 195–216.

Bradham, W.S. and Caruccio, F.T. (1990). A comparative study of tailings analysis using acid/base accounting, cells, columns and soxhlets. Proceedings of the 1990 Mining and Reclamation Conference and Exhibition, April 23–26, Charleston.

Bucknam, C.H. (1999). NMS analytical methods book. Newmont Metallurgical Services.

Castilla-Gomez, J. and Herrera-Herbert, J. (2015). Environmental analysis of mining operations: dynamic tools for impact assessment. Minerals Engineering 76: 87–96.

Coastech Research, Inc. (1989). Investigation of prediction techniques for acid mine drainage. MEND Project 1.16.1a. Canada Centre for Mineral and Energy Technology, Energy, Mines and Resources Canada. Available at http://mend-nedem.org/wp-content/uploads/1161A.pdf [Accessed 10th November 208].

Coastech Research, Inc. (1991) (Revised 2008). Acid rock drainage prediction manual. MEND Project 1.16.1b. Department of Energy, Mines and Resources, Canada. Available at http://mend-nedem.org/mend-report/acid-rock-drainage-prediction-manual/ [Accessed: 8 June 2020].

Dold, B. (2017). Acid rock drainage prediction: a critical review. Journal of Geochemical Exploration 172: 120–132.

Duncan, D.W. and A. Bruynesteyn. (1979). Determination of acid production potential of waste materials. American Institute of Mining, Metallurgical, and Petroleum Engineers (AIME), Littleton, CO.

Elaw (2010), Guidebook for evaluating mining project EIAs. Environmental law alliance worldwide (elaw). Available at https://www.elaw.org/files/mining-eia-guidebook/Full-Guidebook.pdf [Accessed 6th January 2019]

El Haggar, S. (2007). Sustainable industrial design and waste management: cradle-to-cradle for sustainable development. Academic Press, Cambridge, Massachusetts.

Ferguson, K.D. and Erickson, P.M. (1988). Pre-mine prediction of acid mine drainage. In: Salomons, W. and Forstner, U. (Editors), Environmental management of solid waste – dredged material and mine tailings. Springer-Verlag, New York, pp 24–43.

Gordon, A.R. (1994). Environmental consequences of coal mine closure. The Geographical Journal 160 (1): 33–40.

Gray, N.F. (1997). Environmental impact and remediation of acid mine drainage: a management problem. Environmental Geology 30(1/2): 62–71.

Grube, W.E., Smith, R.M., Singh, R.N. and Sobek, A.A. (1973). Characterization of coal overburden materials and mine soils in advance of surface mining. Research and applied technology symposium on mined-land reclamation, 134–151.

Hesketh, A.H., Broadhurst, J.L., Bryan, C.G., van Hille, R.P. and Harrison, S.T.L. (2010). Biokinetic test for the characterisation of AMD generation potential of sulphide mineral wastes. Hydrometallurgy 104: 459–464.

Jones, D., Taylor, J., Pape, S., McCullough, C.D., Brown, P., Garvie, A., Appleyard, S., Miller, S., Unger, C., Laurencont, T., Slater, S., Williams, D., Scott, P., Fawcett, M., Waggitt, P. and Robertson, A. (2016). Preventing acid and metalliferous drainage. Leading practice sustainable development program for the mining industry. Available at https://www.industry.gov.au/sites/default/files/2019-04/lpsdp-preventing-acid-and-metalliferous-drainage-handbook-english.pdf [Accessed: 8 August 2020].

Kania, T. (1998). Laboratory methods for acid-base accounting: An update. In: Brady, K.B.C., Smith, M.W. and Shueck, J. (Editors), Coal mine drainage prediction and pollution prevention in Pennsylvania. The Pennsylvania Department of Environmental Protection, Harrisburg, PA.

Lapakko, K. (1992). Evaluation of Tests for Predicting Mine Waste Drainage pH. Draft Report to the Western Governors' Association, May 1992.

Lapakko, K.A. (1994). Evaluation of neutralization potential determinations for metal mine waste and a proposed alternative. In: Proceedings from the International Land Reclamation and Mine Drainage Conference and Third International Conference on the Abatement of Acidic Drainage, Vol. 1, American Society for Surface Mining and Reclamation, Pittsburgh, PA, April 24–29, 129–137.

Lapakko, K. (2002). Metal mine rock and waste characterisation tools: an overview. Mining, minerals and sustainable development (MMSD). Available at https://pubs.iied.org/pdfs/G00559.pdf [Accessed 27 May 2018].

Lapakko, K.A. (2003). Developments in humidity cell tests and their application. In: Jambor, J.L., Blowes, D.W. and Ritchie, A.I.M. (Editors), Short course: environmental aspects of mine wastes. Mineralogical Association of Canada, Quebec, pp. 147–164.

Lawrence, R.W. and Day, S. (1997). Short course: chemical prediction techniques for ARD. In: Proceedings from the 4th International Conference on Acid Rock Drainage, Vancouver, British Columbia.

Lawrence, R.W., Ritcey, G.M., Poling, G.W. and Marchant, P.B. (1989). Strategies for the prediction of acid mine drainage. In: Proceedings from the 13th Annual British Columbia Mine Reclamation Symposium. The Technical and Research Committee on Reclamation, Vernon, British Columbia, Canada.

Lawrence, R.W. and Wang, Y. (1996). Determination of neutralisation potential for acid rock drainage prediction. MEND/NEDEM Report 1.16.3, Canadian Centre for Mineral and Energy Technology, Ottawa.

Li, J., Smart, R.St.C., Schumann, R.C., Gerson, A.R. and Levay. G. (2007). A simplified method for estimation of jarosites and acid-forming sulphates in acid mine wastes. Science of the Total Environment 373: 391–403.

Lottermoser, B.G. (2010). Mine wastes: characterisation, treatment, environmental impacts. Third edition. Springer, New York.

Lottermoser, B.G. (2011). Recycling, reuse and rehabilitation of mine wastes. Elements 7: 405–41.

Lottermoser, B.G. (2015). Predicting acid mine drainage: past, present, future. Available at https://mining-report.de/english/predicting-acid-mine-drainage-past-present-future/ [Accessed: 20 February 2019].

Maest, A.S., Kuipers, J.R., Travers, C.L. and Atkins, D.A. (2005). Predicting water quality at hardrock mines: methods and models, uncertainties, and state-of-the-art. Available at https://www.waterboards.ca.gov/academy/courses/acid/supporting_material/predictwaterqualityhardrockmines1.pdf [Accessed 13 November 2018].

Miller, S.D., Jeffrey, J.J. and Donohue, T.A. (1994). Developments in predicting and management of acid forming mine wastes in Australia and Southeast Asia. In: Proceedings from the International Land Reclamation and Mine Drainage Conference and Third International Conference on the Abatement of Acidic Drainage, Pittsburgh, PA, pp. 177–184.

Mine Environment Neutral Drainage (MEND). (2000). MEND Manual, Volume 3 – Prediction (MEND 5.4.2c). Available at http://mend-nedem.org/wp-content/uploads/2013/01/5.4.2c.pdf [Accessed: 12 November 2018].

Morgan, R.K. (2012). Environmental impact assessment: the state of the art. Impact assessment and project appraisal 30(1): 5–14.

Morin, K.A. and Hutt, N.M. (1994). Observed preferential depletion of neutralization potential over sulfide minerals in kinetics tests: site-specific criteria for safe NP/AP ratios. In: Proceedings of International Land Reclamation and Mine Drainage Conference and the Third International Conference on the Abatement of Acidic Drainage, April 24–29, 1994. USBM Special Publication, SP 06A-94, Pittsburgh, PA.

Morin, K. A. and Hutt, N.M. (1997). Control of acidic drainage in layered waste rock at the Samatosum minesite: laboratory studies and field monitoring. MEND Project 2.37.3. Available at http://mend-nedem.org/wp-content/uploads/2013/01/2.37.3.pdf [Accessed 7 December, 2017]

Morin, K.A. and Hutt, N.M. (1998). Kinetic test and risk assessment for ARD. 5th Annual BC metal leaching and ARD workshop, Vancouver, Canada. Available at http://www.mdag.com/MDAG%20Paper%20Database/M0003%20-%20Morin%20and%20Hutt%201998%20-%20ARD%20Risk%20%26%20Kinetic%20Tests.pdf [Accessed 23 May 2018].

Morin, K. A. and Hutt, N.M. (2001). Environmental geochemistry of mine site drainage: practical theory and case studies. MDAG Publishing, Vancouver, British Columbia.

Noble, B.F. (2011). Environmental impact assessment. John Wiley & Sons, Ltd, Chichester.

Noble, B.F. (2015). Introduction to environmental impact assessment: a guide to principles and practice. Oxford University Press, Don Mills, Ontario.

Parbhakar-Fox, A.K., Edraki, M., Walters, S., Bradshaw, D. (2011). Development of a textural index for the prediction of acid rock drainage. Minerals Engineering 24: 1277–1287.

Parbhakar-Fox, A. and Lottermoser, B.G. (2015). A critical review of acid rock drainage prediction methods and practices. Minerals Engineering 82: 107–124.

Perry, E.F. (1998). Interpretation of acid-base accounting. In: Brady, K.B.C., Smith, M.W. and Shueck, J. (Editors), Coal mine drainage prediction and pollution prevention in Pennsylvania. The Pennsylvania Department of Environmental Protection, Harrisburg, PA.

Price, W.A. (2009). Prediction Manual for Drainage Chemistry from Sulphidic Geologic Materials, MEND Program, Natural Resources, Canada. Available at http://mend-nedem.org/wp-content/uploads/1.20.1_PredictionManual.pdf [Accessed 17 December 2018].

Rae, I., Taylor, J., Pape, S., Yardi, R., Bennett, J., Brown, P., Currey, N., Jones, D., Miller, S., Mudd, G., Simmonds, S., Slater, S. and Williams, D. (2007). Managing acid and metalliferous drainage. Leading practice sustainable development program for the mining industry, Commonwealth of Australia. Available at https://www.im4dc.org/wp-content/uploads/2014/01/Managing-acid-and-metalliferous-drainage.pdf [Accessed 7 August 2017].

Ridge, T. and Seif, J.M. (1998). Coal mine drainage prediction and pollution prevention in Pennsylvania. The Pennsylvania Department of Environmental Protection, Harrisburg, PA. Available at http://files.dep.state.pa.us/Mining/BureauOfMiningPrograms/BMPPortalFiles/Coal_Mine_Drainage_Prediction_and_Pollution_Prevention_in_Pennsylvania.pdf [Accessed 16 March 2018].

Skousen, J.G., Smith, R.M. and Sencindiver, J.C. (1990). The development of the acid base account. West Virginia mining and reclamation association, Charleston, West Virginia. Green lands, 20(1): 32–37.

Smart, R., Skinner, W.M., Levay, G., Gerson, A.R., Thomas, J.E., Sobieraj, H., Schumann, R., Weisener, C.G., Weber, P.A., Miller, S.D. and Stewart, W.A. (2002). ARD test handbook: project P387, a prediction and kinetic control of acid mine drainage. AMIRA, International Ltd., Melbourne, Australia.

Smith, R.M., Grube, W.E., Arkle, T.A. and Sobek, A.A. (1974). Mine spoil potentials for soil and water quality. U.S. Environmental Protection Agency. Available at https://nepis.epa.gov/Exe/ZyPDF.cgi/9101F1IB.PDF?Dockey=9101F1IB.PDF [Accessed 14 June 2018].

Smith, R.M., Sobek, A.A., Arkle Jr., T., Sencindiver, J.C. and Freeman, J.R. (1976). Extensive overburden potentials for soil and water quality. U.S. Environmental

Protection Agency. Available at https://nepis.epa.gov/Exe/ZyPDF.cgi/ 2000ZQDY.PDF?Dockey=2000ZQDY.PDF [Accessed 4 February 2018].

Sobek, A.A., Schuller, W.A., Freeman, J.R. and Smith, R.M. (1978). Field and laboratory methods applicable to overburdens and minesoils. U.S. Environmental Protection Agency. Available at https://nepis.epa.gov/Exe/ZyPDF.cgi/ 91017FGB.PDF?Dockey=91017FGB.PDF [Accessed 29 January 2020].

The International Association for Impact assessment (IAIA). (1999). Principles of environmental impact assessment best practice. Available at https://www.iaia. org/uploads/pdf/principlesEA_1.pdf [Accessed 29 August 2018].

Tuttle, M.L.W., Briggs, P.H. and Berry, C.J. (2003). A method to separate phases of sulfur in mine-waste piles and natural alteration zones and to use sulfur isotopic compositions to investigate release of metals and acidity to the environment. Proceedings from: 6th International conference on acid rock drainage, Cairns, Queensland, Australia.

USEPA. (1994). Acid mine drainage prediction. U.S. Environmental Protection Agency, Office of Solid Waste, Washington, USA. Available at https://www.epa. gov/sites/production/files/2015-09/documents/amd.pdf [Accessed 9 November 2018].

Weber, P. A, Stewart, W.A., Skinner, W.M., Weisener, C.G., Thomas, J.E. and Smart, R.S.T.C. (2005). A methodology to determine the acid-neutralization capacity of rock samples. Can. Mineral. 43: 1183–1192.

West Virginia University. (1971). Mine spoil potentials for water quality and controlled erosion. USEPA. Available at https://nepis.epa.gov/Exe/ZyPDF.cgi/9100GXFY. PDF?Dockey=9100GXFY.PDF [Accessed 2 October 2019].

White III, W.W., Lapakko, K.A. and Cox, R.L. (1999). Static test methods most commonly used to predict acid mine drainage: practical guidelines for use and interpretation. In: Plumlee, G.S. and Lodgson, M.J. (Editors), The environmental geochemistry of mineral deposits part A: processes, techniques, and health issues. Society of Economic Geologists, Littleton. Reviews of Economic Geology, 6A, pp. 325–338.

ZCCM-IH. (2005). Consolidated environmental management plan, Phase II (CEMP II), Copperbelt environmental project (CEP), Zambia.

3

Chemistry of Acid Mine Drainage Formation

Geoffrey S. Simate

CONTENTS

3.1 Introduction

There is no doubt that acid mine drainage (AMD) is one of the serious environmental problems in the world. Typically, AMD contains high concentrations of sulphate, iron, and many other metallic ions at low pH (Prasad and Henry, 2009). Its formation involves a series of complex chemical, biological, and electrochemical reactions (Zdun, 2001). It is, therefore, important that there is a proper understanding of the formation of AMD by evaluating the biogeochemical interactions and the sequences in these processes (Dold, 2014). In addition, a proper understanding of the chemistry is required if the means of controlling the environmental hazard or enhancing the reaction for commercial exploitation are to be devised (Lowson, 1982). Indeed, understanding the chemistry behind the AMD formation will help create more cost-effective prevention and remediation solutions. Therefore, this chapter discusses the chemistry of AMD formation in detail. First, the chapter focuses on the chemistry of AMD formation with respect to pyrite. Pyrite is chosen amongst many other sulphide minerals because the process of AMD production is made more clear by considering the reactions during the oxidation of pyrite (Simate and Ndlovu, 2014), and that pyrite oxidation is the main process responsible for the generation of AMD (España, 2008). Second, the chapter will briefly discuss the oxidative dissolution of a selected number of other sulphide minerals.

3.2 Oxidation of Sulphide Minerals

The mining and excavation operations have already been discussed in Chapter 1. These operations allow the introduction of oxygen onto the mineral surface and thus initiating the oxidation of minerals which are normally in a reduced state (Banks et al., 1997). The most common family of such minerals is the sulphides. In principle, the oxidation of sulphide minerals occurs when the mineral surface is exposed to an oxidant and water, either in oxygenated or anoxic systems, depending on the oxidant (Blowes et al., 2003). In addition, microorganisms also indirectly play a role in the oxidation of sulphide minerals (Rossi, 1990; Acevedo et al., 2004; Brierley and Brierley, 2001; Nestor et al., 2001; Kodali et al., 2004; Adams et al., 2005; Ndlovu, 2008). Therefore, as a result of various interacting factors, the AMD generation process may be considered as being a complex process involving chemical, biological, and electrochemical reactions that are dependent on the conditions of the environment (Zdun, 2001).

3.2.1 Oxidation of Pyrite

3.2.1.1 Introduction

Pyrite which is found in a wide variety of geological formations is the most widespread and abundant of the sulphide minerals (Craig and Vokes, 1993). It is widely distributed and forms under extremely varied conditions. For example, it can be produced by magmatic (molten rock) segregation, or by hydrothermal solutions, and as stalactite growth. It occurs as an accessory mineral in igneous rocks, in vein deposits with quartz and sulphide minerals, and in sedimentary rocks, such as shale, coal, and limestone (Rafferty, 2020). Pyrite is sometimes called *fool's gold* because of its similarity to gold in colour (brassy-yellow) and shape (King, 2020a; Rafferty, 2020). However, pyrite is quite easy to distinguish from gold, i.e., pyrite is much lighter, but harder than gold and cannot be scratched with a fingernail or knife (TMM, 2014). In addition, pyrite will tarnish when exposed to acid, whereas gold is nonreactive (IGS, 2011).

3.2.1.2 Oxidation Process

Pyrite is stable under anaerobic conditions, but is oxidised and dissolved to release soluble iron species and sulphuric acid when it comes in contact with oxygen and water (Satur et al., 2007). In fact, the oxidative dissolution of pyrite is one of the most extensively studied geochemical processes by many researchers (Lowson, 1982; Nordstrom and Alpers, 1999; Edwards et al., 2000; España, 2008; Simate and Ndlovu, 2014). The pyrite oxidation and the factors affecting the kinetics of oxidation (O_2, Fe^{3+}, temperature, pH, E_h, and the presence or absence of microorganisms) have been the focus of extensive study because of their importance in both environmental remediation and mineral separation (Blowes et al., 2003). As shown in Figure 3.1, the oxidation of pyrite follows a cycle of complex reactions (Stumm and Morgan, 1996; Banks et al., 1997; Ali, 2011; Buzzi et al., 2013) involving surface interactions with dissolved O_2, Fe^{3+}, and other mineral-based catalysts such as MnO_2 (Blowes et al., 2003; Simate and Ndlovu, 2014). Ideally, several products are formed during the oxidation of pyrite including metastable secondary products such as ferrihydrite ($5Fe_2O_3 \cdot 9H_2O$), schwertmannite (between $Fe_8O_8(OH)_6SO_4$ and $Fe_{16}O_{16}(OH)_{10}(SO_4)_3$), and goethite ($FeO(OH)$), as well as the more stable secondary jarosite ($KFe_3(SO_4)_2(OH)_6$), and hematite (Fe_2O_3) depending on the geochemical conditions (Dold, 2010; Dold, 2014).

For pyrite oxidation, oxygen and ferric iron are the two possible available oxidants. However, the initial step and most important reaction is the oxidation of the pyrite (or sulphide) in the presence of atmospheric oxygen and water forming dissolved ferrous iron, sulphate, and hydrogen as shown in Equation 3.1 (Akcil and Koldas, 2006; Dold, 2010). The initial step of pyrite

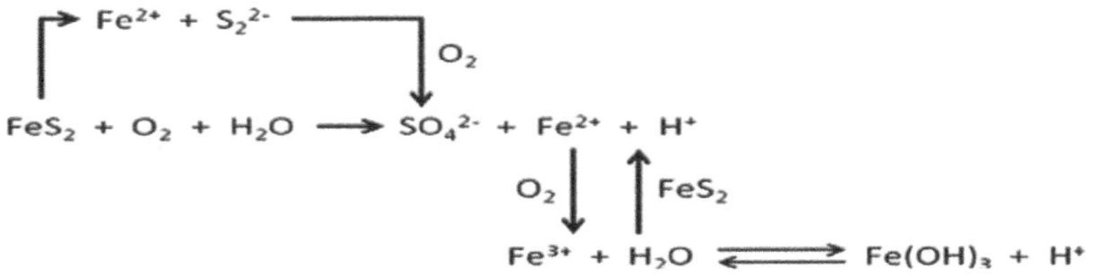

FIGURE 3.1
Model for the oxidation of pyrite. (From Stumm and Morgan, 1996; Ali, 2011; Buzzi et al., 2013.)

oxidation is also illustrated in Figure 3.1 (Stumm and Morgan, 1996; Banks et al., 1997; Ali, 2011; Buzzi et al., 2013). It must be noted that though two oxidisable species are present in pyrite (ferrous iron and sulphidic sulphur), it has been experimentally determined that irrespective of the mechanism (oxygen or ferric mediated), during the initial solubilisation of pyrite only the sulphidic sulphur is oxidised and the iron passes into solution in the ferrous state (Lowson, 1982).

$$2FeS_2 + 7O_2 + 2H_2O \rightarrow 2Fe^{2+} + 4SO_4^{2-} + 4H^+ \tag{3.1}$$

In an environment which is sufficiently oxidizing (dependent on O_2 concentration, pH greater than 3.5 and bacterial activity), the ferrous iron generated as shown in Equation 3.1 may be oxidised to ferric iron according to reaction 3.2 (Blowes et al., 2003; Akcil and Koldas, 2006; Udayabhanu and Prasad, 2010). However, according to Fripp et al. (2000), if the concentration of oxygen is low, reaction 3.2 will not occur until the pH reaches 8.5.

$$4Fe^{2+} + O_2 + 4H^+ \leftrightarrow 4Fe^{3+} + 2H_2O \tag{3.2}$$

Once ferric iron is produced by the oxidation of ferrous iron (reaction 3.2), which is the case at low pH conditions and strongly accelerated by microbiological activities, then ferric iron also becomes an oxidant of pyrite (reaction 3.3) (Dold, 2010). In fact, an important factor in the oxidation of pyrite and the generation of acid mine waters is that Fe^{3+} is able to oxidise pyrite under anoxic subaqueous conditions at a much faster rate than does molecular oxygen (España, 2008). In other words, though oxygen is a primary oxidant, the ferric iron (Fe^{3+}) resulting from the oxidation of ferrous iron is now recognised as a more powerful oxidant than oxygen even at near-neutral pH (Zdun, 2001).

$$FeS_2 + 14Fe^{3+} + 8H_2O \rightarrow 15Fe^{2+} + 2SO_4^{2-} + 16H^+ \tag{3.3}$$

It is noted that at pH < 3.5, reaction 3.2 is several orders of magnitude slower than reaction 3.1 (España, 2008). Therefore, the oxidation of Fe^{2+} by oxygen is usually considered as the rate-limiting step in pyrite oxidation (Singer and Stumm, 1970; Skousen et al., 1998). However, the presence of acidophilic bacteria such as *Acidithiobacillus ferrooxidans* and *Leptospirillum ferrooxidans* greatly accelerates (by a factor of around 10[6]) the abiotic oxidation rate (Singer and Stumm, 1970; Nordstrom and Alpers, 1999; España, 2008), thus maintaining a high concentration of ferric iron in the system (España, 2008).

According to Singer and Stumm (1970), the overall process resulting from the combination of reactions 3.2 and 3.3 is traditionally known as the 'propagation cycle' and along with reaction 3.1 depicts a model by which pyrite oxidation initially starts by reaction 3.1 with oxygen as the oxidant at circum-neutral pH conditions, and as pH decreases to about 4 the oxidation of pyrite proceeds through reaction 3.3. It must be noted, however, that oxygen will always be required to replenish the supply of ferric iron according to reaction 3.2, so that the overall rate of pyrite oxidation is largely dependent on the overall rate of oxygen transport by advection and diffusion (Nordstrom and Alpers, 1999; Ritchie, 1994; Ritchie, 2003; España, 2008).

It is noted further that at pH values between 2.3 and 3.5, ferric iron formed in reaction 3.2 may precipitate as $Fe(OH)_3$ (and to a lesser degree as jarosite, $H_3OFe_3(SO_4)_2(OH)_6$) while simultaneously producing acid as shown in reaction 3.4 (Blowes et al., 2003; Akcil and Koldas, 2006; Dold, 2010; Dold, 2014). The hydrolysis and subsequent precipitation of $Fe(OH)_3$ produces most of the acid in the whole process (Dold, 2010; Dold, 2014). If pH is less than 2, ferric hydrolysis products like $Fe(OH)_3$ are not stable and Fe^{3+} remains in solution (Dold, 2010; Dold, 2014). Therefore, any remaining Fe^{3+} from reaction 3.2 that does not precipitate into $Fe(OH)_3$ (or jarosite) from solution through reaction 3.4 may be used to oxidise additional pyrite, according to reaction 3.3 (Akcil and Koldas, 2006).

$$Fe^{3+} + 3H_2O \leftrightarrow 4Fe(OH)_3 \downarrow + 3H^+ \qquad (3.4)$$

The process of pyrite oxidation relates to all sulphide minerals once exposed to oxidizing conditions (e.g., chalcopyrite, bornite, molybdenite, arsenopyrite, enargite, galena and sphalerite among others). In other words, while the principal sulphide mineral in mine wastes is pyrite, other sulphide minerals are also susceptible to oxidation releasing elements such as aluminium, arsenic, cadmium, cobalt, copper, mercury, nickel, lead, and zinc into the water flowing through the mine waste (Blowes et al., 2003). The oxidation of other sulphide minerals is discussed in the next sections.

3.2.2 Oxidation of Pyrrhotite

3.2.2.1 Introduction

Pyrrhotite group of minerals are considered as all the iron monosulphides of the general formula, $Fe_{(1-x)}S$ (where $0 \leq x \leq 0.125$) (Wang and Salveson, 2005). The pyrrhotite group of minerals often occurs in association with a variety of ore deposits including Ni-Cu, Pb-Zn, and platinum group elements and appears in different crystallographic forms and compositions (Becker et al., 2010). Ideally, the pyrrhotite group of minerals is extremely complex (Wang and Salveson, 2005) with each type exhibiting subtly different physical and chemical properties (Ekmekçi et al., 2010). Pyrrhotite group includes troilite (FeS) which is hexagonal and pyrrhotite ($Fe_{(1-x)}S$) which may be monoclinic or hexagonal (Wang and Salveson, 2005).

3.2.2.2 Oxidation Process

The oxidation of pyrrhotite is not a very well understood process compared to that of pyrite, and the rate of controls of the reactions and the oxidation products are also poorly known (Nicholson and Scharer, 1994; Fox et al., 1997). However, pyrite (FeS_2) and pyrrhotite ($Fe_{(1-x)}S$) are two of the most common iron-sulphide minerals in areas where AMD is prevalent (Fox et al., 1997). It is known that dissolved oxygen and Fe^{3+} are important oxidants of pyrrhotite. Therefore, pyrrhotite dissolution can proceed through oxidative or nonoxidative reactions (Blowes et al, 2003). The overall reaction when oxygen is the primary oxidant may be written as follows according to Blowes et al. (2003):

$$Fe_{(1-x)}S + (2-x/2)O_2 + xH_2O \rightarrow (1-x)Fe^{2+} + 2SO_4^{2-} + 2xH^+ \qquad (3.5)$$

The stoichiometry of pyrrhotite affects the relative production of acid (Dold, 2010). For example, if $x = 0$ and the formula is FeS, no H^+ will be produced in the oxidation reaction; at the other extreme ($x = 0.125$), the maximum amount of acid will be produced by the iron-deficient Fe_7S_8 phase (Dold, 2010).

As for nonoxidative dissolution of pyrrhotite, it occurs when predominant S^{2-} surface species are exposed to acidic solutions; and the reaction occurs as follows (Blowes et al., 2003):

$$Fe_{(1-x)}S + 2H^+ \rightarrow (1-x)Fe^{2+} + H_2S \qquad (3.6)$$

3.2.3 Oxidation of Chalcopyrite

3.2.3.1 Introduction

Chalcopyrite is the most important copper-bearing ore mineral (Vaughan et al., 1995), comprising approximately 70% of copper reserves in the world (Baba et al., 2012). It is a mineral predominantly found in igneous and metamorphic rock and in metalliferous veins (Baba et al. 2012; McGraw-Hill Encyclopedia, 1998). Chalcopyrite is a mineral with a brassy to golden yellow colour (Baba et al., 2012; Mamedov et al., 2012). It contains several minerals including copper, zinc, sulphur, and iron that were produced at different times (Baba et al., 2012).

3.2.3.2 Oxidation Process

Chalcopyrite dissolution is usually suggested to be an electrochemical corrosion activity with oxidants, such as Fe^{3+} or dissolved O_2, being reduced at the mineral surface (Biegler and Swift, 1979; Li et al., 2017). The resulting Fe^{2+} may be re-oxidised either by iron-oxidizing microorganisms or, at a slower rate, by O_2 (Li et al., 2017; Nazari and Asselin, 2009).

In the presence of ferric ions under acidic conditions, the oxidation of chalcopyrite can be represented as shown in reaction 3.7 (Blowes et al., 2003).

$$CuFeS_2 + 4Fe^{3+} \rightarrow Cu^{2+} + 5Fe^{2+} + 2S^0 \tag{3.7}$$

A study by Rimstidt et al. (1994) showed that, with an increase in Fe^{3+} concentration, the oxidation rate of chalcopyrite increases, though with an oxidation rate of 1–2 orders of magnitude less than that of pyrite. Other studies have shown that the combination of ferrous iron oxidation and ferrihydrate hydrolysis is the main acid producing process as shown in reaction 3.8 (Dold, 2010).

$$4CuFeS_2 + 17O_2 + 10H_2O \rightarrow 4Cu^{2+} + 4Fe(OH)_3 + 8SO_4^{2-} + 8H^+ \tag{3.8}$$

Dold (2010) noted that complete oxidation of chalcopyrite without acid production can be represented as reaction 3.9.

$$2CuFeS_2 + 4O_2 \rightarrow 2Cu^{2+} + Fe^{2+} + SO_4^{2-} \tag{3.9}$$

It has also been established that the dissolution of chalcopyrite can be greatly influenced by galvanic effects (Blowes et al., 2003). A study by Dutrizac and MacDonald (1973) reported that the presence of pyrite or molybdenite in association with chalcopyrite can cause accelerated rates of chalcopyrite dissolution, whereas according to Blowes et al. (2003) the presence of iron rich sphalerite and galena can slow the dissolution.

3.2.4 Oxidation of Arsenopyrite

3.2.4.1 Introduction

Arsenopyrite (FeAsS) is the most common arsenic (As) bearing mineral (Saxe et al., 2005). Arsenopyrite and other primary arsenic minerals are formed only under high temperature conditions (Drewniak and Sklodowska, 2013) and are found in a variety of ore deposits, including magmatic, hydrothermal, and porphyry-style systems (Corkhill and Vaughan, 2009; Drewniak and Sklodowska, 2013). It is a common mineral constituent of refractory gold ores (Corkhill and Vaughan, 2009; Andrews and Merkle, 1999) and thus arsenopyrite is often mined, processed to extract the gold and discarded as solid waste (Corkhill and Vaughan, 2009). Other mineral constituents of arsenopyrite are copper and silver (Dos Santos et al., 2017). In addition, natural arsenopyrite samples are always associated with pyrite and are generally found with large domains of pyrite randomly inlaid in its structure (Fleet and Mumin, 1997; Dos Santos et al., 2017).

3.2.4.2 Oxidation Process

Arsenopyrite is stable under reducing conditions (Corkhill and Vaughan, 2009; Nesbitt et al., 1995). However, when arsenopyrite has been mined and exposed to the environment, it oxidises leading to the release of arsenite (As(III)), arsenate (As(V)) in addition to acid and heavy metals (Dos Santos et al., 2017). The oxidation of asernopyrite is a two-step process represented by reactions 3.10 and 3.11 (Drewniak and Sklodowska, 2013).

$$4FeAsS + 11O_2 + 6H_2O \rightarrow 4Fe^{2+} + 4\ SO_4^{2-} + 4H_3AsO_3 \qquad (3.10)$$

$$2H_3AsO_3 + O_2 \rightarrow 2H_2AsO_4^- + 2H^+ \qquad (3.11)$$

Combining reaction 3.10 and reaction 3.11 results in arsenopyrite oxidation process represented by the following reaction path (Dold, 2010; Mok and Wai, 1994; Simate and Ndlovu, 2014):

$$4FeAsS + 13O_2 + 6H_2O \rightarrow 4Fe^{2+} + 4\ SO_4^{2-} + 4H_2AsO_4^- + 4H^+ \quad (3.12)$$

When ferrous iron generated in reaction 3.12 is oxidised forming ferric iron according to reaction 3.2, the ferric iron may hydrolyse at low pH generating ferrihydrate precipitation (reaction 3.4). Therefore, the overall arsenopyrite oxidation reaction can be written as follows:

$$2FeAsS + 7O_2 + 12H_2O \rightarrow 2Fe(OH)_3 + 2SO_4^{2-} + 2H_2AsO_4^- + 6H^+ \quad (3.13)$$

Research studies have shown that the oxidation rate of arsenopyrite is similar to the oxidation rate of pyrite if ferric iron is the oxidant (Dold, 2010) whereas the oxidation rate of arsenopyrite is lower than that of pyrite if oxygen is the oxidant (Mok and Wai, 1994; Dold, 2010).

The galvanic effect between arsenopyrite and pyrite minerals has also been found to influence the dissolution of the two sulphide ores. For example, in the presence of arsenopyrite, the oxidation rate of pyrite is delayed whereas the oxidation rate of arsenopyrite increases (Dos Santos et al., 2017).

3.2.5 Oxidation of Sphalerite

3.2.5.1 Introduction

Sphalerite with a chemical composition of (Zn,Fe)S is a sulphide mineral found in metamorphic, igneous, and sedimentary rocks in many areas globally (King, 2020b). Sphalerite compositions can vary among different sulphide-mineral deposit types (Stanton et al., 2006). For example, it contains variable amounts of iron up to 25% by weight that substitute for zinc in the mineral lattice (King, 2020b), and the structure of sphalerite can accommodate a wide variety of substitutions, such as Cu, Mn, and In (Jambor et al., 2005). Sphalerite also contains trace to minor amounts of rare earth elements such as cadmium, indium, germanium, or gallium, which, if present in large quantities, can be recovered as profitable by-products (King, 2020b).

3.2.5.2 Oxidation Process

Research has shown that the oxidation of sphalerite is dependent on a number of parameters including the concentration of oxidants, such as dissolved O_2 or Fe^{3+} in solution, the temperature, and the pH (Bobeck and Su, 1985; Crundwell, 1988; Rimstidt et al., 1994; Blowes et al., 2003). The overall oxidation reaction for pure sphalerite, assuming that all sulphur is oxidised to sulphate, is given in reaction 3.14 (Blowes et al., 2003).

$$ZnS + 2O_2 \rightarrow Zn^{2+} + SO_4^{2-} \tag{3.14}$$

As can be seen from reaction 3.14, sphalerite falls in the category of non-acid producing sulphide minerals (Dold, 2010). However, if iron substitutes for zinc, sphalerite will be an acid generator in a similar way as pyrrhotite due to the hydrolysis of ferric iron (Blowes et al., 2003).

3.2.6 Oxidation of Galena

3.2.6.1 Introduction

Galena is a sulphide mineral with a chemical composition of PbS. The mineral occurs in igneous and metamorphic rocks in medium to low temperature hydrothermal veins (Ogwata and Onwughalu, 2019). In sedimentary rocks, it exits as veins, breccia cements, isolated grains, and as replacements of limestone and dolostone. Galena is also commonly associated with acid-generating minerals, such as pyrite and pyrrhotite (Blowes et al., 2003).

3.2.6.2 Oxidation Process

Several researchers have studied the oxidation of galena and observed that in natural oxygenated environments, galena weathers to anglesite ($PbSO_4$), which is weakly soluble below the pH of 6 (Lin, 1997; Blowes et al., 2003; Shapter et al., 2000) according to reactions 3.15 and 3.16.

$$PbS + 2O_2 \rightarrow Pb^{2+} + SO_4^{2-} \tag{3.15}$$

$$Pb^{2+} + SO_4^{2-} \rightarrow PbSO_4 \tag{3.16}$$

Jambor and Blowes (1998) found that as the secondary anglesite forms, it creates a layer on galena that can prevent the mineral from direct contact with oxidizing reagents because anglesite has a low solubility. A study by Fornasiero et al. (1994) showed that in the absence of oxygen, both lead and sulphide ions are released into solution in the form of free lead ions and hydrogen sulphide.

Rimstidt et al. (1994) also showed that, under acidic conditions, galena may also be oxidised by Fe^{3+} ions as follows:

$$PbS + 8Fe^{3+} + 4H_2O \rightarrow Pb^{2+} + 8Fe^{2+} + SO_4^{2-} + 8H^+ \qquad (3.17)$$

In general, in the presence of Fe^{3+}, the oxidation of MeS (where Me = divalent metal) produces acidity according to reaction schemes where part of the oxidation capacity of the system is derived from Fe^{3+} as given in reaction 3.18 (Dold, 2010).

$$2MeS + 4Fe^{3+} + 3O_2 + 2H_2O \rightarrow 2Me^{2+} + 4Fe^{2+} + 2SO_4^{2-} + 4H^+ \quad (3.18)$$

Other studies have shown that the oxidation of galena in air may lead to the formation of lead hydroxide and lead oxide (Evans and Raftery, 1982; Buckley and Woods, 1984; Blowes et al., 2003). On the other hand, the exposure of galena to aqueous solutions may result in the formation of lead oxides and lead sulphate surface products (Fornasiero et al., 1994; Kartio et al., 1996; Kim et al., 1995; Blowes et al., 2003).

3.4 Concluding Remarks

This chapter outlined various key reactions in which sulphide minerals are oxidised and dissolved. There is no doubt that the aqueous oxidation and dissolution of sulphide minerals play important roles in the production of environmentally detrimental AMD. The generation of AMD may also be accompanied by release of toxic metals and metalloids to the environment.

From a number of research studies, the dissolution of sulphide minerals is usually proposed to be an electrochemical corrosion process with oxidants, such as Fe^{3+} or dissolved O_2, being reduced at the sulphide mineral surface. Research has also shown that the oxidation of sulphide minerals by aqueous ferric iron generates significantly greater quantities of acid than the oxidation by oxygen. In fact, at low pHs, the oxidation of some sulphide minerals such as pyrite by ferric iron has been found to be in the range of 10–100 times faster than by oxygen, thus making ferric iron a more effective oxidant than oxygen.

References

Acevedo, F., Gentina, J.C. and Valencia P. (2004). Optimisation of pulp density and particle size in the bioxidation of a pyritic gold concentrate by *Sulfolobus metallicus*. World Journal of Microbiology and Biotechnology 20: 865–869.

Adams, J.D., Pennington, P., McLemore, V.T., Wilson, G.W., Tachie-Menson, S. and Gutierrez, L.A.F. (2005). The role of microorganisms in acid rock drainage. Available at https://geoinfo.nmt.edu/staff/mclemore/documents/adams_sme. pdf [Accessed 2 December 2016].

Akcil, A. and Koldas, S. (2006). Acid mine drainage (AMD): causes, treatment and case studies. Journal of Cleaner Production 14: 1139–1145.

Ali, M. S. (2011). Remediation of acid mine waters. In: Rüde, T.R., Freund, A. and Wolkersdorfer, C. (Editors), Mine Water – Managing the Challenges. 11th International Mine Water Association Congress, Aachen, Germany, pp. 253–258.

Andrews, L. and Merkle, R.K.W. (1999). Mineralogical factors affecting arsenopyrite oxidation rate during acid ferric sulphate and bacterial leaching of refractory gold ores. In: Amils, R. and Ballester, A. (Editors), Biohydrometallurgy and the Environment Toward the Mining of the 21st Century Issue Part A. Elsevier, New York, pp. 109–117.

Baba, A.A., Ayinla, K.I., Adekola, F.A., Ghosh, M.K., Ayanda, O.S., Bale, R.B., Sheik, A.R. and Pradhan, S.R. (2012). A review on novel techniques for chalcopyrite ore processing. International Journal of Mining Engineering and Mineral Processing 1(1): 1–16

Banks, D., Younger, P.L., Arnesen, R.T., Iversen, E.R. and Banks, S.B. (1997). Minewater chemistry: the good, the bad and the ugly. Environmental Geology 32: 157–174.

Becker, M., de Villiers, J. and Bradshaw, D. (2010). The flotation of magnetic and nonmagnetic pyrrhotite from selected nickel ore deposits. Minerals Engineering 23: 1045–1052.

Biegler, T. and Swift, D.A. (1979). Anodic electrochemistry of chalcopyrite. Journal of Applied Electrochemistry 9: 545–554.

Blowes, D.W., Ptacek, C.J., Jambor, J.L. and Weisener, C.J. (2003). The geochemistry of acid mine drainage. Treatise on Geochemistry 9: 149–204.

Bobeck, G.E. and Su, H. (1985). The kinetics of dissolution of sphalerite in ferric chloride solution. Metallurgical Transactions B (16b): 413–424.

Brierley, J.A. and Brierley, C.L. (2001). Present and future commercial applications of biohydrometallurgy. Hydrometallurgy 59: 233–239.

Buckley, A.N. and Woods, R.W. (1984). An X-ray photoelectron spectroscopic study of the oxidation of galena. Applied Surface Science 17: 401–414.

Buzzi, D.C., Viegas, L.S., Rodrigues, M.A.S., Bernardes, A.M. and Tenório, J.A.S. (2013). Water recovery from acid mine drainage by electrodialysis. Minerals Engineering 4: 82–89.

Corkhill, C.L. and Vaughan, D.J. (2009). Arsenopyrite oxidation – a review. Applied Geochemistry 24: 2342–2361.

Craig, J.R. and Vokes, F.M. (1993). The metamorphism of pyrite and pyritic ores: an overview. Mineralogical Magazine 57, 3–18.

Crundwell, F.K. (1988). Effect of iron impurity in zinc sulfide concentrates on the rate of dissolution. American Institute of Chemical Engineers Journal 34(7): 1128–1134.

Dold, B. (2010). Basic concepts in environmental geochemistry of sulphide mine waste management. In: Kumar, E.S. (Editor), Waste Management. In-Tech, Rijeka, pp. 173–198.

Dold, B. (2014). Evolution of acid mine drainage formation in sulphidic mine tailings. Minerals 2014 4(3): 621–641.

Dos Santos, E.C., Lourenço, M.P., Pettersson, L.G. M. and Duarte, H.A. (2017). Stability, structure, and electronic properties of the pyrite/arsenopyrite solid–solid interface – a DFT study. Journal of Physical Chemistry C 121: 8042–8051.

Drewniak, L. and Sklodowska, A. (2013). Arsenic-transforming microbes and their role in biomining processes. Environmental Science and Pollution Research 20: 7728–7739.

Dutrizac J. E. and MacDonald R. J. C. (1973). The effect of some impurities on the rate of chalcopyrite dissolution. Canadian Metallurgical Quarterly 12(4): 409–420.

Edwards, K.J., Bond, P.L., Druschel, G.K., McGuire, M.M., Hamers, R.J. and Banfield, J.F. (2000). Geochemical and biological aspects of sulfide mineral dissolution: lessons from Iron Mountain, California. Chemical Geology 169: 383–397.

Ekmekçi, Z., Becker, M. and Tekes, E. (2010). The relationship between the electrochemical, mineralogical and flotation characteristics of pyrrhotite samples from different Ni ores. Journal of Electroanalytical Chemistry 647: 133–143.

España, J.S. (2008). Acid mine drainage in the Iberian pyrite belt: an overview with special emphasis on generation mechanisms, aqueous composition and associated mineral phases. Available at http://www.ehu.eus/sem/macla_pdf/macla10/Macla10_34.pdf [Accessed 4 December 2016].

Evans, S. and Raftery, E. (1982). Electron spectroscopic studies of galena and its oxidation by microwave-generated oxygen species and by air. Journal of the Chemical Society, Faraday Transactions 78: 3545–3560.

Fleet, M.E. and Mumin, A.H. (1997). Gold-bearing arsenian pyrite and marcasite and arsenopyrite from Carlin Trend Gold Deposits and Laboratory Synthesis. American Mineralogist 82: 182–193.

Fornasiero, D., Li F., Ralston J. and Smart, R.S.C. (1994). Oxidation of galena surfaces. Journal of Colloid and Interface Science 164: 333–344.

Fox, D., Robinson, C. and Zentilli, M. (1997). Pyrrhotite and associated sulphides and their relationship to acid rock drainage in the Halifax Formation, Meguma Group, Nova Scotia. Atlantic Geology 33: 87–103.

Fripp, J., Ziemkiewicz, P.F. and Charkavorki, H. (2000). Acid mine drainage treatment. EMRRP-SR-14. Available at http://el.erdc.usace.army.mil/elpubs/pdf/sr14.pdf [Accessed 7 December 2013].

IGS. (2011). Pyrite: Fool's gold. Available at https://igws.indiana.edu/ReferenceDocs/Pyrite_card.pdf [Accessed 12 April 2020].

Jambor, J.L. and Blowes, D.W. (1998). Theory and applications of mineralogy in environmental studies of sulfide-bearing mine waste. In: Cabri, L.J. and Vaughan, D.J. (Editors), Short Course Handbook on Ore and Environmental Mineralogy. Mineralogical Society of Canada, Nepean, pp. 367–401.

Jambor, J.L., Ptacek, C.J., Blowes, D.W. and Moncur, M.C. (2005). Acid drainage from the oxidation of iron sulfides and sphalerite in mine wastes. In: Fujisawa, T. (Editor), Proceedings from: Lead and Zinc '05, Vol. 1, The Mining and Materials Processing Institute of Japan, Japan.

Kartio, I., Laajalehto, K., Kaurila, T. and Suoninen, E. (1996). A study of galena (PbS) surfaces under controlled potential in pH 4.6 solution by synchrotron radiation excited photoelectron spectroscopy. Applied Surface Science 93: 167–177.

Kim, B. S., Hayes, R.A., Prestidge, C.A., Ralston, J. and Smart, R.S.C. (1995). Scanning tunnelling microscopy studies of galena: the mechanisms of oxidation in aqueous solution. Langmuir 11: 2554–2562.

King, H.M. (2020a). Pyrite. Available at https://geology.com/minerals/pyrite.shtml [Accessed 12 April 2020].

King, H.M. (2020b). Sphalerite. Available at https://geology.com/minerals/sphalerite.shtml [Accessed 14 April 2020].

Kodali, B., Rao, M.B., Narasu, M.L. and Pogaku, R. (2004). Effect of biochemical reactions in enhancement of rate of leaching. Chemical Engineering Science 59: 5069–5073.

Li, Y., Qian, G., Brown, P.L. and Gerson, A.R. (2017). Chalcopyrite dissolution: scanning photoelectron microscopy examination of the evolution of sulfur species with and without added iron or pyrite. Geochimica et Cosmochimica Acta 212: 33–47.

Lin, Z. (1997). Mineralogical and chemical characterization of wastes from a sulfuric acid industry in Falun Sweden. Environmental Geology 30: 153–162.

Lowson, R.T. (1982). Aqueous oxidation of pyrite by molecular oxygen. Chemical Reviews 82 (5): 461–497.

Mamedov, E.A., Ahmed, E.I. and Chiragov, M.I. (2012). Mineralogy character and types of the coper-gold-sulphide mineralization of El Samra area, Kid belt, in South Eastern Sina, Egypt. International Journal of Advanced and Technical Research 2(6): 48–61

McGraw-Hill Encylopedia. (1998). Encylopedia of Science and Technology. Longman, Tokyo.

Mok, W.M. and Wai, C.M. (1994). Mobilization of arsenic in contaminated river waters. In: Nriagu, J.O. (Editor), Arsenic in the Environment. Part I Cycling and Characterization. John Wiley Interscience, New York.

Nazari, G. and Asselin, E. (2009) Morphology of chalcopyrite leaching in acidic ferric sulfate media. Hydrometallurgy 96: 183–188.

Ndlovu, S. (2008). Biohydrometallurgy for sustainable development in the African mineral industry. Hydrometallurgy 91: 20–27.

Nesbitt, H.W., Muir, I.J. and Pratt, A.R. (1995). Oxidation of arsenopyrite by air and air-saturated, distilled water and implications for mechanisms of oxidation. Geochimica et Cosmochimica Acta 59, 1773–1786.

Nestor, D., Valdivia, U. and Chaves, A. P. (2001). Mechanisms of bioleaching of a refractory mineral of gold with *Thiobacillus ferrooxidans*. International Journal of Mineral Processing 62: 187–198.

Nicholson, R.V. and Scharer, J.M. (1994). Laboratory studies of pyrrhotite oxidation kinetics. In Environmental Geochemistry of Sulphide Oxidation. In: Alpers, C.N. and Blowes, D.W. (Editors), Environmental Geochemistry of Sulfide Oxidation. American Chemical Society, Washington D.C., pp. 14–30.

Nordstrom, D.K. and Alpers, C.N. (1999). Geochemistry of acid mine waters. In: Plumlee, G.S. and Logsdon, M.J. (Editors), The Environmental Geochemistry of Mineral Deposits. Society of Economic Geologists, Littleton, CO, pp. 133–156.

Ogwata, C. M and Onwughalu, M. K. (2019). Occurrence of galena and its potentials for economic and green energy revolution in Nigeria. Iconic Research and Engineering Journals 3(1): 139–142.

Prasad, D. and Henry, J.G. (2009). Removal of sulphates acidity and iron from acid mine drainage in a bench scale biochemical treatment system. Environmental Technology 30(2): 151–160.

Rafferty, J.P. (2020). Fool's gold, iron pyrite. In: Augustyn, A., Bauer, P., Duignan, B., Eldridge, A., Gregersen, E., McKenna, A., Petruzzello, M., Rafferty, J.P., Ray, M., Rogers, K., Tikkanen, A., Wallenfeldt, J., Zeidan, A. and Zelazko A. (Editors), Encyclopaedia Britannica. Available at https://www.britannica.com/science/acanthite [Accessed 12 April 2020].

Rimstidt, J.D., Chermak, J.A. and Gagen, P.M. (1994). Rates of reaction of galena, spalerite, chalcopyrite, and asenopyrite with Fe(III) in acidic solutions. In: Alpers, C.N. and Blowes, D.W. (Editors), Environmental Geochemistry of Sulfide Oxidation. American Chemical Society, Washington, pp. 2–13.

Ritchie, A.I.M. (1994). Sulfide oxidation mechanisms: controls and rates of oxygen transport. In: Jambor, J.L. and Blowes D.W. (Editors), Short Course Handbook on Environmental Geochemistry of Sulfide Mine-Waste. Mineralogical Association of Canada, Nepean, pp. 201–244.

Ritchie, A.I.M. (2003). Oxidation and gas transport in piles of sulphidic material. In: Jambor, J.L., Blowes, D.W. and Ritchie, A.I.M. (Editors), Environmental Aspects of Mine Wastes. Mineralogical Association of Canada, Short Course Series Volume 31, Vancouver, British Columbia, pp. 73–94.

Rossi, G. (1990). Biohydrometallurgy. McGraw-Hill Book Company, New York.

Satur, J., Hiroyoshi, N., Tsunekawa, M., Ito, M. and Okamoto, H. (2007). Carrier-microencapsulation for preventing pyrite oxidation. International Journal of Mineral Processing 83: 116–124.

Saxe, J. K., Bowers, T. S. and Reid, K. R. (2010). Arsenic. In: Morrison, R. D. and Murphy, L. (Editor), Environmental Forensics: Contaminant Specific Guide. Elsevier, Inc., New York, pp. 279–292.

Shapter, J. G., Brooker M. H. and Skinner W. M. (2000). Observation of oxidation of galena using Raman spectroscopy. International Journal of Mineral Processing 60: 199–211.

Simate, G. S. and Ndlovu, S. (2014). Acid mine drainage: challenges and opportunities. Journal of Environmental Chemical Engineering 2 (3): 1785–1803.

Singer, P.C. and Stumm, W. (1970). Acidic mine drainage: the rate-determining step. Science 167: 1121–1123.

Skousen, J., Rose, A., Geidel, G., Foreman, J., Evans, R. and Hellier, W. (1998). Handbook of Technologies for Avoidance and Remediation of Acid Mine Drainage. The National Mine Land Reclamation Centre, West Virginia University, West Virginia.

Stanton, M.R., Taylor, C.D., Gemery-Hill, P.A. and Shanks III, W.C. (2006). Laboratory studies of sphalerite decomposition: applications to the weathering of mine wastes and potential effects on water quality. Available at https://www.asmr.us/Portals/0/Documents/Conference-Proceedings/2006/2090-Stanton.pdf [Accessed 15 April 2020].

Stumm, W. and Morgan, J.J. (1996). Aquatic Chemistry: Equilibria and Rates in Natural Waters. Willey-Interscience, New York.

TMM. (2014). Pyrite (fool's fold): It's for collectors, not for fools. Available at http://www.treasuremountainmining.com/index.php?route=pavblog/blog&id=56 [Accessed 12 April 2020].

Udayabhanu, S. G. and Prasad, B. (2010). Studies on environmental impact of acid mine drainage generation and treatment: an appraisal. Indian Journal of Environmental Protection 30(11): 953–967

Vaughan, D. J., England, K.E.R., Kelsall, G.H. and Yin, Q. (1995). Electrochemical oxidation of chalcopyrite ($CuFeS_2$) and the related metal-enriched derivatives $Cu_4Fe_5S_8$, $Cu_9Fe_9S_{16}$, and $Cu_9Fe_8S_{16}$. American Mineralogist 80: 725–731.

Wang, H. and Salveson, I. (2005). A review on the mineral chemistry of the non-stoichiometric iron sulphide, $Fe_{(1-x)}$ ($0 \leq x \leq 0.125$): polymorphs, phase relations and transitions, electronic and magnetic structures. Phase Transitions 78(7–8): 547–567.

Zdun, T. (2001). Modelling the hydrodynamics of collie mining void 5B. MSc Dissertation, University of Western Australia, Australia.

4

Legislation and Policies Governing the Management of Acid Mine Drainage

Geoffrey S. Simate

CONTENTS

4.1 Introduction

Mining is one of the oldest industrial sectors in many countries that is of significant economic importance (OECD, 2019). As a result, the emphasis of mining policies in many countries in the world to date has focused mainly on the industrial and economic contributions of the mining industry (Amezaga and Kroll, 2005). However, mining industries create severe environmental problems by discharging waste in the form of solid, liquid and gaseous emissions (Wiertz, 1999; Chan et al., 2008; Ndlovu et al., 2017). Therefore, with a view of protecting the environment due to the production of waste from mining industries, regulations and policies have been enacted to control the disposal and/or recycling of mining and metallurgical wastes (Ndlovu et al., 2017). At the moment, governments have responsibilities to protect human health and the environment from the harmful effects of mining, including mitigating potential adverse impacts of acid mine drainage (AMD) (Commonwealth of Australia, 2016). It must be noted, however, that various countries have developed different approaches for managing AMD, or mining waste, in general, which differ both in the legislative scope and effectiveness (Kumar and Singh, 2013; Ndlovu et al., 2017). This chapter discusses legislation governing the management of AMD for a selected number of countries, but where there is no distinction between AMD and mine waste, in general, only legislation pertaining to mine waste will be discussed.

4.2 Legislation and Policies

As stated in Chapter 3, the formation of AMD involves complex processes including chemical, biological and electrochemical reactions that are dependent on the conditions of the environment. AMD itself is characterised by low values of pH (high acidity), high salinity, high osmotic pressure and high levels of sulphate and heavy metals (Mohan and Chander, 2001; Mohan and Chander, 2006a,b; García et al., 2013; Deloitte, 2013). AMD, by its nature, has given rise to several adverse environmental impacts including toxicity to aquatic organisms, destruction of the ecosystems, corrosion of infrastructure and tainting of water in regions where freshwater is already in short supply (Singh, 1987; Ruihua et al., 2011; Simate and Ndlovu, 2014). The adverse effects of AMD on plant life, human life and aquatic life have been reported in Chapter 5. Indeed, AMD gives rise to a range of environmental problems that will have to be addressed not only by technology, but also by socio-institutional interventions embedded in law and governance. Therefore, the subsections of this section of the chapter discuss guiding policy actions and/or legislations or laws of selected number of countries that are earmarked to help them in the mitigation of environmental impacts caused by AMD.

4.2.1 Analysis of Legislation and Policies of South Africa on Acid Mine Drainage

AMD is a multi-dimensional issue which is multi-scalar at the same time. Therefore, in order to determine the many challenges that law and governance face, from the South African perspective, Feris and Kotze (2014) are of the view that it is important to first understand some of the critical challenges associated with AMD. Two challenges were listed in this respect, namely (1) environmental challenges: this has been addressed extensively in several areas of this book, particularly, in Chapter 5, and (2) challenges for law and governance: this refers mainly to the insufficient legislation and regulatory practices that were in place several decades ago. Such legal frameworks were supposed to ensure proper rehabilitation of a substantial number of mines that have been left abandoned in South Africa after the completion of mining. In other words, prior to current legislations, numerous historical mining operations had been abandoned by their operators with little or no provision for the remediation of the impacts caused by mining (Deloitte, 2013). For example, prior to the promulgation of the current legislations, mining companies were bound only by the Water Act of 1956, which was insufficient in dealing with mine closures (Naidoo, 2017). Thus, historically, it was legally possible for mines that were no longer profitable to be boarded up and abandoned and the land transferred to the government (Mpofu et al., 2018). Confirmation of this is the Fanie Botha Accord of 1975 (Chamber of Mines). An agreement was reached between the Chamber of Mines and government, which stated that government would take over ownership of abandoned mines that existed before 1976 (Flynn and Chirwa, 2005).

The issue of abandoned and now ownerless mines is very problematic from legal perspective because it leads to a situation which shifts liability to the government as stated already, and ultimately taxpayers, who were not responsible for the pollution and who benefited less from the profits of the polluter(s). Several other challenges for law and governance also exist as discussed by Feris and Kotze (2014).

To start with, in addressing the issue of AMD in South Africa, it is important to note that the Constitution of the Republic of South is pivotal in safeguarding the environment, health and well-being of its people (Feris and Kotze, 2014). Furthermore, a number of statutory regulations are also vital in enforcing the objectives of the constitution. These two elements are discussed further in the subheadings that follow. Mpofu et al. (2018), however, argue that even within the framework of the latest environmental legislations, there is very limited regulation by government in real terms.

4.2.1.1 The Constitution of the Republic of South

Two subsections within Section 24 of the Constitution of the Republic of South Africa deal with the issue pertaining to the environment, whereas Section 27 deals with water matters. For example, subsection (a) of Section 24 calls for an environment that is not harmful to human health or well-being and subsection (b) requires that it is more important for the environment to be protected for the good of present and future generations. In the context of environmental protection, subsection (b) mandates the national government to establish appropriate legislative and other measures that are aimed at preventing pollution and ecological degradation, promote conservation and secure the ecologically sustainable development and utilisation of natural resources. Ecologically sustainable development is simply defined as "using, conserving and enhancing the community's resources so that ecological processes, on which life depends, are maintained, and the total quality of life, now and in the future, can be increased" (Australian Government, 1992).

Section 27 of the constitution provides a formal assurance of the right of access to sufficient water by every person. In this regard, the Water Services Act (WSA) of 1997 mandates the water services authorities to supply water-related services in a sustainable manner which implies that water is not only conserved, but water supplied to consumers is also acceptable for use (WSA, 1997). The aspect of providing water of an acceptable quality requires that local authorities also consider and address the impacts of AMD on its water services, sources and infrastructure when fulfilling its task of supplying water of good quality to its customers (Feris and Kotze, 2014). There is no doubt the ecological impacts of AMD also have various socio-economic effects on local authorities across South Africa as they grapple to give people water in sufficient quantity and an acceptable quality.

4.2.1.2 Statutory Frameworks That Regulate Pollution Caused by Acid Mine Drainage

In South Africa, mining activities are regulated by legislations from the mining, water and environmental divisions (Mpofu et al., 2018). However, according to Thomashausen et al. (2016), there are no specific regulations on AMD. Nevertheless, the provisions under National Water Act (NWA), Mineral and Petroleum Resources Development Act (MRPDA) and National Environmental Management Act (NEMA) on environmental damages may apply to AMD liabilities. In other words, the current environmental legislations relevant to AMD in South Africa includes NEMA, NWA and MRPDA. Therefore, in this subsection the statutory frameworks that may be used to regulate pollution caused by AMD are explored.

4.2.1.2.1 National Water Act

The South African NWA (Act 36 of 1998) provides a framework for the protection, use, development, conservation, management and control of water resources in the country as a whole (DWS, 2017). The NWA has been hailed by the international water community as one of the most progressive pieces of water legislation in the world and a major step forward in the translation of the concept of integrated water resources management (IWRM) into legislation (Schreiner, 2013). The act is premised on balancing the three legs of social benefit, economic efficiency and environmental sustainability and sets out the legal framework for the national government to protect, use, develop, conserve, manage and control water resources in the country (Schreiner, 2013). It also incorporates the principle of subsidiarity – management of water resources at the lowest appropriate level – through catchment-management agencies.

According to a study by Naidoo (2014), the NWA is the principal Act that governs water resource management in South Africa. The Act requires that pollution or degradation of the environment must be prevented or resolved. Furthermore, the NWA states that sustainability and equity are identified as central guiding principles in the protection, use, development, conservation, management and control of water resources (NWA, 1998). These principles recognise the basic human needs of present and future generations, the need to protect water resources, the need to promote social and economic development through the use of water and the need to establish suitable institutions in order to achieve the purpose of the Act (NWA, 1998).

The impact of the laws such as NWA and many others is manifested by aspects such as waste discharge costs and the polluter pays principle that have become important focuses of water management in the South African collieries (Postma and Schwab, 2002). Indeed, the "polluter pays" principle supports NWA and has direct implications for the mining industry which is closely related to AMD. The "polluter pays" principle stipulates that those who are responsible for producing, allowing or causing pollution should be held liable for the costs of clean-up and the legal enforcement (Cordato, 2001). For example, according to Postma and Schwab (2002), the implications of NWA that a mine be held responsible for its impact on water resources even after achieving certified formal mine closure from the Department of Minerals and Energy (DME) remain the basis for long-term water management employing a risk-based approach.

One important implementation instrument of the NWA is the Trans Caledon Tunnel Authority (TCTA). The TCTA is a state owned entity that has become directly involved with the AMD issue (Naidoo, 2014). It was established in terms of government Notice No. 2631 in the Government Gazette No. 10545 of December 1986. However, the 1986 notice has been replaced with Government Notice 277 in the Government Gazette No. 21017 of March 2000, which was circulated in terms of the NWA (TCTA, 2011). The TCTA's aim is to provide a sustainable water supply in the Southern African Region and is directly involved in South Africa's Lesotho Highlands Water Project. The TCTA also provides advisory support to the Department of Water Affairs and Forestry (DWAF) on the AMD project that has been implemented since 2010 (Naidoo, 2014).

4.2.1.2.2 National Environmental Management Act

The National Environmental Management Act No. 107 of 1998 serves as a general framework within which environmental management and implementation plans must be formulated. It serves as a guideline by reference to which any organ of state must exercise any function when taking any decisions in terms of NEMA or any statutory provision regarding the protection of the environment, and it serves as principles by reference to which a conciliator under NEMA must make recommendations and, lastly, guide any law concerned with protection or management of the environment (DWAF, 2008).

The NEMA contains certain principles that are applicable throughout the country by the actions of all organs of state that may affect the environment and applies together with all other appropriate and relevant considerations, which include the state's responsibility to respect, protect, promote and fulfil the social and economic rights in Chapter 2 of the Constitution of the Republic of South Africa (NWA, 1998).

The NEMA is administered by the Department of Environmental Affairs and Tourism (DEAT) and addresses AMD and mining impacts through statutory requirements for Environmental Impact Assessments (EIAs) and Environmental Management Programmes (EMPs) (Hobbs et al., 2008). It must also be noted that the National Environmental Management Amendment Act (No. 62 of 2008) and the NWA stipulate that a person(s) or party answerable or responsible for any mining operation shall take all reasonable steps to prevent pollution or degradation from taking place (Mpofu et al., 2018). In other words, NEMA requires that pollution or degradation of the environment must be prevented or rectified. If the landowner or person responsible for the pollution fails to take the required action, DEAT may take legal actions and recover the costs from the polluter (Hobbs et al., 2008).

4.2.1.2.3 Mineral and Petroleum Resources Development Act

The Mineral and Petroleum Resources Development Act 28 of 2002 is intended (1) to make provision for equitable access to and sustainable development of the nation's mineral and petroleum resources; and (2) to provide for matters connected therewith (MPDRA, 2002). In terms of the MPRDA, the principles set out in Section 2 of NEMA apply to all prospecting and mining operations (Mpofu et al., 2018). This implies that any prospecting or mining operation must comply with generally accepted principles geared towards sustainable development by incorporating social, economic and environmental factors into the planning and implementation of such operations (Mpofu et al., 2018). It is recognised that NEMA and the MPRDA have laid down new obligations for the mining and other industries, which include the requirement to monitor and remediate pollution of water resources (NEMA, 1998; MPRDA, 2002). Modern South African law in general also recognises that rehabilitative management of mines needs to continue after extractive operations have ended and that planning for the mine closure phase should always be in place. Furthermore, Section 48 (1)(a) of the MPRDA states that closure objectives and how they relate to the mining operation and its environmental and social setting must be included in the EMP that is developed during the planning stages of the mining operations (MPDRA, 2002).

It is also important to note that after the promulgation of the Minerals Act of 1991, all operational mines are required to provide funds to enable environmental and social rehabilitation after mine closure (Deloitte, 2013). The MPRDA of 2002, which, together with General Notice Regulation 527 of 2004 and associated guidelines, provide a methodology which is allowing for the financial estimation of the closure quantum to be provided by the mine; and this estimation is revised annually to ensure sufficient provision of funds (Deloitte, 2013).

4.2.1.3 Summary

The South African Constitution is the supreme law of the land, and Section 24 entrenches a right to a clean environment whilst the Constitution also enshrines a right to clean water in Section 27. All mining activity regulation including licensing is regulated by the MPRDA. Water licensing occurs under the NWA. Environmental regulation is catered for by the NEMA. Indeed, there is no doubt that the impact of mining on the freshwater sources in the river systems of South Africa is of serious concern (Jacobs and Testa, 2014). Unfortunately, a number of studies, so far, have shown that the government's efforts to finally manage AMD problem are compromised by numerous issues that tend to shift the focus away from it (Mpofu et al., 2018). For example, environmental legislation that enables action to address the AMD problem has been found to be inadequate in terms of enforcement (Naidoo, 2014; Mpofu et al., 2018). Furthermore, current policies that govern mining and water usage are inadequate and appear to be subject to abuse by those who are politically connected (Mpofu et al., 2018). However, it is also noted that prior to enactment of a number of legislations such as the MPRDA, numerous historical mining operations had been abandoned by their operators with little or no provision for the rehabilitation of the impacts caused by mining.

4.2.2 Analysis of Legislation and Policies of the United States of America on Acid Mine Drainage

The United States is a vast country that comprises of 50 states, with diverse culture and laws from state-to-state. Coal, which is one of the major causes of AMD, has been mined in the United States since the 1740s (Jacobs and Testa, 2014). Thereafter, gold mining which is also a cause of AMD gained prominence in the 1960s (Testa and Pompy, 2007; Jacobs and Testa, 2014).

An extensive regulatory system has been developed to govern current mining operations in the United States, as well as to guide the clean-up of historical ones (Hudson et al., 1999). In fact, there are several dozens of federal environmental laws and regulations that cover all aspects of mining in the United States (NMA, 2015). The framework for these regulations is primarily based on federal laws dating back to the late 1960s (Hudson et al., 1999). In addition, each state has laws and regulations that mining companies must follow (Ndlovu et al., 2017). Ideally, regulatory standards established at state levels are commonly equal to or more stringent than federal standards (Hudson et al., 1999).

4.2.2.1 Federal Laws and Policies That Regulate Pollution Caused by Acid Mine Drainage

Some of the major federal laws and regulations governing the mineral industry including AMD in the United States are analysed in this section.

4.2.2.1.1 National Environmental Policy Act

The National Environmental Policy Act (NEPA), passed in 1969, but enacted in 1970, established the basic environmental policies for the nation (Hudson et al., 1999). The NEPA defines processes for evaluating and communicating the environmental consequences of federal decisions and actions, such as the permitting of new mine development on federal lands. With respect to mining, the NEPA applies to mining operations requiring federal approval. The processes established by NEPA are used by concerned parties to ensure that environmental considerations are included in federal decisions. It requires federal agencies to prepare environmental impact statements (EIS) for major federal actions that may significantly affect the environment. These procedures exist to ensure that environmental information is available to public officials and citizens before actions are taken.

4.2.2.1.2 Comprehensive Environmental Response, Compensation and Liability Act

This Comprehensive Environmental Response, Compensation, and Liability Act (CERCLA) or law commonly known as Superfund, because of its funding aspects, was enacted by Congress on 11 December 1980 (Jacobs and Testa, 2014). This law requires operations to report inventory of chemicals handled and releases of hazardous substances to the environment. Hazardous substances are broadly defined under CERCLA and have included mining, milling and smelter wastes that are currently excluded from regulation under Resource Conservation and Recovery Act (RCRA) (Hudson et al., 1999; Ndlovu et al., 2017). It requires clean up of sites where hazardous substances are found.

This law created a tax on the chemical and petroleum industries and provided broad federal authority to respond directly to releases or threatened releases of hazardous substances that may endanger public health or the environment (Jacobs and Testa, 2014). The Superfund program was established to locate, investigate and clean up the worst abandoned hazardous waste sites nationwide and is currently being used by the U.S. Environmental Protection Agency (EPA) to clean up mineral-related contamination at numerous locations (Hudson et al., 1999). Clean-up alternatives range from "no action", at little cost, to removal of the entire mineralised zone, costing millions of dollars (Jacobs and Testa, 2014).

4.2.2.1.3 Resource Conservation and Recovery Act

The *Resource Conservation and Recovery Act* (RCRA), which is an amendment to the Solid Waste Disposal Act, was passed in 1976. The goals of the RCRA law are to conserve energy and natural resources, reduce the amount of waste generated and ensure that wastes are managed to protect human health and the environment. The RCRA gives EPA power to make and enforce regulations for managing many kinds of wastes.

In 1980, Congress specifically, albeit temporarily, excluded oil and gas wastes, mining wastes, waste from the combination of coal and other fossil fuels, and cement kiln dust waste from being regulated under Subtitle C, Hazardous Waste (Bevill Amendment) (Jacobs and Testa, 2014). As stated already, the changes imply that most mining, milling and smelting solid wastes in the category of "high-volume, low-hazard" materials have been exempted from regulation under RCRA (Hudson et al., 1999). The regulation of high-volume, low-hazard mining wastes is now the primary responsibility of the states. Therefore, the RCRA regulations only apply to three kinds of waste management: municipal, solid waste landfills; hazardous waste generators and transporters, and treatment, storage and disposal facilities; and underground tanks that store hazardous materials.

4.2.2.1.4 Surface Mining Control and Reclamation Act

The Surface Mining Control and Reclamation Act of 1977 (SMCRA) is the primary federal law that regulates the environmental effects of coal mining in the United States. SMCRA was basically created to provide a regulatory framework for regulating coal mining and providing a mechanism for remediation activities for lands and waters that had been affected adversely by past coal-mining activities (Jacobs and Testa, 2014). Two programs were created: one for regulating active coal mines and a second for reclaiming abandoned mine lands. SMCRA also created the Office of Surface Mining Reclamation and Enforcement (OSMRE). The OSMRE collects fees on the basis of a tonnage of surface mined coal, coal-mined underground and lignite on all active mining operations. The SMCRA funds are intended for clean-up projects at coal mines that were abandoned prior to 3 August 1977. In some cases, SMCRA can be used to address abandoned hardrock mines provided that the state certifies that the responsible party(ies) have already addressed all of the coal mine problems under their jurisdiction.

4.2.2.1.5 Federal Water Pollution Control Act (Clean Water Act)

According to Hudson et al. (1999), the Federal Water Pollution Control Act commonly referred to as the Clean Water Act (CWA) came into effect in 1977. The CWA's goal is to make all surface waters safe and eventually to stop all harmful discharges. One of the principal tools established by CWA is a permitting system for surface water discharges, known as the National Pollutant Discharge Elimination System (NPDES) (Hudson et al., 1999). The CWA requires mining operations to meet standards for surface water quality and for controlling discharges to surface water. The CWA-based regulations cover such mining-related situations as the disposal of mining-related waters, the pumping or draining of mine water to the surface, storm water runoff in mining operation areas, and control of seeps from mine tailings impoundments (Hudson et al., 1999).

4.2.2.1.6 *National Pollutant Discharge Elimination System Permit Program*

The CWA established the NPDES permit program as stated earlier under CWA in the preceding heading. This program controls water pollution by regulating point sources that discharge pollutants into waters of the United States. Point sources are discrete conveyances such as pipes or human-made ditches. Individual homes that are connected to a municipal system, use a septic system, or do not have surface discharge are not affected by this program, but industrial, municipal and other facilities are affected by this program if their discharges go directly into surface waters. Under this program it is illegal to discharge pollutants from a point source to waters of the United States, except in compliance with an NPDES permit. The US EPA and states with EPA-approved programs are authorised to issue permits.

4.2.2.1.7 *Title 30 of the Code of Federal Regulations Part 876*

This regulation is extracted from the chapter by Jacobs and Testa (2014) and states as follows: Regulations pertaining to AMD treatment and abatement are found under *Title 30 of the Code of Federal Regulations* Part 876 (30 CFR Part 876), Section 876, Acid Mine Drainage Treatment and Abatement Program. This set of federal regulations provides scope, information collection, eligibility, treatment and abatement plan content, and plan approval.

The Acid Mine Drainage Treatment and Abatement Program provides for the comprehensive abatement of the causes and treatment of the effects of AMD within qualified hydrologic units affected by coal-mining practices. AMD plans must adhere to *30 Code of Federal Regulations* Part 876.13 and include:

- Identification of the qualified hydrological unit
- Extent of the sources of AMD within the hydrological unit
- Identification of individual projects and the measures proposed to be undertaken to abate and treat the causes or effects of AMD within the hydrological unit
- Cost of undertaking the proposed sources of funding for such measures
- Analysis of the cost-effectiveness and environmental benefits of abatement and treatment measures

4.2.2.2 Summary

The importance of the environment, in general, and water, in particular, to the United States is seen in the vast number of regulations. These regulations were set up as guidelines on environmental and water issues and most importantly to provide mine operators, communities and regulators with essential information. In other words, they contain information and knowledge earmarked to assist all sectors of the mining industry, within and beyond about the requirements set by legislation. In addition, some of the regulations raise funds via a levy on active mines that is administered by the US Treasury to pay for the reclamation of abandoned mines, and some operations which pay taxes provide broad federal authority to respond directly to releases or threatened releases of hazardous substances that may pose eminent danger to the public as a whole.

4.2.3 Analysis of Legislation and Policies of the European Union on Acid Mine Drainage

Europe has a long tradition of mining activities, dating back as far as records of human settlement in Europe with the development of flint mines in France over 10,000 years ago (Wolkersdorfer and Bowell, 2004). In other words, mining activities have been of interest in the history of the European community from the very beginning (Hámor, 2004). Though Europe has a long and rich history of mining, there are now significantly more abandoned mines than operational mines across the continent (Jarvis et al., 2012). Consequently, considerable attention is focused on addressing the environmental legacy problems associated with these orphan sites (Jarvis et al., 2012). This implies that though the European mining policies have been shaped by the historical importance of mining for industrial development, the relatively recent introduction of environmental concerns in public policy has also changed the discourse (Amezaga and Kroll, 2005). Therefore, in addition to general industrial regulations, the European Union (EU) has adopted some legislations specific to mining (Szczepanski, 2012). In other words, the EU has developed a set of environmental directives that have had a significant effect on the mining industries of member nations (McKinley, 2004); and these rules concern the environmental impact of mining (especially waste and groundwater), as well as occupational health and safety (Szczepanski, 2012).

4.2.3.1 Analysis of Legislative Instruments and Policies of the European Union

This section deals with a selected number of existing EU legislative instruments pertaining to mining, environment and water management. However, according to Jarvis et al. (2012), two key legislative instruments which are influential in driving environmental improvements are the EU Water

Framework Directive (2000/60/EC) and EU Directive on the Management of Waste from the Extractive Industries (2006/21/EC). These two major pieces of legislation are subject to transposition (or transferring) into national regulations across individual states of the EU and are, therefore, subject to slight differences in application between nations. Nevertheless, the underlying principles apply across all EU member states.

4.2.3.1.1 *Water Framework Directive*

The Water Framework Directive (WFD) addresses concerns similar to those of the Clean Water Act (Ndlovu et al., 2017) of the United States. The EU WFD, adopted in 2000, takes a pioneering approach to protecting water based on natural geographical formations: river basins (EU Directive, 2000). The WFD obliges member states to draw up river basin management plans (RBMPs) to safeguard each of the 110 river basin districts. Under the WFD, member states have to hold extensive consultations with the public and interested parties to identify the problems, the solutions and their costs, to be included in river basin management plans. Public support and involvement is a precondition for the protection of waters. Without popular backing, regulatory measures will not succeed. European citizens have a key role to play in the implementation of the WFD and in helping governments to balance the social, environmental and economic questions to be taken into account.

Kroll et al. (2002) argue that the effects of mining on water are not addressed categorically in the WFD. The only specific reference to mine water in WFD is found in Article 11 (j), which allows reinjection of mine-derived water into the same aquifer (Kroll et al., 2002). Thus, despite being theoretically applicable to mine water management, emergent practices relative to implementation of the WFD require clarification.

4.2.3.1.2 *Waste Framework Directive*

Directive 2008/98/EC on waste (Waste Framework Directive) sets the basic concepts and definitions related to waste management, such as definitions of waste, by-products, recycling and recovery (EU Directive, 2008). It explains when waste ceases to be waste and becomes a secondary raw material (so-called end-of waste criteria). The Directive also lays down some basic waste management principles (polluter pays principle, extended producer responsibility).

4.2.3.1.3 *Extractive Waste Directive*

According to EU Directive (2006), in the EU, wastes derived from the extraction and refining industries are regulated under the Extractive Waste Directive 2006/21/EC ("the Directive" or "EWD"). In the Directive, extractive waste is defined as

> waste resulting from the prospecting, extraction, treatment and storage of mineral resources and the working of quarries.

In terms of what is "waste", the EWD makes reference to the definition as provided by the Waste Framework Directive 2008/98/EC as

> any substance or object which the holder discards or intends or is required to discard.

The Directive provides for measures, procedures and guidance to prevent or reduce as far as possible any adverse effects on the environment, in particular, water, air, soil, fauna and flora and landscape, and any resultant risks to human health, brought about as a result of the management of waste from the extractive industries (EU Directive, 2006).

The scope of the Directive covers extractive waste as already defined, including waste rock (unused extraction product), and mine tailings which are defined in the Extractive Waste Directive as:

> waste solids or slurries that remain after the treatment of minerals by separation processes (e.g. crushing, grinding, size-sorting, flotation and other physico-chemical techniques) to remove the valuable minerals from the less valuable rock.

4.2.3.1.4 Other Directives

Amongst other key directives are the Environmental Impact Assessment Directive (similar to the EIS requirements of the United States), and Hazardous Waste, and Landfill Directives (all including the Waste Framework address concerns similar to those of the US RCRA).

4.2.3.2 Summary

Mining operations and the disposal of mining wastes have sometimes caused significant environmental and social damage. Some communities and indigenous groups living near mines in the EU have alleged human rights abuses due to pollution from mining activities (Scannell, 2012). Therefore, appropriate EU legislation and policies should strongly influence approaches to mine waste/water management in Europe. However, because European-wide legislation is subject to national-level transposition and implementation, the overall picture of mine waste/water management in Europe is complex (Jarvis et al., 2012). There are, nevertheless, some research and development themes that are common across Europe and are wide ranging (Jarvis et al., 2012). For example, the partnership for acid drainage remediation in Europe (PADRE) is one of the research strategies whose main aim is to act as a repository for capturing the wide variety of R&D activities. It also ensures that researchers, consultants and mining industry personnel are not only aware of both the historic and current outputs of the R&D, but also avoid replication of effort. The specific objectives of PADRE as stated in its statutes include (1) to provide a network and collaborative platform for European and international research and development into techniques for the prediction, prevention and remediation of acidic drainage in Europe, (2) to promote dissemination of knowledge of current best-practice and innovations relating to acidic drainage prediction, prevention and remediation, with particular reference to European conditions, including the evolving framework of relevant EU legislation, (3) to advance the training of present and future generations of European professionals who will engage in the art and science of acidic drainage prediction, prevention and remediation and (4) to actively collaborate with a Global Alliance of organisations based in other continents which share similar objectives (Jarvis et al., 2012).

As can be seen from the EU directives and PADRE, the efforts of the EU on issues of mining, environment and water management legislations seem to be well coordinated. In fact, each country's environmental laws are derived from the EU directives; and PADRE is endeavouring to ensure that professionals developing solutions to mining pollution are aware of the complimentary work of others across Europe, and that internationally significant work conducted in Europe is effectively rolled out to a global audience. This clearly shows that the EU has adopted advanced regulations to improve environmental standards in mining. For example, the EU documents on best available techniques that are designed to significantly reduce environmental impacts are readily available.

4.2.4 Analysis of Legislation and Policies of Australia on Acid Mine Drainage

For a better understanding, any analysis of Australia's mining laws should also give a brief overview of the nation's system of government (Chambers, 2020). In this regard, Australia's constitution provides for a federal system of government. This results in a division of jurisdiction over mineral resources between two levels: Federal (or Commonwealth) and the State (there are seven states and two territories).

According to Cunsolo and McKenzie (2019), the regulatory framework for the mineral extraction process is divided mainly into the two levels of government, but the local government, which has literally no jurisdiction over mineral resources at all except a few responsibilities, may also be included:

1. *Australian Federal Government* – Federal involvement in mining regulation is not extensive. It involves indirect policy formulations, such as: taxation, foreign investment law, competition policy, trade and customs, corporation law, international agreements, native title, and national environmental laws. Ideally, the only key interaction with the Australian federal government is in relation to foreign investment approval (through the Foreign Investment Review Board)

2. *State and Territories* – State/Territory governments are constitutionally vested with the primary responsibility for: land administration,

granting mineral exploration and mining titles, regulating mine operations (including environmental and occupational health and safety) and collecting royalties on minerals produced.

3. *Local Governments* – Operating at city, town or shire/district level, the local governments are responsible for handling community needs. Interaction with local governments is usually in relation to the upkeep of roads and town services for mine staff and their families.

4.2.4.1 Regulatory Frameworks for Assessing and Managing Acid Mine Drainage in Australia

According to the Commonwealth of Australia (2007) and Commonwealth of Australia (2016), the federal (or commonwealth), state and local governments have legislation and guidelines in place that are relevant to mine site AMD management. The aim is to protect environmental aspects such as biodiversity, water resources (quantity and quality), landforms, existing and potential future land uses, and cultural and environmental heritage.

The state and territory governments have primary responsibility for oversight and regulation of mining operations. This is often administered through a mining resources agency, a natural resource management agency and/or a statutory environment authority. The Commonwealth is primarily involved where issues of national environmental significance have been established or where there are agreed national frameworks for managing certain environmental aspects.

The following are key Commonwealth regulations that are relevant to AMD: (1) Environment Protection and Biodiversity Conservation Act, (2) National Environment Protection Measures (NEPM), (3) Australian and New Zealand Mineral and Energy Council (ANZMEC), and (4) ANZECC/ARMCANZ Water Quality Guidelines. Basically, the ANZECC/ARMCANZ Water Quality Guidelines provide a risk-based approach to the development of site discharge standards.

4.2.4.1.1 *Environment Protection and Biodiversity Conservation Act*

The Environment Protection and Biodiversity Conservation Act 1999 (EPBC Act) is the Australian government's key piece of environmental legislation which commenced on 16 July 2000 (EPBC, 2000). The EPBC Act enables the Australian government to create ties with the states and territories in providing a truly national scheme of environment and heritage protection and biodiversity conservation. In addition, the EPBC Act focuses the interests of the Australian government on the protection of matters of national environmental significance, with the states and territories having responsibility for matters of state and local significance.

Amendments to the EPBC Act became law on 22 June 2013, making water resources a matter of national environmental significance, in relation to coal seam gas and large coal mining development (EPBC, 2013).

4.2.4.1.2 *National Environment Protection Council*

The National Environment Protection Council is a statutory body with law-making powers established under the National Environment Protection Council Act 1994 (Commonwealth) and corresponding legislation in other Australian jurisdictions (NEPC, 2015). The National Environment Protection Council Act 1994 (NEPC Act) recognises the importance of communities and businesses in protecting Australia's environment, and that national outcomes are best achieved through regionally tailored approaches.

The NEPC has two primary functions (NEPC, 2015):

- to make National Environment Protection Measures (NEPMs)
- to assess and report on the implementation and effectiveness of NEPMs in participating jurisdictions

National Environment Protection Measures (NEPMs), created under the NEPC Act, can be used to establish nationally consistent environmental standards, goals, guidelines or protocols in relation to air, water, noise, site contamination, hazardous waste and recycling. A NEPM is a Commonwealth legislative instrument. Once a NEPM is made or varied, its implementation is the prerogative of each jurisdiction. The regulation is just one of a suite of implementation tools a jurisdiction may use.

The NEPMs provide a single national framework that addresses one or more environmental issues, with the flexibility for local implementation to take into account variability between jurisdictions. This provides certainty and consistency for business and the community in the management of these environmental issues, while reducing the need for regulation.

Currently, there are seven NEPMs as follows (NEPC, 2015):

1. *Air Toxics* – Sets out a nationally consistent approach to collection of data on toxic air pollutants (such as benzene) in order to deliver a comprehensive information base from which standards can be developed to manage the air pollutants in order to protect human health.

2. *Ambient Air Quality* – Establishes a nationally consistent framework for monitoring and reporting on air quality, including the presence of pollutants such as carbon monoxide, lead and particulates. Work commenced in 2013–2014, towards making a variation to this NEPM, which included a public consultation. The final variation of the framework was completed in 2015–2016, with the compilation prepared on 25 February 2016 (NEPM, 2016).

3. *Assessment of Site Contamination* – Provides a nationally consistent approach to the assessment of site contamination so as to ensure sound environmental management practices by regulators, site assessors, environmental auditors, landowners, developers and

industry. It has been highly effective in providing authoritative guidance to practitioners in this field.

4. *Diesel Vehicle Emissions* – Supports reducing pollution from diesel vehicles. Several jurisdictions operate a suite of programs to reduce exhaust emissions from diesel vehicles.

5. *Movement of Controlled Waste* – Operates to minimise potential environmental and human health impacts related to the movement of certain waste materials, by ensuring that waste that is being moved between states and territories is properly identified, transported and handled in ways consistent with environmentally sound management practices.

6. *National Pollutant Inventory* – Provides a framework for collection and dissemination of information to improve ambient air and water quality, minimise environmental impacts associated with hazardous wastes and improve the sustainable use of resources.

7. *Used Packaging Materials* – Operates to minimise environmental impacts of packaging materials, through design (optimizing packaging to use resources more efficiently), recycling (efficiently collecting and recycling packaging) and product stewardship (demonstrating commitment by stakeholders).

4.2.4.1.3 Australian and New Zealand Mineral and Energy Council

In 1995, the Australian and New Zealand Mineral and Energy Council (ANZMEC) published a baseline environmental guideline for operating mines in Australia, which required the need for acid generation to be predicted and incorporated in the mine closure plan (ANZMEC, 1995, 2000). To better understand the impact of AMD in Australia and to provide the basis for assessing long-term management options, the Office of the Supervising Scientist and the Australian Centre for Minesite Rehabilitation Research initiated the preparation of a status report on AMD. Results from the survey (Harries, 1997) suggested that about 54 sites in Australia were managing significant amounts of potentially acid-generating wastes, where "significant amounts" means that more than 10% of the waste is potentially acid generating or there is more than 10 million metric ton (mt) of potentially acid-generating wastes. About 62 additional sites were managing some potentially acid-generating wastes, but less than 10% of the total wastes and less than 10 mt.

4.2.4.1.4 Water Quality Guidelines

The new ANZECC/ARMCANZ Water Quality Guidelines were introduced in 2000, and represent a major shift in the way surface water quality is managed in Australia (Batley et al., 2003). The guideline package consists of several large volumes of information and provides a complete outline of how the guidelines should be applied, together with a lengthy discussion of the underpinning science. The new guidelines represent world's best practice in water quality management, and adherence to their principles and philosophy is clearly desirable. It is important to note that, whereas the role of regulatory agencies is to protect or improve ambient water quality through managing effluents, they also need to consider cumulative impacts, total loads and contaminant cycling through sediments, biota and water, as well as temporary or permanent storages of contaminants and their ultimate fate and impacts. This means that both a spatial and temporal view is required along with recognition that pollutants can and will move downstream and downwind and change over time. This means that regulatory issues have a broader, whole of catchment focus, rather than the more localised concerns that industries might have.

The new guidelines provide guidelines for water quality in relation to a number of environmental values (previously termed "beneficial uses"). These are somewhat altered from those in the 1992 guidelines and now comprise the following (Batley et al., 2003):

- *Aquatic Ecosystems* – This is the primary focus of the new guidelines.
- *Primary Industries* – The new guidelines have amalgamated agriculture, aquaculture and human consumption of aquatic foods into one environmental value.
- *Recreation and Aesthetics* – New guidelines are still under review. Until these are revised and released, the ANZECC 1992 Guidelines still apply.
- *Drinking Water* – The Australian NHMRC and ARMCANZ Drinking Water Guidelines (NHMRC/ARMCANZ 1996) apply.
- *Industrial Water* – No specific guidance is given because the requirements for industrial waters are so varied.
- *Cultural and Spiritual Values* – No specific guidance is given in the new guidelines, but it is recommended that they should be considered during the planning phase of water resource management.

Note: According to Batley et al. (2003), guidelines for aquatic ecosystem protection are usually the most stringent. For the most part, these are the major focus of management actions, the greatest source of contention amongst the stakeholders and, therefore, the values that determine the acceptability of water quality. As a consequence, the guideline volumes devote the greatest effort dealing with aquatic ecosystem protection.

4.2.4.2 Summary

In Australia, the state and territory have primary responsibilities for oversight and regulation of mining operations (Commonwealth of Australia, 2007), with an overlay of Commonwealth (federal) regulation (Herbert Smith Freehills, 2012). In other words, mining activities in each of the states are governed by their respective Mining Acts and Mining Regulations. The primary means by which state and territory governments regulate AMD is through the standard authorisations required for a mining project, including mining leases, EIAs and water resources (Commonwealth of Australia, 2007). Although the exact structure, legislation and regulatory regime applicable to AMD vary somewhat between jurisdictions, in general, they all seek to minimise environmental impacts during operations and achieve sustainable landforms following rehabilitation through the minimisation of pollutant release.

According to Commonwealth of Australia (2007), the key considerations under state and territory legislation include: (1) identification and assessment of AMD risks in the environment and social impact assessment, (2) determination of financial bonds based on adequate management of AMD issues post closure, (3) management of compliance with national water quality guidelines and (4) availability, quality and use of local and regional water resources.

4.3 Concluding Remarks

Society's need for metals is undisputed. For example, the demand for metal usage has risen due to a continuously growing population, an increase in standard of living in emerging economies, and on going consumption in industrialised countries. Technological innovations, such as newer digital technologies, also require an increase in the variety of metals as well. However, despite the importance of the mining industry in fulfilling the need for minerals and its contributions to economic and social development globally, concerns about aspects of its relationship to the degradation of the environment have been lingering. There is no doubt mining is also increasingly influenced by other competing land uses, such as urban development, agriculture and nature conservation. However, it is inevitable that a balanced consideration of economic, environmental and social aspects with a view of ensuring sustainable development of the mining industry is included in the frameworks underlying the development of coherent legislations and policies. What is clear from this chapter is that though different countries have different regulations and policies from one another, they all agree in many aspects about the importance of protecting the environment and water resources. It must also be noted that a better understanding of the activities leading to AMD generation could enable countries to develop legislation and policies that are likely to achieve the desired results of combating pollution of the environment and water resources at an early stage. Indeed, there are many examples and case studies globally that provide evidence that sound planning and management can significantly reduce environmental risks of AMD. For example, the EU and other industrialised and developed nations have also continued to fund various research projects on environmental technologies and policy strategies towards responsible mining.

References

Amezaga, J.M. and Kroll, A. (2005). European Union policies and mine water management. In: Wolkersdorfer, C. and Bowell, R. (Editors), Contemporary reviews of mine water studies in Europe. Springer-Verlag, Berlin.

Australian and New Zealand Mineral and Energy Council (ANZMEC). (1995). Baseline environmental guidelines for new and existing mines (Report 95-02). Available at https://portals.iucn.org/library/node/25728 [Accessed 7 July 2020].

Australian and New Zealand Mineral and Energy Council (ANZMEC). (2000). Strategic framework for mine closure. ANZMEC Secretariat, Canberra.

Australian Government. (1992). National strategy for ecologically sustainable development – part 1 introduction. Available at https://www.industry.gov.au/sites/default/files/2019-04/lpsdp-preventing-acid-and-metalliferous-drainage-handbook-english.pdf [Accessed 1 July 2020].

Batley, G.E., Humphrey, C.L., Apte, S.C. and Stauber, J.L. (2003). A guide to the application of the ANZECC/ARMCANZ water quality guidelines in the Minerals industry. Australian Centre for Mining Environmental Research, Kenmore, Queensland.

Chambers, R.H. (2020). An overview of the Australian legal framework for mining projects in Australia. Chambers and Company International Lawyers, New York.

Chan, B.K.C., Bouzalakos, S. and Dedeney, A.W.L. (2008). Integrated waste and water management in mining and metallurgical industries. Transactions of Nonferrous Metals Society of China 18: 1497–1505.

Commonwealth of Australia. (2007). Managing acid and metalliferous drainage: leading practice sustainable development program for the mining industry. Available at https://www.im4dc.org/wp-content/uploads/2014/01/Managing-acid-and-metalliferous-drainage.pdf [Accessed 7 May 2020].

Commonwealth of Australia. (2016). Preventing acid and metalliferous drainage: leading practice sustainable development program for the mining industry. Available at https://www.industry.gov.au/sites/default/files/2019-04/lpsdp-preventing-acid-and-metalliferous-drainage-handbook-english.pdf [Accessed 3 May 2020].

Cordato, R.E. (2001). The polluter pays principle: a proper guide for environmental. The Institute for Research on the Economics of Taxation Policy, Washington, D.C.

Cunsolo, A. and McKenzie, B. (2019). Mining in Australia: overview. Available at https://uk.practicallaw.thomsonreuters.com/8-576-7530?transitionType=Default&contextData=(sc.Default)&firstPage=true&bhcp=1 [Accessed 7 July 2020].

Deloitte. (2013). Acid mine water drainage: debating a sustainable solution to a serious issue. Available at https://www.industry.gov.au/sites/default/files/2019-04/lpsdp-preventing-acid-and-metalliferous-drainage-handbook-english.pdf [Accessed 3 July 2020].

Department of Water Affairs and Forestry (DWAF). (2008). Best practice guideline G5: water management aspects for mine closure. Available at https://www. mineralscouncil.org.za/work/environment/environmental-resources/send/ 26-environmental-resources/348-g5-water-management-aspects-for-mine-closure [Accessed 3 July 2020].

Department of Water and Sanitation (DWS). (2017). Mine water management policy position draft for external consultation and discussion. Available at https://cisp. cachefly.net/assets/articles/attachments/69904_40965_gon657.pdf [Accessed 2 July 2020].

EPBC. (2000). The Environment Protection and Biodiversity Conservation Act 1999 (EPBC Act). Available at https://www.environment.gov.au/epbc/about [Accessed 8 July 2020].

EPBC. (2013). Water resources – 2013 EPBC Act Amendment – Water trigger. Available at https://www.environment.gov.au/epbc/what-is-protected/water-resources [Accessed 8 July 2020].

EU Directive. (2000). The EU Water Framework Directive. Available at https:// ec.europa.eu/environment/pubs/pdf/factsheets/wfd/en.pdf [Accessed 7 July 2020].

EU Directive. (2006). Extractive Waste Directive 2006/21/EC. Available at https:// ec.europa.eu/environment/waste/studies/mining/waste_extractive_industries. pdf [Accessed 7 July 2020].

EU Directive. (2008). The EU Waste Framework Directive. Available at https:// ec.europa.eu/environment/eco-innovation/files/docs/publi/waste-directive-ecoi-projects.pdf [Accessed 7 July 2020].

Feris, L. and Kotze, L.J. (2014). The regulation of acid mine drainage in South Africa: law and governance perspectives. Potchefstroom Electronic Law Journal 17 (5): 2105–2116.

Flynn, S. and Chirwa, D.M. (2005). The constitutional implications of commercializing water in South Africa. In: McDonald, D. and Ruiters, G (Editors), The age of commodity: water privatization in Southern Africa. Taylor & Francis, Oxfordshire, pp. 59–76.

García, V., Häyrynen, P., Landaburu-Aguirre, J., Pirilä, M., Keiski, R.L. and Urtiag, A. (2013). Purification techniques for the recovery of valuable compounds from acid mine drainage and cyanide tailings: application of green engineering principles. Journal of Chemical Technology and Biotechnology 89: 803–813.

Hámor, T. (2004). Sustainable mining in the European Union: the legislative aspect. Environmental Management 33 (2): 252–261.

Harries, J. (1997). Acid mine drainage in Australia: its extent and potential future liability. Available at http://www.environment.gov.au/science/supervising-scientist/publications/ssr/acid-mine-drainage-australia-its-extent-and-potential-future-liability [Accessed 7 July 2020].

Herbert Smith Freehills. (2012). Mining law in Asia. Available at http://files. chinagoabroad.com/Public/uploads/v2/guide/attachment/Mining% 20law%20in%20Asia_Herbert%20Smith%20Freehills%20Lawyers.pdf.PDF [Accessed 27 July 2020].

Hobbs, P., Oelofse, S.H.H. and Rascher, J. (2008). Management of environmental impacts from coal mining in the upper Olifants River catchment as a function 150 of age and scale. Water Resources Development 24(3): 417–43.

Hudson, T.L, Fox, F.D. and Plumlee, G.S. (1999). Metal mining and the environment. American Geological Institute, Alexandria, VA.

Jacobs, J.A. and Testa, S.M. (2014). Mine reclamation policy and regulations of selected jurisdictions. In: Jacobs, J.A., Lehr, J.H. and Testa, S.M. (Editors), Acid mine drainage, rock drainage, and acid sulfate soils: causes, assessment, prediction, prevention, and remediation. John Wiley & Sons, London, pp. 319–324.

Jarvis, A.P., Alakangas, L., Azzie, B., Lindahl, L., Loredo, J., Madai, F., Walder, I.F. and Wolkersdorfer, C. (2012). Developments and challenges in the management of mining wastes and waters in Europe. In: Proceedings of the 9th International Conference on Acid Rock Drainage, Ottawa, Canada, 20–26 May, pp 769–780.

Kroll, A., Amézaga, J.M., Younger, P.L. and Wolkersdorfer, C. (2002). Regulation of mine waters in the European Union: the contribution of scientific research to policy development. Mine Water and the Environment 21: 193–200.

Kumar, U. and Singh, D.N. (2013). E-waste management through regulations. International Journal of Engineering Inventions 3 (2): 6–14.

McKinley, M.J. (2004). Mining. Available at http://www.pollutionissues.com/Li-Na/Mining.html [Accessed 6 July 2020].

Mineral and Petroleum Resources Development Act 28 of 2002 (MPDRA). Government Gazette. Available at https://www.gov.za/sites/default/files/gcis_document/201409/a28-020.pdf [Accessed 4 July 2020].

Mohan, D. and Chander, S. (2001). Single component and multi-component adsorption of metal ions by activated carbons. Colloids and surfaces a: physicochemical and engineering aspects 177: 183–196.

Mohan, D. and Chander, S. (2006a). Single, binary and multicomponent sorption of iron and manganese on lignite. Journal of Colloid and Interface Science 299: 76–87.

Mohan, D. and Chander, S. (2006b). Removal and recovery of metal ions from acid mine drainage using lignite – a low cost sorbent. Journal of Hazardous Materials B137: 1545–1553

Mpofu, C., Morodi, T.J. and Hattingh, J.P. (2018). Governance and socio-political issues in management of acid mine drainage in South Africa. Water Policy 20: 77–89.

Naidoo, S. (2017). Acid mine drainage in South Africa: development actors, policy impacts, and broader implications. Springer, Cham, Switzerland

Naidoo, S. (2014). Development actors and the issues of acid mine drainage in the Vaal river system. Dissertation, University of South Africa

National Environment Protection (Ambient Air Quality) Measure (NEPM, 2016). Available at https://www.legislation.gov.au/Details/F2016C00215 [Accessed 8 July 2020].

National Environment Protection Council (NEPC). (2015). National Environment Protection Council 2014–2015 Annual Report. Available at http://www.nepc.gov.au/system/files/resources/e3da1ed8-68f0-48e5-937a-5de0045feb62/files/nepc-annual-report-2014-15.pdf [Accessed 8 July 2020].

National Environmental Management Act, 1998 (Act 107 of 1998) (NEMA). Government Gazette. Available at https://www.gov.za/sites/default/files/gcis_document/201409/a107-98.pdf [Accessed 3 July 2020].

National Mining Association (NMA). (2015). Federal environmental laws that govern U.S. mining. Available at https://nma.org/wp-content/uploads/2016/10/NMA-Fact-Sheet-Federal-Laws-that-Govern-Mining-v2.pdf [Accessed 4 July 2020].

National Water Act, 1998 (Act 36 of 1998) (NWA). Government Gazette. Available at https://www.gov.za/sites/default/files/gcis_document/201409/a36-98.pdf [Accessed 3 July 2020].

Ndlovu, S., Simate, G.S. and Matinde, E. (2017). Waste production and utilization in the metal extraction industry. CRC Press, Boca Raton, FL.

OECD. (2019). Mining and Green Growth in the Eastern Europe, Caucasus, and Central Asia region. Available at https://www.oecd.org/environment/outreach/20190413_Mining%20and%20Green%20Growth%20Final.pdf [Accessed 6 July 2020].

Postma, B. and Schwab, R. (2002). Mine closure: the way forward from DWAF's perspective. Paper presented at the WISA Mine Water Division, Mine Closure Conference, Randfontein, 23–24 October.

Ruihua, L., Lin, Z., Tao, T. and Bo, L. (2011). Phosphorus removal performance of acid mine drainage from wastewater. Journal of Hazardous Materials 190: 669–676.

Scannell, Y. (2012). The regulation of mining and mining waste in the European Union. Washington and Lee Journal of Energy, Climate and the Environment 3(2): 177–267

Schreiner, B. (2013). Viewpoint – why has the South African National Water Act been so difficult to implement? Water Alternatives 6(2): 239–245.

Simate, G.S. and Ndlovu, S. (2014). Acid mine drainage: challenges and opportunities. Journal of Environmental Chemical Engineering 2(3): 1785–1803.

Singh, G. (1987). Mine water quality deterioration due to acid mine drainage. International Journal of Mine Water 6(1): 49–61.

Szczepanski, M. (2012). Mining in the EU: regulation and the way forward. Available at https://www.europarl.europa.eu/RegData/bibliotheque/briefing/2012/120376/LDM_BRI(2012)120376_REV1_EN.pdf [Accessed 6 July 2020].

Testa, S.M. and Pompy, J.S. (2007). Report on backfilling of open-pit metallic mines in California. Available at https://www.conservation.ca.gov/smgb/reports/Documents/Information_Reports/SMGB%20IR%202007-02.pdf [Accessed 4 July 2020].

Thomashausen, S., Capone, J.E. and Langalanga, A. (2016). Water risks in the mingsector South Africa1. Available at http://ccsi.columbia.edu/files/2016/06/Water-Template-South-Africa.pdf [Accessed 2 June 2020].

Trans-Caledon Tunnel Authority (TCTA). (2011). AMD Witwatersrand basin due diligence. Presentation to the Parliamentary Committee on Water and Environmental Affairs, Cape Town, 7 September 2011.

Wiertz, J.V. (1999). Mining and metallurgical waste management in Chilean copper industry. IWA Proceedings, Sevilla, Spain, pp. 403–408.

Wolkersdorfer, C. and Bowell, R. (2004). Contemporary reviews of mine water studies in Europe. Mine Water and the Environment 23: 161.

WSA. (1997). Water Services Act, 1997. Available at https://www.gov.za/sites/default/files/gcis_document/201409/a108-97.pdf [Accessed 2 July 2020].

5

Environmental and Health Effects of Acid Mine Drainage

Geoffrey S. Simate

CONTENTS

5.1 Introduction

As discussed in Chapter 3, acid mine drainage (AMD) which, without doubt, is one of the serious pollutants in the world, is extremely acidic (pH as low as 2) and is enriched with iron, aluminium, sulphate and many other heavy metals (Evangelou, 2001). More specifically, AMD is most often characterised by one or more of the four major components: (1) low pH (high acidity), (2) high metal concentrations (with iron being the most common), (3) elevated sulphate levels and (4) excessive suspended solids and/or siltation (AMRC, 2020).

It is obvious that the seriousness of AMD, as a pollutant, arises from its nature, extent and difficulty of resolution, as well as the economic costs of its prevention and remediation. Research has shown that AMD has the potential to affect the ecosystem and the environment negatively through the release of acidic water that contains high concentrations of metals and metalloids (CSIR 2013; Simate and Ndlovu, 2014). It is very clear from such studies that AMD effluents are associated with numerous environmental impacts such as the increase in the acidity of soils and water sources, in addition to the content of dissolved salts (Foureaux et al., 2020). AMD is also responsible for various human and animal health problems if it is disposed without proper treatment in water bodies intended for consumption (Foureaux et al., 2020). The most unfortunate fact is that AMD which is one of the most prominent environmental issues currently facing the mining industry is regarded as second only to global warming and stratospheric ozone depletion in terms of global ecological risk (Crane and Sapsford, 2018). This chapter discusses the pertinent findings of the impacts of AMD on the ecosystem, in particular, humans, plants and aquatic species based on the available documented literature. The chapter will specifically focus on the effects of acidity and heavy metals on plants, aquatic species and human health.

5.2 Effect of High Acidity

Several studies have characterised AMD, but for the purpose of this chapter AMD is basically highly acidic water having pH of less than 5 (or even as low as 2) and also contains several contaminants including heavy metals (Evangelou, 2001; RoyChowdhury et al., 2015; John et al., 2017). The acidity of water is a function of the concentration or, more correctly, the activity of free hydrogen ions (H^+). Since the molar concentration of these ions is normally very small, it is usually expressed as pH, the negative base-10 logarithm of the hydrogen ion concentration in mol L^{-1} as shown in Equation 5.1:

$$pH = -\log_{10}\left[H^+\right] \tag{5.1}$$

The pH ranges from 1 (highly acidic) through 7 (neutral) to 14 (highly basic). An understanding of the acidity in AMD is significant. Acidity, for example, has the ability to solubilise metals and increase their toxicity because metals in a dissolved state tend to be more harmful to aquatic species, humans and plants (Hunter 1980; Spyra, 2017).

5.2.1 Effect of High Acidity on Plants

Since the eighteenth century botanists have been aware that the chemical nature of soil affects plant growth and distribution (Rorison, 1973). For example, factors associated with soil acidity have been found to be limiting the growth of plants in many parts of the world (Kidd and Proctor, 2001). In fact, research on the effects of acidic soils on plant growth has been performed in the world; however, it is still a serious problem, especially in mine site where revegetation is necessary for environmental reclamation (Matsumoto et al., 2017).

In the natural environment, soil pH influences myriads of soil biological, chemical and physical properties and processes that affect plant growth and biomass yield (Neina, 2019). For example, the soil pH affects the abundance and activity of soil organisms (from microorganisms to arthropods) responsible for transformations of nutrients (De Boer and Kowalchuk, 2001; Nicol et al., 2008; Soti et al., 2015). However, the diversity of plant species has been found to be low in most acidic soils (Dupré et al., 2002; Soti et al., 2015) as essential nutrients (such as Ca, Mg, K, PO4 and Mo) exist in unavailable forms to plants thus causing nutrient deficiency (Larcher, 2003; Soti et al., 2015). Additionally, acidic soils have high cation exchange capacity and promote leaching of nutrients which results in soil being unfavourable for plant growth (Johnson, 2002; Soti et al., 2015).

There is no doubt that plants need a proper balance of macro and micronutrients in the soil and thus the soil pH, as stated already, has a significant role that it plays on the availability of nutrients in the soil (Halcomb and Fare, 2002; Larcher, 2003; Soti et al., 2015) and on the growth of different kinds of plants (Halcomb and Fare, 2002; Simate and Ndlovu, 2014). For example, when the soil pH is low, phosphorus, potassium and nitrogen are tied up in the soil and not available to plants (Halcomb and Fare, 2002). Calcium and magnesium, which are essential plant nutrients, may also be absent or deficient in low pH soils (Halcomb and Fare, 2002; Simate and Ndlovu, 2014). At low pH, toxic elements such as aluminium, iron and manganese are also released from soil particles, thus increasing their toxicity (Schrock et al., 2001; Halcomb and Fare, 2002). Furthermore, if soil pH is low, the activity of soil organisms that break down organic matter is reduced (Halcomb and Fare, 2002). Proper soil pH increases microorganism activity which produces improved soil tilth, aeration and drainage. This is turn allows for better use of nutrients, increased root development and drought tolerance (Halcomb and Fare, 2002; Simate and Ndlovu, 2014).

Other effects of soil acidity include the fact that important productive plants such as lucerne, phalaris, canola and barley are difficult to establish and grow in acidic soils (Teshome, 2017). Soil acidity is also able to alter root length and architecture, root hair development, and deforms the root tip in some plant species (Haling et al., 2011). Basically, soil acidity affects root development, leading to reduced nutrient and water uptake and deficiency in essential plant nutrients, such as K, Ca and Mg (Abdenna et al., 2007).

The two most important toxicities in acidic soils are those of aluminium and manganese (Slattery et al., 1999; Osman, 2014). For example, in strongly acidic soils (pH < 4.3) aluminium and manganese become more available in the soil solution and are harmful to plant roots (Osman, 2014). In other words, at low pH, toxic elements such as aluminium, iron and manganese are also released from soil particles, thus increasing their toxicity (Halcomb and Fare, 2002; Schrock et al., 2001). Aluminium toxicity is the most common plant symptom in acidic soils and causes root stunting (Slattery et al. 2000;

Osman, 2014). Reduced root growth impedes nutrient and water uptake and results in decreased production (Osman, 2014).

5.2.2 Effect of High Acidity on Aquatic Species

The pH of waters is important to aquatic life because pH affects the ability of aquatic organisms to regulate basic life-sustaining processes, particularly the exchanges of respiratory gasses and salts with the water in which they live (Rober-Bryan, Inc., 2004). In other words, pH (or more appropriately H⁺ activity) has a large influence on many important chemical reactions such as dissociation of organic acids and concentration and speciation of potentially toxic aluminium (Sullivan, 2000). More specifically, the pH of the water is a major physical factor that determines the distributions of organisms in aquatic habitats (Clark et al., 2004). It is noted that in many instances important physiological processes operate normally in most aquatic biota under a relatively wide pH range (e.g., 6–9 pH) (Rober-Bryan, Inc., 2004). In fact, most of the freshwater lakes, streams and ponds have a natural pH in the range of 6 to 8 (Lenntech, 2014). It must be noted, however, that the acceptable range of pH to aquatic life, particularly fish, depends on numerous other factors, including prior pH acclimatisation, water temperature, dissolved oxygen concentration and the concentrations and ratios of various cations and anions (Rober-Bryan, Inc., 2004).

Unfortunately, human activities present aquatic species with numerous environmental challenges including altered pH regimes (freshwater acidification) (Isaza et al., 2020). It must also be noted that the increased acidity caused by AMD has a range of negative effects on aquatic species depending on the severity of the pH change (Coil et al., 2013). Nevertheless, aquatic organisms are affected by acidic water at all trophic levels, resulting in changes in productivity and biomass accumulation, and the extermination of sensitive species (Lacoul et al., 2011). Low pH often stunts the growth of frogs, toads and salamanders (Thoreau, 2002). Table 5.1 is a summary of the effects of a wide range of pH (3–11.5) on aquatic life (Thoreau, 2002; Simate and Ndlovu, 2014). Table 5.1 clearly demonstrates that there is no definitive pH range within which all freshwater aquatic life are unharmed (Rober-Bryan, Inc., 2004).

To sum up, research has shown that the acid streams resulting from mining activities from certain types of mineral deposits are highly toxic to the aquatic environment (Cotter and Brigden, 2006). In extreme cases, many river systems and former mine sites are totally inhospitable to aquatic life, with the exception of "extremophile" bacteria (Coil et al., 2013). The extreme acidity is toxic to most aquatic life and even after neutralisation, the precipitate formed continues to affect aquatic organisms (Cotter and Brigden, 2006). Toxic elements, such as copper, cadmium and zinc that are often associated with AMD also substantially contribute to the devastating ecological effects of AMD (Cotter and Brigden, 2006). Additionally, heightened acidity

TABLE 5.1

Effects of pH on Aquatic Life

pH	Effect
3.5–3.0	Toxic to most fish; some plants and invertebrates can survive such as the waterbug, water boatmen and white mosses
4.0–3.5	Lethal to salmonids
4.5–4.0	Harmful to salmonids, tench, bream, roach, goldfish and the common carp; all stock of fish disappear because embryos fail to mature at this level
5.0–4.5	Harmful to salmonid eggs, fry and the common carp; the lake is usually considered dead and a "wet desert"; it is unable to support a variety of life
6.0–5.0	Critical pH level, when the ecology of the lake changes greatly; the number and variety of species begin to change; salmon, roach and minnow begin to become less diverse; less diversity in algae, zooplankton, aquatic insects, insect larvae; rainbow trout do not occur and molluscs become rare; there is a great decline in salmonid fishing; the fungi and bacteria that are important in organic matter decomposition are not tolerant so the organic matter degrades more slowly and valuable nutrients are trapped at the bed and are not released back into the ecosystem; most of the green algae and diatoms (siliceous phytoplankton) that are normally present disappear. The reduction in green plants allows light to penetrate further so acid lakes seem crystal clear and blue; snails and phytoplankton disappear
9.0–6.5	Harmless to most fish
9.5–9.0	Harmful to salmonids, harmful to perch if persistent
10.0–9.5	Slowly lethal to salmonids
11.0–10.5	Lethal to salmonids, carp, tench, goldfish and pike
11.5–11.0	Lethal to all fish

Sources: Thoreau, 2002; Simate and Ndlovu, 2014.

reduces the ability of streams to buffer against further chemical changes (Coil et al., 2013).

5.2.3 Effect of High Acidity on Human Health

The polluted AMD water has serious impact on both the ecosystem and on any humans who may come into contact with it (Garland, 2011). Whilst there is research documenting the effects of AMD on the ecosystem, less is known about the potential effects of AMD on human health (Garland, 2011). However, the acidity of AMD indirectly affects human health mainly due to the mobilisation, transport and even chemical transformation of toxic metals (Goyer et al., 1985). Through the food chain, the intake of toxic as well as essential elements may be altered in man (Oskarsson et al., 1996). For example, acidification increases bioconversion of mercury to methylmercury, which accumulates in fish, thus increasing the risk of toxicity to people who eat fish (Goyer et al., 1985).

Another indirect, but devastating effect of high acidity in AMD concerns the vivid orange colour which forms when AMD is neutralised because of the precipitation of iron oxides and hydroxides (Cotter and Brigden, 2006). This precipitate, often called ochre, is very fine and can deposit and imbed on the river, stream or ocean bed thus cementing substrates for small animals (Cotter and Brigden, 2006). As a result, the small animals that used to feed on the bottom of the river or stream or ocean (benthic organisms) can no longer feed and die due to starvation and thus get depleted (Cotter and Brigden, 2006). In view of the fact that such small aquatic animals are at the bottom of the aquatic food chain, their depletion has an impact higher up the food chain all the way to the fish that feed on them (Cotter and Brigden, 2006). Therefore, even if the acidity and heavy metals are neutralised, AMD still affects humans and wildlife a long way down stream because of the indirect effects (Cotter and Brigden, 2006).

5.3 Effect of Heavy Metals

AMD contains a variety of dissolved heavy metals from abandoned and active mining areas (Gaikwad and Gupta, 2008). Appenroth (2010) argues that the definition of heavy metals should be more unequivocal and that it should be based on the periodic system of elements than how it is commonly defined. In this context, Appenroth (2010) defines heavy metals by the following groups of the periodic table of elements: (1) all transition elements, except La and Ac (=Transition metals), (2) rare earth elements, subdivided in the series of lanthanides and the series of actinides including La and Ac themselves (=Rare earth metals), and (3) a heterogenous group of elements including the metal Bi, the amphoteric oxides forming elements Al, Ga, In, Tl, Sn, Pb, Sb and Po, and the metalloids Ge, As and Te (= Lead-group elements as termed by Appenroth, 2010).

It is noted also that many other scholars define heavy metals as elements with atomic density greater than 6 g/cm^3 (Gardea-Torresdey et al., 2005; Akpor and Muchie, 2010), i.e., density of five times or more greater than that of water (Asati et al., 2016) or are conventionally defined as elements with metallic properties and an atomic number greater than 20 (Tangahu et al., 2011) and have serious implications on human life due to their acute and long-term toxicity (Ndlovu et al., 2013). In fact any toxic metal may be called a heavy metal, irrespective of its atomic mass or density (Singh et al., 2011). However, Appenroth (2010) also argues that heavy metals are not toxic per se, but only when a certain threshold of internal concentrations is exceeded. In fact, heavy metals are intrinsic components of the environment which are either essential or non-essential (Asati et al., 2016). The eight most common heavy metal pollutants listed by the Environment Protection Agency (EPA) are: As, Cd, Cr, Cu, Hg, Ni, Pb and Zn (Athar and Vohora, 2001; Khayatzadeh and Abbasi, 2010).

5.3.1 Effect of Heavy Metals on Plants

The influence of heavy metals on plants and their metabolic activities has been studied extensively (Cheng, 2003; Jiwan and Ajay, 2011; Asati et al., 2016). The toxicity of heavy metals to plants varies with plant species, specific metal, concentration of the metal, chemical form of the metal, soil composition and prevailing pH (Asati et al., 2016). In other words, high concentrations and/or certain mixtures of heavy metals in plant tissues can affect plant growth in different manners (Gardea-Torresdey et al., 2005). However, many heavy metals are considered to be essential for plant growth (Asati et al., 2016). For example, Cu and Zn can either serve as cofactor and activators of enzyme reactions (Mildvan, 1970; Asati et al., 2016). Some of the heavy metals such as As, Cd, Hg, Pb or Se are not essential for plant growth, since they do not perform any known physiological function in plants (Jiwan and Ajay, 2011). Other heavy metals such as Co, Fe, Mn, Mo, Ni and V are essential in minute quantities, but become harmful in excessive amounts (Asati et al., 2016). Specifically, if plentiful amounts of heavy metals are accumulated in plants, they adversely affect the absorption and transport of essential elements, disturb the metabolism and have an impact on growth and reproduction (Xu and Shi 2000; Cheng, 2003). In general, plants experience oxidative stress upon exposure to heavy metals that leads to cellular damage and disturbance of cellular ionic homeostasis (Yadav, 2010; Singh et al., 2011), thus disrupting the physiology and morphology of plants (Gardea-Torresdey et al., 2005). Table 5.2 is a summary of the main undesirable effects of some of the heavy metals on plants (Loneragan, 1988; Elamin and Wilcox, 1986; Marin et al., 1993; Wu, 1994; Barrachina et al., 1995; Cox et al., 1996; Sinha et al., 1997; Abedin et al., 2002; Gardea-Torresdey et al., 2005; Jayakumar et al., 2007; Solomon, 2008; Li et al., 2009; Akpor and Muchie, 2010; Yadav, 2010; Asati et al., 2016).

5.3.2 Effect of Heavy Metals on Aquatic Species

The contamination of aquatic ecosystems with heavy metal has long been recognised as a serious pollution problem (Ayandiran et al., 2009). This is because heavy metals are highly persistent, and research has shown that they are toxic even in trace amounts such that they can potentially induce severe oxidative stress in aquatic organisms (Jiwan and Ajay, 2011).

Aquatic organisms, such as fish, accumulate heavy metals directly from contaminated water and indirectly via the food chain (Khayatzadeh and Abbasi, 2010). In addition, Jennings et al. (2008) also state that fish are exposed indirectly to metals through ingestion of contaminated sediments and food items. Ayandiran et al. (2009) contend that there are in fact five potential routes for a water pollutant to enter a fish, namely, through the food, non-food particles, gills, oral consumption of water and the skin. However, the presence of metals and their subsequent toxicity vary between fish species

TABLE 5.2

Main Undesirable Effects of Heavy Metals on Plants

Heavy Metal	Effects
Manganese	Not readily remobilised through phloem to other organs after reaching the leave; necrotic brown spotting on leaves, petioles and stems is a common symptom of Mn toxicity; crinkle leaf leading to chlorosis and browning of leaf, stem and petiole
Cadmium	Decreases seed germination, lipid content and plant growth; induces phytochelatins production; impaired aquatic plant growth
Iron	Iron toxicity leads to reduction of plant photosynthesis and yield an increase in oxidative stress and ascorbate peroxidise activity in tobacco, canola, soybean and hydrilla verticillata; secretion of acids from the roots which lowers soil pH
Lead	Reduces chlorophyll production and plant growth; increases superoxide dismutase
Nickel	Reduces seed germination, dry mass accumulation, protein production, chlorophylls and enzymes; increases free amino acids
Mercury	Decreases photosynthetic activity, water uptake and antioxidant enzymes; accumulates phenol and proline
Zinc	Reduces Ni toxicity and seed germination; increases plant growth and ATP/chlorophyll ratio
Chromuim	Decreases enzyme activity and plant growth; produces membrane damage, chlorosis and root damage
Copper	Inhibits photosynthesis, plant growth and reproductive process; decreases thylakoid surface area
Cobalt	Phytotoxicity study has shown the adverse effect on shoot growth and biomass; excess of Co restricts the concentration of Fe, chlorophyll, protein and catalase activity in leaves of cauliflower; other effects include reduction in shoot length, root length and total leaf area; decrease in chlorophyll content; reduction in plant nutrient content and antioxidant enzyme activity; decrease in plant sugar, amino acid, and protein content has been noticed
Arsenic	Reduces fruit yield, decreases the leaf fresh weight in tomatoes; stunted growth, chlorosis and wilting in canola; reduces seed germination, decrease in seedling height, reduces leaf area and dry matter production in rice

Sources: Loneragan, 1988; Elamin and Wilcox, 1986; Marin et al., 1993; Wu, 1994; Barrachina et al., 1995; Cox et al., 1996; Sinha et al., 1997; Abedin et al., 2002; Gardea-Torresdey et al., 2005; Jayakumar et al., 2007; Solomon, 2008; Li et al., 2009; Akpor and Muchie, 2010; Yadav, 2010; Asati et al., 2016.

and also depend on age, developmental stage and other physiological factors (Khayatzadeh and Abbasi, 2010). There is no doubt also that the toxicity of heavy metals to aquatic species differs depending on a specific metal, concentration of the metal and chemical form of the metal. Heavy metals such as cadmium, copper, lead and zinc are of particular concern because of their severe toxicity to aquatic life (Lewis and Clark, 1997). In fact, acute exposure (short-term, high concentration) of these metals can kill organisms directly,

TABLE 5.3

Aquatic Life Protection Standards

Heavy Metal	Permissible Level (ppb)
Aluminium	5 if pH < 6.5, 100 if pH > 6.5
Arsenic	5 (FW), 12.5 (SW)
Cadmium	0.017 (FW), 0.12 (SW)
Lead	1–7 depending on water hardness
Nickel	25–150 depending on water hardness
Manganese	None
Mercury	0.1
Zinc	30 FW
Chromium	Cr^{6+}: 1 (FW), 1.5 (SW); Cr^{3+}: 8.9 (FW), 56 (SW)
Copper	2–4 depending on water hardness
Selenium	1

Source: Solomon, 2008.
FW = freshwater; SW = saltwater.

whereas chronic exposure (long-term, low concentration) can result in either mortality or nonlethal effects such as stunted growth, reduced reproduction, deformities or lesions (Lewis and Clark, 1997). With respect to AMD, Jennings et al. (2008) argue that when fish are exposed directly to metals and H^+ ions in water through their gills, impaired respiration may result from chronic and acute toxicity.

A major concern is that once heavy metals are accumulated by aquatic organisms, they can be transferred through the upper classes of the food chain (Ayandiran et al., 2009). Carnivores, which are at the top of the food chain including humans, obtain most of their heavy metal load from the aquatic ecosystem through their food, especially where fish are present and so there exists the potential for considerable biomagnification (Cumbie, 1975; Mance, 1987; Ayandiran et al., 2009). However, some studies have indicated that not all the metals taken up by fish are accumulated because fish have the ability, to a certain extent, of regulating the concentration of metals in their bodies (Romanenko et al., 1986; Ayandiran et al., 2009). The regulation of the concentrations of metals in fish may occur through the gills, bile, via feces, kidney and the skin (Heath, 1995; Ayandiran et al., 2009). Table 5.3 shows the levels of metals recommended for the protection of aquatic life (Solomon, 2008).

5.3.3 Effect of Heavy Metals on Human Health

A plethora of statements concerning the hazards of heavy metals in AMD on human health have been made by natural scientists, social scientists and activists over decades. However, Garland (2011) argues that whilst there is evidence that shows the effects of AMD on the ecosystem, very little is known about the potential effects of AMD on human health. On the other hand, what is known is that many of the components and pollutants in AMD are dangerous to humans (Garland, 2011). For example, Pb, Cd, Hg and As are widely dispersed in AMD (Peng et al., 2009; Simate and Ndlovu, 2014), and these elements have no beneficial effects in humans, and there is no known homeostasis mechanism for them (Vieira et al., 2011; Morais et al., 2012). However, the metals are considered to be the most toxic to humans and animals; the adverse human health effects associated with exposure to them, even at low concentrations, are diverse and include, but not limited to, neurotoxic and carcinogenic actions (Castro-González and Méndez-Armenta, 2008; Jomova and Valko, 2011; Morais et al., 2012). In addition, many other studies have also indicated that any exposure to heavy metals is capable of causing a myriad of human health effects, ranging from cardio-vascular and pulmonary inflammation to cancer and damage of vital organs (Geiger and Cooper, 2010; Alissa and Ferns, 2011; Jaishankar et al., 2014); and many major human physiological systems including the skeletal, nervous, respiratory, excretory and digestive systems may be affected (Ngole-Jeme and Fantke, 2017).

According to Garland (2011), in order for humans to be affected by the pollutants in AMD, they need to be exposed to the pollutants. Besides, there are a number of ways that people can potentially be exposed to the AMD pollutants (Garland, 2011; Ngole-Jeme and Fantke, 2017). The AMD enters the environment as polluted water, either through surface waters such as rivers or through groundwater. As the water moves through the ecosystem, the pollutants in AMD can then enter other parts of the ecosystem. For example, AMD pollutants in a river could be deposited and end up in the river beds where some aquatic species get their food. As discussed already in Section 5.3.2, what is of a major concern is that once heavy metals are accumulated by aquatic organisms, they can be transferred through the upper classes of the food chain (Ayandiran et al., 2009). Carnivores at the top of the food chain including humans obtain most of their heavy metal load from the aquatic ecosystem through their food, especially where fish are present and so there exists the potential for considerable biomagnification (Cumbie, 1975; Mance, 1987; Ayandiran et al., 2009). The depletion of aquatic organisms affected by AMD either through heavy metals or acidity was emphasised by Cotter and Brigden (2006) as it leads to the domino effects up the ecosystem by reducing the food sources available for animals at the top of the marine and human food chain.

Another AMD exposure pathway to humans occurs if AMD-contaminated water is used for irrigating crops, then there is a risk that the soils and plants can become polluted. In addition, if an animal drinks the polluted water or eats the polluted plants, then there is a risk that the animal can take up the pollution as well. Other exposure pathways have been described by Ngole-Jeme and Fantke (2017) and Jaishankar et al. (2014).

Akpor and Muchie (2010) contend that the danger of heavy metal pollutants in water lies in two aspects of their impact. First, heavy metals have the ability to persist in natural ecosystems for an extended period. Second, they have the ability to accumulate in successive levels of the biological chain, thereby causing acute and chronic diseases. In general, the toxicity or poisoning of heavy metals results from the disruption of metabolic functions. According to Singh et al. (2011), heavy metals disrupt the metabolic functions in two ways: (1) they accumulate in vital organs and glands such as the heart, brain, kidneys, bone and liver where they disrupt their important functions and (2) they inhibit the absorption, interfere with or displace the vital nutritional minerals from their original places, thereby, hindering their biological functions. Ideally, heavy metals are toxic because they have cumulative deleterious effects that can cause chronic degenerative changes (Ibrahim et al., 2006; Alissa and Ferns, 2011), especially to the nervous system, liver and kidneys, and, in some cases, they also have teratogenic and carcinogenic effects (International Agency for Research on Cancer, 1987; Alissa and Ferns, 2011). A summary of some of the heavy metals and their effects on human health together with permissible limits is presented in Table 5.4 (Solomon, 2008; Singh et al., 2011; Monachese et al., 2012).

TABLE 5.4

Heavy Metals and Their Effects on Human Health Together with Permissible Limits

Heavy Metal	Effect	Permissible Level (mg/L)
Arsenic	Bronchitis, dermatitis, poisoning	0.02
Cadmium	Renal dysfunction, lung disease, lung cancer, bone defects, increased blood pressure, kidney damage, bronchitis, bone marrow cancer, gastrointestinal disorder	0.06
Lead	Mental retardation in children, developmental delay, fatal infant encephalopathy, congenital paralysis, sensor neural deafness, liver, kidney, and gastrointestinal damage, acute or chronic damage to the nervous system, epilepticus	0.10
Manganese	Inhalation or contact causes damage to nervous central system	0.26
Mercury	Damage to the nervous system, protoplasm poisoning, spontaneous abortion, minor physiological changes, tremors, gingivitis, acrodynia characterised by pink hands and feet	0.01
Zinc	Damage to nervous membrane	15.0
Chromuim	Damage to the nervous system, fatigue, irritability	0.05
Copper	Anemia, liver and kidney damage, stomach and intestinal irritation	0.10

Sources: Solomon, 2008; Singh et al., 2011; Monachese et al., 2012; Simate and Ndlovu, 2014.

5.4 Concluding Remarks

It is very clear from this chapter that AMD effluents deserve special attention given that AMD is associated with numerous undesirable components such as low pH (high acidity), (2) high metal concentration, (3) elevated sulphate levels and (4) excessive suspended solids and/or siltation (AMRC, 2020). This chapter focused particularly on the effects of heavy metals and acidity on plants, aquatic species and human health. A collection of documented studies in the chapter has shown that heavy metals are toxic to soil, plants, aquatic life and human health if their concentrations are high. Heavy metals exhibit toxic effects towards soil biota, for example, by affecting key microbial processes and decrease the number and activity of soil microorganisms. Several reports have indicated that even low concentrations of heavy metals may inhibit the physiological metabolism of plants. Furthermore, the uptake of heavy metals by plants and aquatic species such as fish and the subsequent accumulation along the food chain is a potential threat to animals and humans. More specifically, contaminations of the aquatic systems by heavy metals can stimulate the production of reactive oxygen species (ROS) that can damage fishes and other aquatic organisms. Heavy metals toxicity to humans has proven to be a major threat and there are several health risks associated with them including interference with metabolic processes (Jaishankar et al., 2014).

The need for proper balance of the soil pH was also emphasised in the chapter since pH is influential on the availability of nutrients and on the growth of different kinds of plants. The knowledge of pH associated changes in aquatic ecosystems is also important. For example, acidity affects the structure and functioning of ecosystems, causes damages to forests and causes the extinction of aquatic organisms (Spyra, 2017). Much of the effect of AMD acidity on humans is indirect. For example, the depletion of aquatic organisms affected by AMD acidity has domino effects up the ecosystem by reducing the food sources available for animals at the top of the marine and human food chain (Cotter and Brigden, 2006).

References

Abandoned Mine Reclamation Clearinghouse (AMRC). (2020). What is AMD? Available at http://www.amrclearinghouse.org/Sub/AMDbasics/WhatIsAMD. htm. [Accessed 11 July 2020.]

Abdenna, D., Negassa, C.W. and Tilahun, G. (2007). Inventory of soil acidity status in crop lands of central and Western Ethiopia. Proceedings of the symposium of utilization of diversity in land use systems: Sustainable and organic approaches to meet human needs. Tropentag, October 9–1, Witzenhausen.

Abedin, M.J., Cotter-Howells, J. and Meharg, A.A. (2002). Arsenic uptake and accumulation in rice (*Oryza sativa* L.) irrigated with contaminated water. Plant and Soil 240 (2): 311–319.

Akpor, O.B. and Muchie, M. (2010). Remediation of heavy metals in drinking water and wastewater treatment systems: Processes and applications. International Journal of the Physical Sciences 5 (12): 1807–1817.

Alissa, E.M. and Ferns, G.A. (2011). Heavy metal poisoning and cardiovascular disease. Journal of Toxicology 2011: 1–20.

Appenroth, K.J. (2010). What are "heavy metals" in plant sciences? Acta Physiologiae Plantarum 32: 615–619.

Asati, A., Pichhode, M. and Nikhil, K. (2016). Effect of heavy metals on plants: An overview. International Journal of Application or Innovation in Engineering and Management 5 (3): 56–66.

Athar, M. and Vohora, S.B. (2001). Heavy metals and environment. New Age International Publisher, New Delhi.

Ayandiran, T.A., Fawole, O.O., Adewoye, S.O. and Ogundiran M.A. (2009). Bioconcentration of metals in the body muscle and gut of *Clarias gariepinus* exposed to sublethal concentrations of soap and detergent effluent. Journal of Cell and Animal Biology 3 (8): 113–118.

Barrachina, A.C., Carbonell, F.B., and Beneyto, J.M. (1995). Arsenic uptake, distribution, and accumulation in tomato plants: Effect of arsenite on plant growth and yield. Journal of Plant Nutrition 18 (6): 1237– 1250.

Castro-González, M.I. and Méndez-Armenta, M. (2008). Heavy metals: Implications associated to fish consumption. Environmental Toxicology and Pharmacology 26: 263–271.

Cheng, S. (2003). Effects of heavy metals on plants and resistance mechanisms. Environmental Science and Pollution Research 10 (4): 256–264.

Clark, T.M., Flis, B.J. and Remold, S.K. (2004). pH tolerances and regulatory abilities of freshwater and euryhaline Aedine mosquito larvae. The Journal of Experimental Biology 207: 2297–2304

Coil, D., McKittrick, E.M.S., Mattox, M., Hoagland, N., Higman, B. and Zamzow, K. (2013). Acid mine drainage. Available at http://yukonconservation.org/docs/AcidMineDrainage.pdf. [Accessed 10 July 2020.]

Cotter, J. and Brigden, K. (2006). Acid mine drainage: the case of the Lafayette mine, Rapu Rapu (Philippines), Greenpeace Research Laboratories Technical Note 09/2006. Available at https://www.greenpeace.to/publications/acid-mine-drainage.pdf. [Accessed 10 July 2020.]

Cox, M.S., Bell, P.F. and Kovar, J.L. (1996). Differential tolerance of canola to arsenic when grown hydroponically or in soil. Journal of Plant Nutrition 19 (12): 1599–1610.

Crane, R.A. and Sapsford, D.J. (2018). Selective formation of copper nanoparticles from acid mine drainage using nanoscale zerovalent iron particles. Journal of Hazardous Materials 347: 252–265.

CSIR. (2013). Characterising the risk of human exposure and health impacts from acid mine drainage in South Africa. Available at https://mhsc.org.za/sites/default/files/public/research_documents/SIM110901%20Report.pdf. [Accessed 2 May 2020.]

Cumbie, P.M. (1975). Mercury levels in Georgia otter, mink and freshwater fish. Bulletin of Environmental Contamination and Toxicology 14 (2): 193–196.

De Boer, W. and Kowalchuk, G. (2001). Nitrification in acid soils: Microorganisms and mechanisms. Soil Biology and Biochemistry 33: 853–866.

Dupré, C., Wessberg, C. and Diekmann, M. (2002). Species richness in deciduous forests: Effects of species pools and environmental variables. Journal of Vegetation Science 13 (4): 505–516.

Elamin, O.M. and Wilcox, G.E. (1986). Effect of magnesium and manganese nutrition on muskmelon growth and manganese toxicity. Journal of the American Society for Horticultural Science 111: 582–587.

Evangelou, V.P. (2001). Pyrite microencapsulation technologies: Principles and potential field application. Ecological Engineering 17: 165–178.

Foureaux, A.F.S., Moreira, V.R., Lebron, Y.A.R., Santos, L.V.S. and Amaral, M.C.S. (2020). Direct contact membrane distillation as an alternative to the conventional methods for value-added compounds recovery from acidic effluents: A review. Separation and Purification Technology 236: 116251–116264.

Gaikwad, R.W. and Gupta, D.V. (2008). Review on removal of heavy metals from acid mine drainage. Applied Ecology and Environmental Research 6 (3): 81–98.

Gardea-Torresdey, J.L., Peralta-Videa, J.R., Rosa, G.D. and Parsons, J.G. (2005). Phytoremediation of heavy metals and study of the metal coordination by X-ray absorption spectroscopy. Coordination Chemistry Reviews 249: 1797–1810.

Garland, R. (2011). Acid mine drainage – Can it affect human health? Quest 7 (4): 46–47.

Geiger, A. and Cooper, J. (2010). Overview of airborne metals regulations, exposure limits, health effects, and contemporary research. Available at http://www.green-resource.com/wp-content/themes/greenresource/uploads/resource_files/resource_lawn/Soil%20pH%20Explained.pdf. [Accessed 12 July 2020.]

Goyer, R.A., Bachmann, J., Clarkson, T.W., Ferris, B.G., Graham, J., Mushak, P., Perl, D.P., Rall, D.P., Schlesinger, R., Sharpe, W. and Wood, J.M. (1985). Potential human health-effects of acid-rain – Report of a workshop. Environmental Health Perspectives 60: 355–368.

Halcomb, M. and Fare, D. (2002). Soil pH explained. Available at http://www.green-resource.com/wp-content/themes/greenresource/uploads/resource_files/resource_lawn/Soil%20pH%20Explained.pdf. [Accessed 9 July 2020.]

Haling, R.E., Simpson, R.J., Culvenor, R.A., Lambers, H. and Richardson, A.E. (2011). Effect of soil acidity, soil strength and macropores on root growth and morphology of perennial grass species differing in acid-soil resistance. Plant, Cell and Environment 34: 444–456.

Heath, A.G. (1995). Water pollution and fish physiology. CRC Press, Boca Raton, FL.

Hunter, R.D. (1980). Effects of low pH and low calcium concentration on the pulmonate snail *Planorbella trivolvis*: A laboratory study. Canadian Journal of Zoology 68: 1578–1583.

Ibrahim, D., Froberg, B., Wolf, A. and Rusyniak, D.E. (2006). Heavy metal poisoning: Clinical presentations and pathophysiology. Clinics in Laboratory Medicine 26 (1): 67–97.

International Agency for Research on Cancer (IARC) (1987). Monographs on the evaluation of the carcinogenic risk of chemicals to humans: Arsenic and arsenic compounds (Group 1), volume 7. World Health Organization, Lyon, France, pp. 100–106.

Isaza, D.F.G., Cramp, R.L. and Franklin, C.E. (2020). Simultaneous exposure to nitrate and low pH reduces the blood oxygen-carrying capacity and functional performance of a freshwater fish. Conservation Physiology 8 (1): 1–15.

Jaishankar, M., Tseten, T., Anbalagan, N., Mathew, B.B. and Beeregowda, K.N. (2014). Toxicity, mechanism, and health effects of some heavy metals. Interdisciplinary Toxicology 7 (2): 60–72.

Jayakumar, K., Jaleel, C.A. and Vijayarengan, P. (2007). Changes in growth, biochemical constituents, and antioxidant potentials in radish (*Raphanus sativus* L.) under cobalt stress. Turkish Journal of Biology 31 (3): 127–136.

Jennings, S.R., Neuman, D.R. and Blicker, P.S. (2008). Acid mine drainage and effects on fish health and ecology: A review. Reclamation Research Group Publication, Bozeman, MT.

Jiwan, S. and Ajay, K.S. (2011). Effects of heavy metals on soil, plants, human health and aquatic life. International Journal of Research in Chemistry and Environment 1 (2): 15–21.

John, R., Singla, V., Goyal, M., Kumar, P. and Singla, A. (2017). Acid mine drainage: Sources, impacts and prevention. Available at https://www.biotecharticles.com/Environmental-Biotechnology-Article/Acid-Mine-Drainage-Sources-Impacts-and-Prevention-3801.html. [Accessed 9 July 2020.]

Johnson, C. (2002). Cation exchange properties of acid forest soils of the northeastern USA. European Journal of Soil Science 53: 271–282.

Jomova, K. and Valko, M. (2011). Advances in metal-induced oxidative stress and human disease. Toxicology 283: 65–87.

Khayatzadeh J., and Abbasi E. (2010). The effects of heavy metals on aquatic animals. The 1st international applied geological congress, Department of Geology, Islamic Azad University – Mashad Branch, Iran, 26–28 April 2010, pp. 688–694. Available at http://conference.khuisf.ac.ir/DorsaPax/userfiles/file/pazhohesh/zamin%20mashad/125.pdf. [Accessed 11 July 2020.]

Kidd, P.S. and Proctor, J. (2001). Why plants grow poorly on very acid soils: Are ecologists missing the obvious? Journal of Experimental Botany 52 (357): 791–799.

Lacoul, P., Freedman B. and Clair, T. (2011). Effects of acidification on aquatic biota in Atlantic Canada. Environmental Reviews 19: 429–460.

Larcher, W. (2003). Physiological plant ecology: Ecophysiology and stress physiology of functional groups. Springer, Berlin.

Lenntech. (2014). Acids and alkalis in freshwater: Effects of changes in pH on freshwater ecosystems. Available at https://www.lenntech.com/aquatic/acids-alkalis.htm. [Accessed 10 July 2020.]

Lewis, M.E. and Clark, M.L. (1997). How does streamflow affect metals in the upper Arkansas River. US Department of the Interior, US Geological Survey, Reston, VA.

Li, H.F., Gray, C., Mico, C., Zhao, F.J. and McGrath, S.P. (2009). Phytotoxicity and bioavailability of cobalt to plants in a range of soils. Chemosphere 75: 979–986.

Loneragan, J.F. (1988). Distribution and movement of manganese in plants. In: Graham, R.D., Hannam, R.J. and Uren, N.C. (Editors), Manganese in soils and plants. Kluwer, Dordrecht, pp. 113–124.

Mance, G. (1987). Pollution threat of heavy metals in aquatic environments. Springer, Netherlands.

Marin, A.R., Pezeshki, S.R., Masscheleyn, P.H. and Choi, H.S. (1993). Effect of dimethylarsinic acid (DMAA) on growth, tissue arsenic and photosynthesis of rice plants. Journal of Plant Nutrition 16 (5): 865–880.

Matsumoto, S., Shimada, H., Sasaoka, T., Miyajima, I., Kusuma, G.J. and Gautama, R.S. (2017). Effects of acid soils on plant growth and successful Revegetation in the case of mine. In: Oshunsanya, S. (Editor), Soil pH for nutrient availability and crop performance. IntechOpen Limited, London, pp. 9–27.

Mildvan, A.S. (1970). Metal in enzymes catalysis. In: Boyer P.D. (Editor), The enzymes: Kinetics and mechanism, Volume 2, Third Edition. Academic Press, London, pp. 445–536.

Monachese, M., Burton, J.P. and Reid, G. (2012). Bioremediation and tolerance of humans to heavy metals through microbial processes: A potential role for probiotics? Applied and Environmental Microbiology 78 (18), 6397–6404.

Morais, S., Costa, F.G. and Pereira, M.L. (2012). Heavy metals and human health. In: Oosthuizen, J. (Editor), Environmental health – Emerging issues and practice, pp. 227–246. Intech, Rijeka, Croatia.

Ndlovu, S., Simate, G.S., Seepe, L., Shemi, A., Sibanda, V. and van Dyk, L. (2013). The removal of Co^{2+}, V^{3+} and Cr^{3+} from waste effluents using cassava waste. South African Journal of Chemical Engineering 18 (1): 1–19.

Neina, D. (2019). The role of soil pH in plant nutrition and soil remediation. Applied and Environmental Soil Science 2019, 1–9.

Ngole-Jeme, V.M. and Fantke, P. (2017). Ecological and human health risks associated with abandoned gold mine tailings contaminated soil. PLoS One 12 (2): e0172517.

Nicol, G.W., Leininger, S., Schleper, C. and Prosser, J.I. (2008). The influence of soil pH on the diversity, abundance and transcriptional activity of ammonia oxidizing archaea and bacteria. Environmental Microbiology 10(11): 2966–2978.

Oskarsson, A., Nordberg, G., Block, M., Rasmussen, F., Vahter, M., Petterson, R., Skerfving, S., Wicklund, G.A., Oeborn, I., Heikensten, M.L. and Thuvander, A. (1996). Adverse health effects due to soil and water acidification: A Swedish research program. Available at https://thewaternetwork.com/_/hydrogeology-and-groundwater-remediation/storage/TFX%5CDocumentBundle%5CEntity%5CDocument-VMn1IX525Aby2soG9Rl7kA/0OoysQD3JM-E38mFcLlClQ/file/ph_turbidity_04phreq.pdf. [Accessed January 2014.]

Osman, K.T. (2014). Soil degradation, conservation and remediation. Springer, Heidelberg.

Peng, B., Tang, X.Y., Yu, C.X., Xie, S.R., Xiao, M.L., Song, Z. and Tu, X. (2009). Heavy metal geochemistry of the acid mine drainage discharged from the Hejiacun uranium mine in central Hunan, China. Environmental Geology 57 (2): 421–434.

Rober-Bryan, Inc. (RBI). (2004). pH requirements of freshwater aquatic life: Technical memorandum. Available at https://thewaternetwork.com/_/hydrogeology-and-groundwater-remediation/storage/TFX%5CDocumentBundle%5CEntity%5CDocument-VMn1IX525Aby2soG9Rl7kA/0OoysQD3JM-E38mFcLlClQ/file/ph_turbidity_04phreq.pdf. [Accessed 9 July 2020.]

Romanenko, V.D., Malyzheva, T.D. and Yevtushenko, N.Y.U. (1986). The role of various organs in regulating zinc metabolism in fish. Hydrobiological Journal 21 (3): 7–12.

Rorison, I.H. (1973). The effect-of extreme soil acidity on the nutrient uptake and physiology of plants. In: Dost, H. (Editor), Proceedings of the international symposium on acid sulphate soils. Publication 18, Volume 1. International Institute for Land Reclamation and Improvement, Wageningen.

RoyChowdhury, A., Sarkar1, D. and Datta, R. (2015). Remediation of acid mine drainage-impacted water. Current Pollution Reports 1: 131–141.

Schrock, S., Vallar, A. and Weaver, J. (2001). The effect of acidic conditions on photosynthesis in two aquatic plants. Journal of Honors Lab Investigations 1 (1): 22–26.

Simate, G.S. and Ndlovu, S. (2014). Acid mine drainage: Challenges and opportunities. Journal of Environmental Chemical Engineering 2 (3): 1785–1803.

Singh, R., Gautam, N., Mishra, A. and Gupta, R. (2011). Heavy metals and living systems: An overview. Indian Journal of Pharmacolology 43 (3): 246–253.

Sinha, S., Guptha, M. and Chandra, P. (1997). Oxidative stress induced by iron in *Hydrilla verticillata* (i.f) Royle: Response of antioxidants. Ecotoxicology and Environmental Safety 38: 286–291.

Slattery, M.L., Benson, J., Curtin, K., Ma, K.N., Schaeffer, D. and Potter, J.D. (2000). Carotenoids and colon cancer. American Journal of Clinical Nutrition 71 (2): 575–582.

Slattery, W.J., Conyers, M.K. and Aitken, R.L. (1999). Soil, pH, aluminium, manganese and line requirement. In: Peverill, K.I., Sparrow, L.A. and Reuters, D.J. (Editors), Soil analysis: An interpretation manual. CSIRO Publishing, Collingwood, pp. 103–128.

Solomon, F. (2008). Impacts of metals on aquatic ecosystems and human health. Available at https://pdfs.semanticscholar.org/6572/7277c6270165b2329e363ff6 45f3bec8c586.pdf. [Accessed 11 July 2020.]

Soti, P.G., Jayachandran, K., Koptur, S. and Volin, J.C. (2015). Effect of soil pH on growth, nutrient uptake, and mycorrhizal colonization in exotic invasive *Lygodium microphyllum*. Plant Ecology 216(7): 989–998.

Spyra, A. (2017). Acidic, neutral and alkaline forest ponds as a landscape element affecting the biodiversity of freshwater snails. Science of Nature 104 (73): 1–12.

Sullivan, T.J. (2000). Aquatic effects of acidic deposition. Lewis Publishers, Boca Raton, FL.

Tangahu, B.V., Abdullah, S.R.S., Basri, H, Idris, M, Anuar, N. and Mukhlisin, M. (2011). A review on heavy metals (As, Pb, and Hg) uptake by plants through phytoremediation. International Journal of Chemical Engineering 2011, 1–31.

Teshome, N. (2017). Influence of potassium fertilization and liming on growth, grain yield, and quality of soybean (*Glycine max* L. (Merrill) on acidic soil in Gobu Sayo District, Western Ethiopia. Dissertation, Jimma University College of Agriculture and Veterinary Medicine, Ethiopia.

Thoreau, H.D. (2002). Effects of acid rain on aquatic species. Available at https:// www.rst2.org/ties/acidrain/PDF/3effects/ef6.pdf. [Accessed 10 July 2020.]

Vieira, C., Morais, S., Ramos, S., Delerue-Matos, C. and Oliveira, M.B.P.P. (2011). Mercury, cadmium, lead and arsenic levels in three pelagic fish species from the Atlantic Ocean: Intra- and inter-specific variability and human health risks for consumption. Food and Chemical Toxicology 49: 923–932.

Wu, S. (1994). Effect of manganese excess on the soybean plant cultivated under various growth conditions. Journal of Plant Nutrition 17: 993–1003.

Xu, Q. and Shi, G. (2000). The toxic effects of single Cd and interaction of Cd with Zn on some physiological index of [*Oenanthe javanica* (Blume) DC]. Journal of Nanjing Normal University (Natural Science Edition) 23 (4): 97–100.

Yadav, S.K. (2010). Heavy metals toxicity in plants: An overview on the role of glutathione and phytochelatins in heavy metal stress tolerance of plants. South African Journal of Botany 76: 167–179.

Part II

Prevention and Remediation Processes of Acid Mine Drainage

It is unanimously agreed that 'prevention is better than cure', hence it is better to prevent the occurrence of AMD in the first place. In contrast to preventive techniques, the remediation and/or corrective methods are those that treat AMD in such a way that it ceases to be a threat to the environment and/or the threat to the environment is minimized. The chapters in this part of the book discuss techniques and processes that are aimed at preventing the formation of AMD, including the number of remediation methods, which are generally categorized as conventional and non-conventional methods, are also critically analysed and discussed. Life-cycle assessment of AMD treatment processes is also covered.

6

Prevention Processes for Acid Mine Drainage

Geoffrey S. Simate

CONTENTS

6.1 Introduction

As discussed in Chapter 5, acid mine drainage (AMD) poses serious dangers on the environment and human health. Doubtlessly, prevention of AMD generation is an important task in order to protect the environmental and health risks associated with it (Kefeni et al., 2017). Therefore, over the years considerable focus has been directed towards the development of technologies for the prevention of AMD. Currently, there are a number of approaches to preventing and/or minimizing the generation of AMD. Actually, prevention of acid production is preferred over treatment of AMD (Brown, 1996). Park et al. (2019) also argue that since prevention techniques do not require continuous maintenance, they are more sustainable than traditional remediation or treatment techniques. Moreover, AMD treatment costs can negatively influence the economic performance or even compromise the viability of a project; thus, AMD formation should be prevented (Pozo-Antonio et al., 2014). Therefore, only the most serious and unforeseen cases of AMD pollution should employ treatment techniques (Pozo-Antonio et al., 2014).

Taking into account that the oxidation of sulphide-bearing minerals and the subsequent AMD generation occurs in the presence of water and oxygen, and is highly accelerated by microorganisms, the general approach for AMD prevention techniques has been to eliminate or reduce one or more of these essential factors (Kuyucak, 2002; Pozo-Antonio et al., 2014; Park et al., 2019). This chapter focuses on a number of such techniques. In particular, the chapter discusses the following techniques that have been explored to prevent or minimise the generation of AMD: (1) control of water movement, (2) selective separation and blending of sulphide-rich materials, (3) physical separation barriers for water and oxygen, (4) inhibition of iron and sulphur-oxidizing microorganisms and (5) other methods, namely, sulphide passivation or micro-encapsulation, desulphurisation and electrochemical protection systems.

6.2 Preventive Techniques for Acid Mine Drainage

This section of the chapter discusses and evaluates, in detail, the preventive techniques for AMD formation. Typically, these techniques are designed to exclude any of the three components that are essential in the formation of AMD: oxygen, water and microorganisms (Kuyucak, 2002; Pozo-Antonio et al., 2014; Park et al., 2019). In other words, prevention of AMD generation mainly requires protection of sulphide-bearing minerals from air, water and microorganisms (Kefeni et al., 2017). It must be noted, however, that the nature of the sulphide-waste materials and the conditions of the mining site dictate the choice and implementation strategy of the preventive techniques (Kuyucak, 2002).

6.2.1 Control of Water Movement

Water is the most important media responsible for the transportation of contaminants and thus all measures aimed at curtailing AMD generation should be concerned with the control of the flow of water (Akcil and Koldas, 2006). The techniques of controlling the flow of water restrict the movement of water through potential acid-producing sulphide materials. In other words, the control technologies rely on the prevention of the entry of water into the AMD source (Barton-Bridges and Robertson, 1989). In this regard, a number of measures can be employed to control the movement of water. Firstly, surface water can be diverted from flowing through potential acid-producing sulphide materials. This can be achieved through the use of well-designed and waterproofed pipes, ditches and/or channels (Kuyucak, 2002; Pozo-Antonio et al., 2014). In addition, surface water diversion may involve the construction of drainage ditches or impervious channels that are able to move surface water quickly across acid-producing sulphide materials (Skousen et al., 2019). Secondly, groundwater can be prevented from infiltrating into sites that contain acid-generating materials using intercepting structures such as grout barriers (or curtains) and reinforced concrete walls (i.e., slurry walls) or even diversion ditches (Skousen et al., 2000; Kuyucak, 2002; Akcil and Koldas, 2006; RoyChowdhury et al., 2015). The use of grouts to separate acid-producing rocks and groundwater has also been recommended by Skousen et al. (2019) who states that "injection of grout barriers or curtains may significantly reduce the volume of groundwater moving through backfills". Thirdly, using underdrains and sealing layers, hydrological water seepage into acid-generating waste can be prevented (Akcil and Koldas, 2006). Fourthly, for open pit mines, it is appropriate to flush or drain water rapidly from spoil heaps before it accumulates to high levels (Akcil and Koldas, 2006). Finally, it is also important to use properly designed slopes and steps in order to minimise the infiltration of water into the mining waste (Kuyucak, 2002; Pozo-Antonio et al., 2014). Despite all the measures to control the flow of water, Sahoo et al. (2013) state that "surface barriers can achieve substantial reductions in water flow through piles, but generally do not control AMD completely".

6.2.2 Selective Separation and Blending of Sulphide-Rich Materials

In this technique, sulphide-rich waste materials are separated from the rest of other materials and disposed into specifically designed and prepared storage areas (Kuyucak, 2002). The primary strategy is to segregate and place the material in the storage areas in order to limit its exposure to water and air (Skousen et al., 1987; Skousen et al., 1998). It is also helpful to compact the material once placed in the storage areas (Skousen et al., 1998) and depending on the neutralisation potential, the wastes are mixed with alkaline amendments such as limestone in order to minimise acidity of the overall system (Skousen et al., 1998; Chowdhury et al., 2015). Indeed, co-disposal and/or mixing of sulphide materials along with some materials (waste rock, limestone) which are either potentially acid consuming or alkaline producing is the most common practice to reduce AMD production from mine waste (Skousen et al., 2000; Kuyucak, 2002; Johnson and Hallberg, 2005; RoyChowdhury et al., 2015; Park et al., 2019). In fact, research has shown that complete mixing of the benign and acid-producing mine wastes is more effective than placing the two materials separately in layers (Kuyucak, 2002). Ideally, this technique is meant to ensure that AMD-generating potential of a reactive mine waste is minimised by either reducing the net acid-producing potential of the reactive waste or by increasing the net alkaline neutralizing potential of the benign waste (Kuyucak, 2002).

As noted already, a number of materials are used for co-disposal or blending of reactive mine wastes. The most frequently used materials include lime, limestone and phosphate minerals (Renton et al., 1988; Mylona et al., 2000; Hakkou et al., 2009; Kastyuchik et al., 2016; Li et al., 2018). Other materials that include industrial by-products and residues have also become popular blending materials for preventing the generation of AMD due to their high neutralisation capabilities (Xenidis et al., 2002; Doye and Duchesne, 2003; Yeheyis et al., 2009; Alakangas et al., 2013;Park et al., 2019). These materials and many others, for example, are known to consume H^+ ions generated by sulphidic waste materials, a property that does not only contribute to the formation of near-neutral AMD (reactions 6.1 and 6.2), but also immobilises soluble metals and metalloids via precipitation (Mylona et al., 2000; Hakkou et al., 2009; Lottermoser, 2003; Park et al., 2019).

$$CaCO_3 + H^+ \rightarrow Ca^{2+} + HCO_3^- \tag{6.1}$$

$$Ca_5(PO_4)_3F + 3H^+ \rightarrow 5Ca^{2+} + 3HPO_4^{2-} + F^- \tag{6.2}$$

It is also important to analyse the chemistry of how some of the blending materials such as limestone, in particular, and the rest, in general, control the oxidation of sulphide waste materials. A number of studies have shown that blending (e.g., by limestone) control the acid production of sulphide minerals through four mechanisms. The first mechanism involves precipitation of ferric iron into the hydroxide form, thus its further participation as an oxidizing agent in the oxidation of sulphide wastes is inhibited (Kelley and Tuovinen, 1988; Mylona et al., 2000; Park et al., 2019). The second mechanism involves the raising of the pH (pH 6.1–8.4) that significantly weakens the activities of the sulphide-oxidizing microorganisms (Nicholson et al., 1988; Mylona et al., 2000; Park et al., 2019). The third mechanism involves the precipitation of oxidised compounds and the formation of a protective layer such as ferric-oxy-hydroxide on the surface of sulphide minerals that reduce their reactive surface areas thus impairing their dissolution further (Nicholson et al., 1990; Mylona et al., 2000; Park et al., 2019). The fourth mechanism also involves the formation of a cemented layer (hardpan) consisting of ferric-oxy-hydroxide and gypsum which has very low permeability that limits the diffusion of O_2 and infiltration of water into sulphide materials (Blowes et al., 1991; Lin, 1997; Mylona et al., 2000; Park et al., 2019). Hallberg et al. (2005) also argue that the formation of hardpans reduces permeability and oxygen diffusion and thus lowers the oxidation rate of sulphide waste materials. Besides, acting as a barrier, the cement-like hard substance also behaves as a stabilizing material (Stehouwer et al., 1995; Skousen et al., 2000; RoyChowdhury et al., 2015).

Other studies have also found that large acid-producing waste rocks can be mixed with fine materials that contain high moisture content such as tailings and disposed together (Kuyucak, 2002; RoyChowdhury et al., 2015). This process results in the filling of large pores of waste rock and can help in minimizing oxygen penetration into the acid-producing mine wastes thereby preventing the oxidation process (Kuyucak, 2002; RoyChowdhury et al., 2015). The effectiveness of tailings as oxygen barriers is influenced by the moisture content maintained in the tailings (Barton-Bridges and Robertson, 1989).

6.2.3 Physical Separation Barriers for Water and Oxygen

The most prevalent and conventional approach to limiting the entry of either oxygen or water, or both to sulphide-bearing waste, is to use physical barriers that consist of wet or dry covers (Kuyucak, 2002; Sahoo et al., 2013; Skousen et al., 2019). Argunhan-Atalay and Yazicigil (2018) also emphasise that barriers, particularly, dry cover barriers, retard the movement of water or oxygen into areas containing acid-producing rocks. It is noted, however, that though the exclusion of oxygen from waste materials is a very effective means of preventing sulphide oxidation, it is equally difficult to achieve than excluding water (Kuyucak, 2002). The sections that follow these opening remarks discuss in detail the dry and water covers.

6.2.3.1 Dry Covers

Pozo-Antonio et al. (2014) argue that dry covers have a number of roles: (1) stabilisation of mining waste so as to prevent erosion by wind and water, (2) improvement of aesthetic appearance and (3) prevention and/or inhibition of the release of AMD pollutants. The prevention and/or inhibition of the release of AMD pollutants is the focus of this section. The dry covers designed to inhibit AMD generation primarily comprise of a sealing layer with low hydraulic conductivity (Figure 6.1) which restricts the supply of

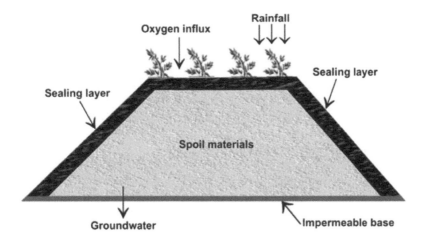

FIGURE 6.1
Diagram of a typical dry cover designed to minimise the production of acidic effluents from sulphide-bearing wastes. (From Sahoo et al., 2013.)

oxygen and limits the percolation of water into the sulphide waste materials, thereby reducing the rate of sulphide oxidation (Kleinmann, 1990; Hallberg et al., 2005; Sahoo et al., 2013; Pozo-Antonio et al., 2014; Skousen et al., 2019). Indeed, the dry covers also referred to as oxygen barriers slow down the movement of water or oxygen into areas containing acid-producing rocks (Argunhan-Atalay and Yazicigil, 2018; Skousen et al. 2019). Furthermore, Vila et al. (2008) acknowledge that the properties of dry covers, which include low permeability and increased moisture content, enable the covers to be used as a barrier to oxygen diffusion.

There is a large number of dry cover designs, but most importantly, the cover should be stable and provide long-term protection (Pozo-Antonio et al., 2014). Fine-grained soil is one of the materials that has been used as effective dry covers (Sahoo et al., 2013). However, according to Vila et al. (2008), the dry covers are usually made out of clay materials. It is also noted that soil covers vary depending on climate, types and volume of waste material, size and geometry of the waste storage facility, available cover materials in the field, etc, (Argunhan-Atalay and Yazicigil, 2018), and thus the choice of appropriate soil covers is important. For several years, some studies indicated that soil covers decreased the generation of acid remarkably compared to uncovered waste materials (Payant et al., 1995). Other studies that used soil cover on the surface of waste rock found that, over time, the effectiveness of the soil cover reduced (Harries and Ritchie, 1985; Yanful and Orlandea, 2000; Wang et al., 2006). Some of the shortcomings of soil covers that lead to their ineffectiveness include sidewall passage of oxygen and water (Yanful and Orlandea, 2000), precipitate infiltration (Wang et al., 2006) and high maintenance cost (Yanful, 1993; Yanful et al., 1999). Most importantly, Swanson and O'Kane (1999) state that "soil cover design is climate specific and a common misconception in soil cover design is the use of compacted clay covers". According to Swanson et al. (1997), low-permeability barriers such as clay are not necessarily the most effective covers in dry climates. This is due to the high potential for drying and cracking, which results in water bypassing the soil matrix (Morris et al., 1992; Daniel and Wu, 1993). Therefore, the ultimate result is a failed cover system (Swanson and O'Kane, 1999).

For decades, researchers have embarked on finding other solutions to minimizing and/or eliminating the generation of AMD. For example, synthetic materials such as plastic liners and polyethylene have been studied and subsequently used to control AMD generation (Sahoo et al., 2013). In a study by Caruccio and Geidel (1983), polyvinyl chloride (PVC) covers were used to completely cover a waste site and the study found that there was a substantial decrease in total acid loads. However, the use of synthetic plastic or polymer liners has a number of limitations such as (1) being too expensive for covering a large volume, (2) being susceptible to cracking and (3) repair costs being exorbitant (Sahoo et al., 2013).

On the other hand, organic materials have been found to be a good replacement for clay as long as the layer is thick enough (Pozo-Antonio et al., 2014).

Several studies have actually indicated that organic carbon-rich materials are able to remove oxygen from sulphide waste materials and thus limit AMD generation (Tremblay, 1994; Peppas et al., 2000; Sahoo et al., 2013; Park et al., 2019). Ideally, the organic covering has proven effective in preventing oxygen from reaching the tailings (Tremblay, 1994). Examples of organic carbon-rich waste materials that can be used as dry covers include wood waste, wood chips, sawdust, municipal sludge, sewage sludge, composted municipal wastes, manure, peat, paper mill sludge and vegetation (Backes et al., 1987; Sahoo et al., 2013; Park et al., 2019). As can be seen from reaction 6.3, the carbon-rich materials consume oxygen that subsequently maintains very low dissolved oxygen within the waste material thus suppressing the oxidation of sulphide waste materials and limit AMD generation (Sahoo et al., 2013; Park et al., 2019).

$$C_6H_{12}O_6 + 6O_2 \rightarrow 6CO_2 + 6H_2O \qquad (6.3)$$

Sahoo et al. (2013) also argue that organic waste provides a pH buffer that neutralises acids. In addition, INAP (2014) states that when organic materials are mixed with sulphide-bearing wastes the organic materials consume oxygen and promote metal reduction in an anoxic environment by naturally occurring bacteria. Bacteria can reduce available sulphate and create insoluble metal sulphide precipitates in the presence of suitable organic substrates. Some studies have indicated that the complexation of free Fe (III) by soluble microbial growth products (SMPs) that are produced by the microorganisms growing in waste rock in the presence of organic matter can reduce the effectiveness of ferric iron as an oxidant (Pandey et al., 2011; Sahoo et al., 2013).

Although the organic carbon-rich waste materials have a number of advantages in terms of cost and are effective in a short term, they also possess some downsides (Sahoo et al., 2013; Park et al., 2019). For example, the materials may contain organic acids that may infiltrate the waste areas and leach out toxic metals such as arsenic, copper, nickel and zinc (Pond et al., 2005; Sahoo et al., 2013). Similarly, organic cover may induce the reductive dissolution of secondary minerals like Fe (III)-oxyhydroxides, leading to the release of toxic elements (e.g., arsenic, cadmium, copper, lead and selenium) previously adsorbed onto or co-precipitated with these phases (Ribet et al., 1995; Park et al., 2019).

A significant number of materials have also been tested as dry covers and were found to reduce the generation of AMD because the materials were able to maintain the high degree of water saturation in the overlying/covering layers (Ribet et al., 1995; Bellaloui et al., 1999; Peppas et al., 2000; Demers et al., 2008; Park et al., 2019). These materials included non-reactive fine mine residue and natural till, low-sulphide tailings, desulphurised tailings, silty materials, alkaline waste, and industrial or municipal wastes such as fly ash, bottom ash, cement kiln dust, red mud bauxite, paper mill waste, pulp/paper residue and organic wastes (Ribet et al., 1995; Bellaloui et al., 1999; Peppas et al., 2000; Bussière et al., 2004; Demers et al., 2008; Park et al., 2019). Demers et al. (2017) also evaluated the possibility of using AMD neutralisation sludge as an alternative for natural soil used as covering materials. The study results, after about almost one and half years, showed that a cover layer composed of sludge-soil mixture (25%–75% by weight) over either waste rock or tailings effectively suppressed AMD generation.

Many other dry covers have been tested and applied, and these include impervious membranes, dry seals, hydraulic mine seals and grout curtains/walls. Some of these have already been discussed in Section 6.2.1.

6.2.3.2 Water Covers

Research has shown that the disposal of sulphide-bearing materials under a water cover such as lakes, oceans, fjord and ponds (Fraser and Robertson, 1994; Skousen et al., 2019) is a suitable technique for excluding oxygen from sulphides and thus limits the generation of AMD (Kuyucak, 2002; Sahoo et al., 2013; Park et al., 2019; Skousen et al., 2019). In fact, water covers are an economical alternative to dry covers because oxygen has a very low solubility in water and its diffusion rate in water is almost four orders of magnitude less than in air (Kuyucak, 2002). Park et al. (2019) also argue that water is widely used as an oxygen barrier because of oxygen's slow diffusion in water $(1.90 \times 10^{-9} \text{ m}^2/\text{s})$ compared to its diffusion in the air $(1.98 \times 10^{-5} \text{ m}^2/\text{s})$ as illustrated in Figure 6.2. Therefore, by having less amount of oxygen under water, highly anoxic conditions are created thus inhibiting the oxidation of sulphide-bearing wastes (Sahoo et al., 2013). In addition, Kleinmann and Crerar (1979)

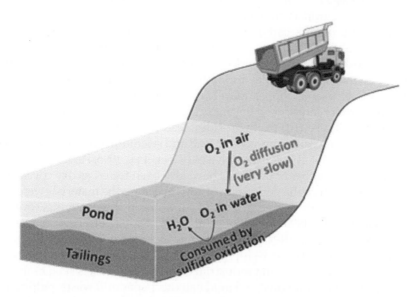

FIGURE 6.2
Sub-aqueous disposal. (From Park et al., 2019.)

state that "microbial catalysis, in association with other mechanisms such as metal hydroxide precipitation and development of sediment barriers between sulphide-bearing wastes and overlying waters also inhibit oxidation".

Water covers have been extensively utilised for decades to reduce the rate of oxygen contact with the sulphide-bearing wastes since it is comparatively less expensive than other options (Sahoo et al., 2013). However, the technique has a number of drawbacks. This is in addition to the practice of disposing the sulphide-bearing wastes into lakes and ocean having been banned by most countries (Skousen et al., 2019). For example, the water cover technique is not applicable in arid and semi-arid regions where annual evaporation exceeds precipitation because the drying out of water cover exposes the sulphide-bearing waste to atmospheric conditions, thereby generating AMD (Lottermoser, 2003; Park et al., 2019). In other words, the technique is limited to sites that can be flooded, or where the water table may be permanently altered to cover sulphide-bearing wastes (Sahoo et al., 2013). Research has also found that water cover is not suitable at sites where the influx of oxygen-containing water occurs, or where mines are only partially flooded (Johnson and Hallberg, 2005). It is also observed that despite water covers minimizing acid generation, there could be a slow release of some metals thus increasing the metal concentration that may increase the recommended water standards (Aube et al., 1995; Kuyucak, 2002). Most importantly, the actual flooding of mine wastes requires a rigorous engineering design and proper maintenance in order to eliminate or minimise the risk of any failure in the embankment that is definitely costly when it occurs (Sahoo et al., 2013).

6.2.4 Inhibition of Iron and Sulphur-Oxidizing Microorganisms

The solubilisation of metals due to the action of microorganisms has been extensively studied (Rawlings, 2005; Simate and Ndlovu, 2008; Simate et al., 2009; Simate et al., 2010). One shared characteristic amongst the microorganisms particularly the acidophilic microorganisms is their ability to produce the ferric iron and sulphuric acid required to degrade the sulphide-bearing minerals and facilitate metal recovery (Rawlings, 2005). Indeed, many other researchers also argue that microorganisms like acidophilic iron- and sulphur-oxidizing microorganisms (e.g., *Thiobacillus ferrooxidans* and *Thiobacillus thiooxidans*) are known to accelerate sulphide oxidation (Sasaki et al., 1998; Bacelar-Nicolau and Johnson, 1999) once the system is in a highly acid state (Kleinmann, 1990). This means that the absence of microorganisms in sulphidic mine wastes could slow down and limit the formation of AMD (Sahoo et al., 2013; Park et al., 2019). It is on the basis of this notion that bactericides such as anionic surfactants, cleaning detergents, organic acids and food preservatives have been used to inhibit the growth of these microorganisms (Kleinmann and Erickson, 1981; Kleinmann, 1982; Kleinmann and Erickson, 1983; Backes et al., 1987; Evangelou, 1995; Lottermoser, 2003; Zhang and Wang, 2017; Park et al., 2019).

According to a number of researchers, in general, bactericides change the protective and greasy coating that allow the internal enzymes in the micro-organisms to maintain a circumneutral pH in order to function normally in an acid environment, and/or bactericides may also disrupt the contact between the microorganism and the sulphide mineral surface (Langworthy 1978; Ingledew 1982; Kleinmann 1998; Sahoo et al., 2013). More specifically, anionic surfactants allow protons to penetrate into the microorganism's cell membranes freely thus causing disruptions to its enzymatic functions at low concentrations and eventual death of the cell at high concentrations occurs (Evangelou, 1995; Lottermoser, 2003; Zhang and Wang, 2017; Park et al., 2019). Similarly, organic acids are harmful to acidophiles because they uncouple the respiratory chain of the microorganisms under acidic conditions via the penetration of their protonated forms through the cell membrane which then deprotonates while inside the cells releasing harmful H^+ ions (Baker-Austin and Dopson, 2007; Park et al., 2019).

Despite bactericides being cost effective compared to dry or wet cover techniques, they have a number of limitations. Bactericides cannot permanently inhibit microbial activity because they are water-soluble and thus are easily washed away from the sulphidic wastes-microorganism interface (Park et al., 2019); and, therefore, repetitive applications of the bactericide are required (Kleinmann, 1998) so as to maintain their effectiveness in controlling AMD generation (Kuyucak, 2002; Skousen et al., 2000; RoyChowdhury et al., 2015). One critical disadvantage of the water-soluble bactericides such as anionic surfactants is that once they are carried away from the sulphidic wastes-micro-organism interface and find themselves into waterbodies, they become toxic to aquatic species (Liwarska et al. 2005; Hodges et al. 2006; Sahoo et al., 2013). Some studies have also found that anionic surfactants work best on fresh and unoxidised sulphides (Johnson and Hallberg, 2005; Sahoo et al., 2013).

6.2.5 Other Methods

6.2.5.1 Sulphide Passivation or Microencapsulation

Sulphide passivation or microencapsulation is amongst the new techniques of controlling AMD at source that falls under the category of chemical barriers (Sahoo et al., 2013; Nyström, 2018). Passivation is a phenomenon that refers to the treatment of reactive sulphide-bearing waste materials and it results in the creation of a chemically inert and protective surface layer that prevents the material from oxygen and ferric iron attack (Zhang and Evangelou, 1998; Sahoo et al., 2013; Zhang et al., 2003; INAP, 2014; Nyström, 2018). In other words, passivation or microencapsulation is a technique wherein a sulphide-bearing material is coated with oxidatively stable materials that prevent the reaction of the sulphide mineral with water and oxygen or an attack by ferric iron (Evangelou, 1995; Evangelou, 2001; Park et al., 2018a). In addition, Bessho et al. (2011) also argue that the new technique inhibits the oxidation of sulphide-bearing materials through the formation of a coating on the surface of the sulphide, which, ideally, prevents both oxygen and ferric iron from oxidizing the sulphide-bearing waste. Research has shown that the existence of a chemically inert and protective surface coating resulting from the passivation treatment process inhibits the oxidation of sulphide-bearing waste materials and subsequent generation of AMD for much longer period (Zhang and Evangelou, 1998; Zhang et al., 2003) than other techniques.

Several additives that could enhance the inactivity of sulphide surfaces that include organic and inorganic materials have been studied (Sahoo et al., 2013; Nyström, 2018; Park et al., 2019). These additives inhibit oxygen, water and ferric iron to diffuse to the surface of sulphide materials (Satur et al., 2007) and thus disrupt the oxidation of sulphide-bearing materials and generation of AMD through various mechanisms (Kleinmann, 1990).

Examples of organic materials that have been used as coatings on the surface of sulphide minerals to suppress oxidation of sulphide minerals include sodium oleate fatty acids (Jiang et al., 2000), natural organic matter (NOM) like humic acid and lignin (Lalvani et al., 1990; Khummalai and Boonamnuayvitaya, 2005; Ačai et al., 2009), lipids and phospholipids (Elsetinow et al., 2003; Zhang et al., 2003; Hao et al., 2006; Hao et al., 2009), oxalic acids (Sasaki et al., 1996) and many others such as diethylenetriamine (DETA), triethylenetetramine (TETA), sodium triethylenetetramine-bisdithiocarbamate (DTC-TETA), methyl ethyl ketone formaldehyde resin modified carbazoles and 8-hydroxyquinoline as discussed in a review article by Park et al. (2019).

A number of inorganic materials are also available as passivating reagents. Readers are referred to Sahoo et al. (2013) for detailed information on research that directly used inorganic coatings for passivating sulphide-bearing waste materials. In summary some of the inorganic materials include phosphate (Nyavor and Egiebor, 1995; Evangelou, 2001), silica (Evangelou, 1996; Evangelou, 2001; Bessho et al., 2011) and permanganate (Ji et al., 2012). Alkaline materials such as sodium hydroxide (NaOH), sodium carbonate (Na_2CO_3), sodium bicarbonate ($NaHCO_3$), lime (CaO, Ca(OH)$_2$) and limestone ($CaCO_3$) have also been successfully used to prevent AMD (Nicholson et al., 1988; Nicholson et al., 1990; Evangelou, 1995; Sahoo et al., 2013). More recently, fly ash (from coal combustion) has been identified as a suitable alternative alkaline material because of its low cost, local availability and self-healing capacity (Pérez-López et al., 2007; Sahoo et al., 2013; Park et al., 2019). Ideally, fly ash has the potential to encapsulate sulphide-bearing materials with iron oxyhydroxide coatings by supplying a sufficient amount of alkalinity (Pérez-López et al., 2007). Pérez-López et al. (2007) also observed that the potential use of fly ash to attenuate the sulphide-bearing oxidation is not limited to its capacity to encapsulate the sulphide grains in a short term, but it can also promote a hardpan on the interface between the fly ash and the sulphide in a medium term that ensures the total neutralisation of the mine residues.

Other studies have shown that under suitable conditions such as near neutral pH in the presence of sufficient alkalinity, some sulphide materials such as pyrite can passivate on their own due to the formation of ferric oxyhydroxide coatings on the surfaces of such sulphide materials (Park et al., 2019). Ideally, the ferric oxyhydroxide coatings lower the rate of oxidation of sulphide materials (Nicholson et al., 1990). Huminicki and Rimstidt (2009) proposed the mechanism of iron oxyhydroxide coating formation on the sulphide as follows: (1) colloidal iron oxyhydroxide precipitates are deposited on the sulphide surface driven by electrostatic attraction between negatively charged sulphide material and positively charged iron oxyhydroxide and (2) the densification and inward propagation of the coating, i.e., the coating becomes thicker and denser making it impermeable and thus a more effective barrier to dissolved oxygen transport.

Unfortunately, the passivation or microencapsulation techniques do not have the ability to target specific problematic acid-generating sulphide-bearing minerals in complex systems (Park et al., 2019). As a result, microencapsulation techniques have resulted in unnecessary consumption of large quantities of expensive reagents. A great improvement to the microencapsulation technique proposed by Satur et al. (2007) that specifically target potential acid-generating minerals is termed carrier-microencapsulation (CME). Since CME can target pyrite and arsenopyrite, for example, even in complex systems containing other minerals like silicates and aluminosilicates, unwanted consumption of chemicals during treatment is drastically reduced (Park et al., 2018a).

Figure 6.3 is a schematic representation of CME. According to Park et al. (2018b), in CME, redox-sensitive organic compounds (e.g., catechol, 1,2-dihydroxybenzene) are used to transform relatively insoluble metal(loid) ions (e.g., Si^{4+} or Ti^{4+}) into soluble metal(loid)-organic complexes (e.g., $[Si(cat)_3]^{2-}$

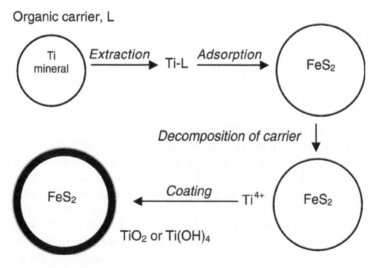

FIGURE 6.3
Schematic illustration of CME. (From Satur et al., 2007.)

and [Ti(cat)$_3$]$^{2-}$), which are stable in solution, but decompose selectively on the surfaces of pyrite and arsenopyrite, for example. The decomposition of metal(loid)-organic complex frees the insoluble metal(loid) ion, which is rapidly precipitated to form stable metal(loid)-oxyhydroxide coatings on the sulphide-bearing material. A number of researchers argue that the unique targeting ability of CME could be attributed to the decomposition of metal(loid)-organic complexes only on surfaces of minerals that dissolve electrochemically like pyrite and arsenopyrite, for example (Crundwell, 1988; Tabelin et al., 2017; Park et al., 2018b).

If successful, passivation is considered as a low-cost prevention technique, especially compared to traditional AMD treatments using alkaline additives (Sahoo et al., 2013; Nyström, 2018). However, most of the materials used for passivation are either too expensive or potentially harmful to the environment (Sahoo et al., 2013; Nyström, 2018). Thus, there is a need to find cost-effective materials able to passivate sulphide surfaces in a long-term perspective.

6.2.5.2 Desulphurisation

Environmental desulphurisation is one technology that has gained popularity in the last couple of decades for the management of acid-generating sulphide materials (Bois et al., 2004; Sahoo et al., 2013; Nadeif et al., 2019). This technique consists of separating sulphide components from non-sulphide mine tailings using the principle of froth flotation (Nadeif et al., 2019). The success of the method relies on how well the sulphide minerals are isolated from the non-sulphide ones (Nyström, 2018). Ideally, two different fractions are generated, the desulphurised tailings that do not generate AMD and a quantity of sulphide-enriched tailings which is acid generating (Bois et al., 2004). In principle, the low-sulphur tailings can be used as dry covers for existing high-sulphur materials (Bussière et al., 2004; Demers et al., 2008; Fyfe and Martin, 2011; Sahoo et al., 2013); however, the low sulphur-bearing tailings should be evaluated to ensure that the cover material is not of an acid-generating type (Sahoo et al., 2013). Alternatively, the resulting desulphurised fraction can be managed separately (Leppinen et al., 1997; Benzaazoua et al., 2000), for example, by backfilling it in underground mines (Benzaazoua and Kongolo, 2003). In view of the aforementioned examples, desulphurisation has demonstrated to be an economically and environmentally effective technique for decreasing the acid-generation potential of the sulphide-bearing materials (Bois et al., 2004; Demers et al., 2008; Nadeif et al., 2019). Furthermore, another advantage of desulphurisation is that, if successful, it can limit the amount of mine waste that needs treatment (Nyström, 2018).

6.2.5.3 Electrochemical Protection Systems

It is established that sulphide minerals exhibit semi-conductor properties (Koch, 1975) and consequently are amenable to electrochemical manipulation (Lin et al., 2001). Therefore, the electrochemical protection systems are

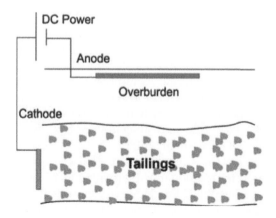

FIGURE 6.4.
Schematic illustration of an electrochemical protection system. (From Lin et al., 2001.)

premised on the fact that electronegative polarisation can be used to prevent the oxidation of sulphide-rich materials in a manner similar to that used in cathodic protection of steel structures (Lin et al., 2001). In this method, the cathode of the electrochemical cell is the tailings pile (or an exposed seam of sulphide-bearing rock), which must be partially submerged. A steel rod forms the electrical contact to the sulphide-bearing rock in the case of a seam, or a grid of metal mesh or graphite in the case of a pile of tailings. An external circuit connects the cathode to the anode (scrap iron), which is submerged in the water body to be protected, and the water also acts as its own supporting electrolyte (Bejan and Bunce, 2015).

Figure 6.4 is a schematic representation of the electrochemical protection system utilizing tailings as the cathode (Lin et al., 2001). The key to this technique involves negatively polarizing the tailing-electrolyte interface so that dissolved oxygen is reduced at the surface of the tailings (Sahoo et al., 2013). In this regard, electrons are transferred to the tailings through the electrical circuit, thus negatively polarizing the tailings/electrolyte interface. As a result, oxygen is reduced, in situ, at the tailings/electrolyte interface through the electrochemical reduction of oxygen as described by reaction 6.4 (Lin et al., 2001).

$$2H_2O + 4e^- + O_2 \rightarrow 4OH^- \tag{6.4}$$

Lin et al. (2001) state that the rate at which the electrochemical reduction reaction proceeds is controlled predominantly by the applied current density that is related to the rate at which oxygen diffuses through the overburden (or soil cover) to the tailings. In addition to removing dissolved oxygen, hydroxyl ions are generated at the tailings, thereby inhibiting the oxidation of sulphide minerals and the associated release of metal ions (Lin et al., 2001). According to Lin et al. (2001), the negative polarisation of tailings also changes the thermodynamic properties of the iron sulphide minerals in the tailings to more stable conditions.

6.3 Concluding Remarks

The ability to reduce oxidation of sulphide-bearing waste materials after mine closure or in existing mines is a widely researched area. Unfortunately, the concept of AMD prevention is a difficult one because it is dependent on several interrelated parameters. However, despite existing difficulties, a number of AMD prevention technologies are in operation globally. Actually, these techniques have shown great promise in delaying and/or the prevention of AMD formation by limiting oxygen availability, microbial activities or reactions between acid-producing minerals and oxygenated water. Unfortunately, the major disadvantage of the preventive technologies is their ineffectiveness in the long term. Moreover, most of these techniques have failed to protect the environment against long and persistent AMD pollution.

References

Ačai, P., Sorrenti, E., Gorner, T., Polakovič, M., Kongolo, M. and de Donato, P. (2009). Pyrite passivation by humic acid investigated by inverse liquid chromatography. Colloids and Surfaces A: Physicochemical and Engineering Aspects 337: 39–46.

Akcil, A. and Koldas, S. (2006). Acid mine drainage (AMD): causes, treatment and case studies. Journal of Cleaner Production 14: 1139–1145.

Alakangas, L., Andersson, E. and Mueller, S. (2013). Neutralization/prevention of acid rock drainage using mixtures of alkaline by-products and sulfidic mine wastes. Environmental Science and Pollution Research 20: 7907–7916.

Argunhan-Atalay, C. and Yazicigil, H. (2018). Modeling and performance assessment of alternative cover systems on a waste rock storage area. Mine Water and the Environment 37: 106–118.

Aube, B.C., St-Arnaud, L.C., Payant, S.C. and Yanful, E.K. (1995). Laboratory evaluation of the effectiveness of water covers for preventing acid generation from pyritic rock. In: Proceedings of Sudbury 1995: Mining and the Environment, May 1995, Sudbury, Ontario, Canada, Vol. 2, pp. 495–504.

Bacelar-Nicolau, P. and Johnson, D.B. (1999). Leaching of pyrite by acidophilic heterotrophic ironoxidizing bacteria in pure and mixed cultures. Applied and Environmental Microbiology 65(2): 585–590.

Backes, C.A., Pulford, I.D. and Duncan, H.J. (1987). Studies on the oxidation of pyrite in colliery spoil. II. Inhibition of the oxidation by amendment treatments. Reclamation and Revegetation Research 6: 1–11.

Baker-Austin, C. and Dopson, M. (2007). Life in acid: pH homeostasis in acidophiles. Trends in Microbiology 15(4): 165–171.

Barton-Bridges, J.P and Robertson, A.M. (1989). Design and reclamation of mine waste facilities to control acid mine drainage. Available at https://www.asmr. us/Portals/0/Documents/Conference-Proceedings/1989/0717-Barton-Bridges. pdf [Accessed 22 April 2020].

Bejan, D. and Bunce, N. (2015). Acid mine drainage: electrochemical approaches to prevention and remediation of acidity and toxic metals. Journal of Applied Electrochemistry 45: 1239–1254.

Bellaloui, A., Chtaini, A., Ballivy, G. and Narasiah, S. (1999). Laboratory investigation of the control of acid mine drainage using alkaline paper mill waste. Water, Air, & Soil Pollution 111: 57–73.

Benzaazoua, M., Bussière, B., Kongolo, M., McLaughlin, J. and Marion, P. (2000). Environmental desulphurization of four Canadian mine tailings using froth flotation. International Journal of Mineral Processing 60: 57–74.

Benzaazoua, M. and Kongolo, M. (2003). Physico-chemical properties of tailing slurries during environmental desulphurization by froth flotation. International Journal of Mineral Processing 69: 221–234.

Bessho, M., Wajima, T., Ida, T. and Nishiyama, T. (2011). Experimental study on prevention of acid mine drainage by silica coating of pyrite waste rocks with amorphous silica solution. Environmental Earth Sciences 64: 311–318.

Blowes, D.W., Reardon E.J., Jambor, J.L. and Cherry, A. (1991). The formation and potential importance of cemented layers in inactive sulphide mine railings. Geochimica et Cosmochimica Acta 55: 965–978.

Bois, D., Poirier, P., Benzaazoua, M. and Bussière, B. (2004). A feasibility study on the use of desulphurized tailings to control acid mine drainage. Available at https://www.onemine.org/document/document.cfm?docid=232136&docor gid=15 [Accessed 30 April 2020].

Brown, T. (1996). Acid mine drainage prevention, control and treatment technology development for the Stockett/Sand Coulee area. Available at https://inis.iaea. org/collection/NCLCollectionStore/_Public/30/022/30022678.pdf [Accessed 17 April 2020].

Bussière, B., Benzaazoua, M., Aubertic, M. and Mbonimpa, M. (2004). A laboratory study of covers made of low-sulfide tailings to prevent acid mine drainage. Environmental Geology 45: 609–622.

Caruccio, F.T. and Geidel, G. (1983). The effect of plastic liner on acid loads/DLM Site, W.V. Available at https://wvmdtaskforce.files.wordpress.com/2015/12/83-caruccio2.pdf [Accessed 25 April 2020].

Crundwell, F.K. (1988). The influence of the electronic structure of solids on the anodic dissolution and leaching of semiconducting sulphide minerals. Hydrometallurgy 21(2): 155–190.

Daniel, D.E. and Wu, Y.K. (1993). Compacted clay liners and covers for arid sites. The Journal of Geotechnical Engineering 119(2): 223–237.

Demers, I., Bussière, B., Benzaazoua, M., Mbonimpa, M. and Blier, A. (2008). Column test investigation on the performance of monolayer covers made of desulphurized tailings to prevent acid mine drainage. Minerals Engineering 21: 317–329.

Demers, I., Mbonimpa, M., Benzaazoua, M., Bouda, M., Awoh, S., Lortie, S. and Gagnon, M. (2017). Use of acid mine drainage treatment sludge by combination with a natural soil as an oxygen barrier cover for mine waste reclamation:

laboratory column tests and intermediate scale field tests. Minerals Engineering 107: 43–52.

Doye, I. and Duchesne, J. (2003). Neutralisation of acid mine drainage with alkaline industrial residues: laboratory investigation using batch-leaching tests. Applied Geochemistry18: 1197–1213.

Elsetinow, A.R., Borda, M.J., Schoonen, M.A.A. and Strongin, D.R. (2003). Suppression of pyrite oxidation in acidic aqueous environments using lipids having two hydrophobic tails. Advances in Environmental Research 7: 969–974.

Evangelou, V.P. (1995). Pyrite Oxidation and Its Control. CRC Press, Boca Raton, FL.

Evangelou, V.P. (1996). Oxidation proof silicate surface coating on iron sulfides. U.S. Patent No. 5, 494, 703.

Evangelou, V.P. (2001). Pyrite microencapsulation technologies: principles and potential field application. Ecological Engineering 17: 165–178.

Fraser, W. and Robertson, J. (1994). Subaqueous disposal of reactive mine waste: an overview and update of case studies-MEND/Canada. Available at https://www.asmr.us/Portals/0/Documents/Conference-Proceedings/1994-Volume-1/0250-Fraser.pdf [Accessed 26 April 2020].

Fyfe, J. and Martin, J. (2011). Innovative closure concepts for Xstrata nickel onaping operations. Available at http://bc-mlard.ca/files/presentations/2007-16-FYFE-MARTIN-innovative-closure-concepts-xstrata.pdf [Accessed 1 May 2020].

Hakkou, R., Benzaazoua, M. and Bussière, B. (2009). Laboratory evaluation of the use of alkaline phosphate wastes for the control of acidic mine drainage. Mine Water and the Environment 28: 206–218.

Hallberg, R.O., Granhagen. J.R. and Liljemark. A. (2005). A fly ash/biosludge dry cover for the mitigation of AMD at the Falun mine. Chemie der Erde 65: 43–63.

Hao, J., Cleveland, C., Lim, E., Strongin, D.R. and Schoonen, M.A.A. (2006). The effect of adsorbed lipid on pyrite oxidation under biotic conditions. Geochemical Transactions 7(8): 1–9.

Hao, J., Murphy, R., Lim, E., Schoonen, M.A.A. and Strongin, D.R. (2009). Effects of phospholipid on pyrite oxidation in the presence of autotrophic and heterotrophic bacteria. Geochimica et Cosmochimica Acta 73: 4111–4123.

Harries, J. R. and Ritchie, A.I.M. (1985). Pore gas composition in waste-rock dumps undergoing pyrite oxidation. Soil Science 140: 143–152.

Hodges, G., Roberts, D.W., Marshall, S.J. and Dearden, J.C. (2006). The aquatic toxicity of anionic surfactants to Dophnia magna: a comparative QSAR study of linear alkylbenzene sulphonates and ester sulphonates. Chemosphere 63: 1443–1450.

Huminicki, D.M.C. and Rimstidt, J.D. (2009). Iron oxyhydroxide coating of pyrite for acid mine drainage control. Applied Geochemistry 24: 1626–1634.

INAP. (2014). Global acid rock drainage (GARD) guide. Available at http://www.gardguide.com/images/5/5f/TheGlobalAcidRockDrainageGuide.pdf [Accessed 28 April 2020].

Ingledew, W.J. (1982). Thiobacillus ferrooxidans the bioenergetics of an acidophilic chemolithotroph. Biochimica et Biophysica Acta 683: 89–117.

Ji, M., Gee, E., Yun, H., Lee, W., Park, Y., Khan, M.A. and Choi, J. (2012). Inhibition of sulfide mineral oxidation by surface coating agents: batch and field studies. Journal of Hazardous Materials 229/230: 298–306.

Jiang, C.L., Wang, X.H. and Parekh, B.K. (2000). Effect of sodium oleate on inhibiting pyrite oxidation. International Journal of Mineral Processing 58: 305–318.

Johnson, D.B. and Hallberg, K.B. (2005). Acid mine drainage remediation options: a review. Science of the Total Environment 338: 3–14.

Kastyuchik, A., Karam, A. and Aïder, M. (2016). Effectiveness of alkaline amendments in acid mine drainage remediation. Environmental Technology and Innovation 6: 49–59.

Kefeni, K.K., Msagati, T.A.M. and Mamba, B.B. (2017). Acid mine drainage: prevention, treatment options, and resource recovery: a review. Journal of Cleaner Production 151: 475–493.

Kelley, B. C. and Tuovinen, O. H. (1988). Microbiological oxidations of minerals in mine tailings. In: Salomons, W. and Forstner, U. (Editors), Chemistry and Biology of Solid Waste: Dredged Material and Mine Tailings. Springer-Verlag, Berlin, FRG, pp. 33–53.

Khummalai, N. and Boonamnuayvitaya, V. (2005). Suppression of arsenopyrite surface oxidation by solgel coatings. Journal of Bioscience and Bioengineering 99: 277–284.

Kleinmann, R. (1982). Method of control of acid drainage from exposed pyritic materials. U.S. Patent No. 4, 314, 966.

Kleinmann, R.L.P. (1990). At-source of acid mine drainage. Mine Water and the Environment 9: 85–96.

Kleinmann, R.L.P. (1998). Bactericidal control of acidic drainage. In: Brady, K.C., Smith, M.W. and Schueck, J. (Editors), Coal Mine Drainage Prediction and Pollution Prevention in Pennsylvania. Pennsylvania Department of Environmental Protection (DEP), Harrisburg, PA, Chapter 15, pp. 15-1–15-6.

Kleinmann, R.L.P. and Crerar, D.A. (1979). Thiobacillus ferrooxidans and the formation of acidity in simulated coal mine environments. Geomicrobiology 1: 373–388

Kleinmann R.L.P. and Erickson, P.M. (1981). Field evaluation of a bactericidal treatment to control acid drainage. In: Graves, D.H. (Editor), Proceedings of the Symposium on Surface Mining Hydrology, Sedimentology and Reclamation. University of Kentucky, Lexington, KY, pp. 325–329.

Kleinmann, R.L.P. and Erickson, P.M. (1983). Control of Acid Mine Drainage from Coal Refuse Using Anionic Surfactants. U.S. Bureau of Mines, Washington, DC, p. 8847.

Koch, D.F.A. (1975). Electrochemistry of sulphide minerals. In: Bockris, J.O.M. and Conway, B.E. (Editors), Modern Aspects of Electrochemistry, No. 10. Plenum Press, New York, pp. 211–237.

Kuyucak, N. (2002). Acid mine drainage prevention and control options. Canadian Institute of Mining and Metallurgy Bulletin 95 (1060): 96–102.

Lalvani, S.B., DeNeve, B.A. and Weston, A. (1990). Passivation of pyrite due to surface treatment. Fuel 69: 1567–1569.

Langworthy, T.A. (1978). Microbial life in extreme pH values. In: Kuschner, D.J (Editor), Microbial Life in Extreme Environments. Academic, New York, pp. 279–315.

Leppinen, J., Salonsaari, P. and Palosaari, V. (1997). Flotation in acid mine drainage control: beneficiation of concentrate. Canadian Metallurgical Quarterly 36: 225–230.

Li, Y., Li, W., Xiao, Q., Song, S., Liu, Y. and Naidu, R. (2018). Acid mine drainage remediation strategies: a review on migration and source controls. Minerals and Metallurgical Processing 35(3): 148–158.

Lin, Z. (1997). Mobilization and retention of heavy metals in mill-tailings from Garpenberg sulfide mines, Sweden. Science of the Total Environment 198: 13–31.

Lin, M., Seed, L., Yetman, D., Fyfe, J., Chesworth, W. and Shelp, G. (2001). Electrochemical cover technology to prevent the formation of acid mine drainage. Available at https://open.library.ubc.ca/cIRcle/collections/59367/items/1.0042382.. [Accessed 1 May 2020].

Liwarska, B.E., Miksch, K., Malachowska, J.A. and Kalka, J. (2005). Acute toxicity and genotoxicity of five selected anionic and nonionic surfactants. Chemosphere 58: 1249–1253.

Lottermoser, B.G. (2003). Mine Wastes: Characterization, Treatment and Environmental Impacts. Springer-Verlag, Berlin Heidelberg, Germany.

Morris, P.H., Graham, J. and Williams, D.J. (1992). Cracking in drying soils. Canadian Geotechnical Journal 29: 263–277.

Mylona, E., Xenidis, A. and Paspaliaris, I. (2000). Inhibition of acid generation from sulphidic wastes by the addition of small amounts of limestone. Minerals Engineering 13(10–11): 1161–1175.

Nadeif, A., Taha, Y., Bouzahzah, H., Hakkou, R. and Benzaazoua, M. (2019). Desulfurization of the old tailings at the Au-Ag-Cu Tiouit Mine (anti-atlas Morocco). Minerals 9(7): 401.

Nicholson, R.V., Gillham, R.W. and Reardon, E.J. (1988). Pyrite oxidation in carbonate-buffered solutions: 1. Experimental kinetics. Geochimica et Cosmochimica Acta 52: 1077–1085.

Nicholson, R.V., Gillham, R.W. and Reardon, E.J. (1990). Pyrite oxidation in carbonate-buffered solutions: 2. Rate control by oxide coatings. Geochimica et Cosmochimica Acta 54: 395–402.

Nyavor, K. and Egiebor, N.O. (1995). Control of pyrite oxidation by phosphate coating. Science of the Total Environment 162: 225–237.

Nyström, E. (2018). Suitability of industrial residues for preventing acid rock drainage generation from waste rock. PhD Thesis, Luleå University of Technology, Sweden.

Pandey, S., Yacob, T.W., Silverstein, J., Rajaram, H., Minchow, K. and Basta, J. (2011). Prevention of acid mine drainage through complexation of ferric iron by soluble microbial growth products. Available at https://ui.adsabs.harvard.edu/abs/2011AGUFM.H43J1370P/abstract [Accessed 25 April 2020].

Park, I., Tabelin, C. B., Jeon, S., Li, X., Seno, K., Ito, M. and Hiroyoshi, N. (2019). A review of recent strategies for acid mine drainage prevention and mine tailings recycling. Chemosphere 219: 588–606.

Park, I., Tabelin, C.B., Magaribuchi, K., Seno, K., Ito, M. and Hiroyoshi, N. (2018a). Suppression of the release of arsenic from arsenopyrite by carrier-microencapsulation using Ti-catechol complex. Journal of Hazardous Materials 344: 322–332.

Park, I., Tabelin, C.B., Seno, K., Jeon, S., Ito, M. and Hiroyoshi, N. (2018b). Simultaneous suppression of acid mine drainage formation and arsenic release by carrier-microencapsulation using aluminum-catecholate complexes. Chemosphere 205: 414–425.

Payant, S.C., St. Amaud, L.C. and Yanful, E.K. (1995). Evaluation of techniques for preventing acidic rock drainage. Available at http://pdf.library.laurentian.ca/medb/conf/Sudbury95/AcidMineDrainage/AMD7.PDF [Accessed 24 April 2020].

Peppas, A., Komnitsas, K. and Halikia, I. (2000). Use of organic covers for acid mine drainage control. Minerals Engineering 13(5): 563–574.

Pérez-López, R., Cama, J., Nieto, J.M. and Ayora, C. (2007). The iron-coating role on the oxidation kinetics of a pyritic sludge doped with fly ash. Geochimica et Cosmochimica Acta 71: 1921–1934.

Pond, A.P., White, S.A., Milczarek, M. and Thompson, T.L. (2005). Accelerated weathering of biosolid-amended copper mine tailings. Journal of Environmental Quality 34: 1293–1301.

Pozo-Antonio, S., Puente-Luna, I., Lagüela-López, S. and Veiga-Ríos, M. (2014). Techniques to correct and prevent acid mine drainage: a review. Dyna 81(186): 73–80.

Rawlings, D.E. (2005). Characteristics and adaptability of iron- and sulfur-oxidizing microorganisms used for the recovery of metals from minerals and their concentrates. Microbial Cell Factories 4(13): 1–15.

Renton, J.J., Stiller, A.H. and Rymer, T.E. (1988). The use of phosphate materials as ameliorants for acid mine drainage. Available at https://www.asmr.us/Portals/0/Documents/Conference-Proceedings/1988-Volume-1/0067-Renton.pdf [Accessed 22 April 2020].

Ribet, I., Ptacek, C.J., Blowes, D.W. and Jambor, J.L. (1995). The potential for metal release by reductive dissolution of weathered mine tailings. Journal of Contaminant Hydrology17: 239–273.

RoyChowdhury, A., Sarkar, D. and Datta, R. (2015). Remediation of acid mine drainage-impacted water. Current Pollution Reports 1: 131–141.

Sahoo, P.K., Kim, K., Equeenuddin, S. M. and Powell, M.A. (2013). Current approaches for mitigating acid mine drainage. In: Whitacre, D.M. (Editor), Reviews of Environmental Contamination and Toxicology, Volume 226. Springer Science, Berlin, pp. 1–32.

Sasaki, K., Tsunekawa, M., Ohtsuka, T. and Konno, H. (1998). The role of sulfur-oxidizing bacteria Thiobacillus thiooxidans in pyrite weathering. Colloids and Surfaces A: Physicochemical and Engineering Aspects 133: 269–278.

Sasaki, K., Tsunekawa, M., Tanaka, S. and Konno, H. (1996). Supression of microbially mediated dissolution of pyrite by originally isolated fulvic acids and related compounds. Colloids and Surfaces A: Physicochemical and Engineering Aspects 119: 241–253.

Satur, J., Hiroyoshi, N., Tsunekawa, M., Ito, M. and Okamoto, H. (2007). Carrier-microencapsulation for preventing pyrite oxidation. International Journal of Mineral Processing 83: 116–124.

Simate, G.S. and Ndlovu, S. (2008). Bacterial leaching of nickel laterites using chemolithotrophic microorganisms: identifying influential factors using statistical design of experiments. International Journal of Mineral Processing 88: 31–36.

Simate, G. S., Ndlovu, S. and Gericke, M. (2009). Bacterial leaching of nickel laterites using chemolithotrophic microorganisms: process optimisation using response surface methodology and central composite rotatable design. Hydrometallurgy 98: 241–246.

Simate, G. S., Ndlovu, S. and Walubita, L.F. (2010). The fungal and chemolithotrophic leaching of nickel laterites – challenges and opportunities. Hydrometallurgy 103: 150–157.

Skousen, J., Rose, A., Geidel, G., Foreman, J., Evans, R. and Hellier, W. (1998). Handbook of Technologies for Avoidance and Remediation of Acid Mine

Drainage. The National Mine Land Reclamation Centre, West Virginia University, Morgantown, WV.

Skousen, J. G., Sencindiver, J. C. and Smith, R. M. (1987). A review of procedures for surface mining and reclamation in areas with acid producing materials. Energy and Water Research Center 871, West Virginia University, Morgantown, WV.

Skousen, J.G., Sexstone, A. and Ziemkiewicz, P.F. (2000). Acid mine drainage control and treatment. Agronomy 41: 31–68.

Skousen, J.G., Ziemkiewicz, P.F. and McDonald, L.M. (2019). Acid mine drainage formation, control and treatment: approaches and strategies. Extractive Industries and Society 6(1): 241–249.

Stehouwer, R., Sutton, P., Fowler, R. and Dick, W. (1995). Minespoil amendment with dry flue gas desulfurization by-products: element solubility and mobility. Journal of Environmental Quality 24: 165–174.

Swanson, D. A. and O'Kane, M. (1999). Application of unsaturated zone hydrology at waste rock facilities: Design of soil covers and prediction of seepage. Available at https://pdfs.semanticscholar.org/84f1/484ffd6c745b092c22da9df89d659220119b.pdf [Accessed 14 October 2020].

Swanson, D.A., Barbour, S.L. and Wilson, G.W. (1997). Dry-site versus wet-site cover design. In: Proceedings of the Fourth International Conference on Acid Rock Drainage. Mine Environment Neutral Drainage (MEND) Program, Vancouver, B.C., May 1997.

Tabelin, C.B., Veerawattananun, S., Ito, M., Hiroyoshi, N. and Igarashi, T. (2017). Pyrite oxidation in the presence of hematite and alumina: II. Effects on the cathodic and anodic half-cell reactions. Science of the Total Environment 581/582, 126–135.

Tremblay, R.L. (1994). Controlling acid mine drainage using an organic cover: the case of the East Sullivan Mine, Abitibi, Quebec. Available at https://www.asmr.us/Portals/0/Documents/Conference-Proceedings/1994-Volume-2/0122-Tremblay.pdf [Accessed 25 April 2020].

Vila, M.C., de Carvalho, S.J., A. da Silva, F. and Fi'uza, A. (2008). Preventing acid mine drainage from mine tailings. WIT Transactions on Ecology and the Environment 109: 729–738.

Wang, H.L., Shang. J.Q., Kovac, V. and Ho, K.S. (2006). Utilization of Atikokan coal fly ash in acid rock drainage from Musselwhite mine tailings. Canadian Geotechnical Journal 43: 229–243.

Xenidis, A., Mylona, E. and Paspaliaris, I. (2002). Potential use of lignite fly ash for the control of acid generation from sulphidic wastes. Waste Management 22: 631–641.

Yanful, E.K. (1993). Oxygen diffusion through soil cover on sulphidic mine tailings. Journal of Geotechnical Engineering 119: 1207–1228.

Yanful, E. and Orlandea, M. (2000). Controlling acid drainage in a pyritic mine waste rock. Part II: geochemistry of drainage. Water, Air, and Soil Pollution 124: 259–284.

Yanful, E.K., Simms, P.H. and Payant, S.C. (1999). Soil cover for controlling acid generation in mine tailings: a laboratory evaluation of the physics and the geochemistry. Water, Air, and Soil Pollution 114: 347–375.

Yeheyis, M.B., Shang, J.Q. and Yanful, E.K. (2009). Long-term evaluation of coal fly ash and mine tailings co-placement: a site-specific study. Journal of Environmental Management 91: 237–244.

Zhang, X., Borda, M.J., Schoonen, M.A.A. and Strongin, D.R. (2003). Adsorption of phospholipids on pyrite and their effect on surface oxidation. Langmuir 19: 8787–8792.

Zhang, Y.L. and Evangelou, V.P. (1998). Formation of ferric hydroxide-silica coatings on pyrite and its oxidation behavior. Soil Science 163: 53–62.

Zhang, M. and Wang, H. (2017). Utilization of bactericide technology for pollution control of acidic coal mine waste. Advances in Engineering Research 129: 667–670.

7

Remediation Processes
for Acid Mine Drainage

Sehliselo Ndlovu and Geoffrey S. Simate

CONTENTS

7.1 Introduction

The prevalence of acid mine drainage (AMD) globally in the mining and metal extraction sector has led to the development of a number of treatment technologies aimed at minimizing and mitigating its impact on the environment. The major aim of these treatment technologies is to reduce the acidity and sulphate content of the water. Most importantly, the treatment methods should not give rise to new high volume and unstable waste products that could impact further on the environment. This means that each method must be aligned with a near zero waste ethos, generate waste that is safe and not a source of environmental pollution or if possible generate products of economic value that can be used or sold to offset some of the treatment costs. As a result, this would reduce the problems associated with waste disposal. The stability of the waste products, as well as the volumes produced should, therefore, be a major criterion in the development of a long-term solution for the treatment of AMD. The treatment methods should also allow for the generation of potable water that can be reused in the mining and metal extraction processes or used in the communities where these industries are located. This could include activities such as agricultural use and other recreational activities. Thus, the treated AMD should be viewed as a resource and, in general, processes with the most likelihood of producing recyclable products, and of minimizing the waste products so that safe disposal can be sustained, should be the main focus for research and development in AMD treatment. This approach which implies the reuse, recycle, and recovery of valuable products from AMD is widely discussed in Chapter 9. This particular chapter (i.e., Chapter 7) will look at some of the technologies that are currently used to remediate AMD.

7.2 Classification of Acid Mine Drainage Treatment Technologies

The AMD technologies can be classified according to three levels of development, i.e., laboratory scale, pilot scale, and proven technologies (DWA, 2013). Laboratory scale technologies include all technologies that have only been tested at a theoretical laboratory scale (DWA, 2013). There is still, however, insufficient data available to attempt to test or implement the technology at full scale. Quite a significant number of recently developed technologies fall under this category as the transition to full-scale development is associated with very high risk and the current state of the mining industry has made most companies very risk averse. Pilot scale technologies are technologies that have been simulated in pilot plants to prove the chemical, physical or biological principles on a larger scale. The risks associated with technology that has been implemented on a pilot plant scale may be less than the risks associated with laboratory scale technology (DWA, 2013). A proven technology is one that has been in operation at a scale comparable with the scale required for the application under consideration (DWA, 2013). The implementation of any novel technologies in a financial tight market requires that any new technology be completely proven in order to achieve its full market deployment. Whilst massive research in the AMD treatment options has resulted in the generation of a significant number of laboratory scale technologies, there are relatively few that have been proven and implemented on a commercial scale.

AMD treatment technologies are commonly categorised as either 'passive' or 'active', both potentially combining physical, biological and chemical approaches. The main purpose of both types is to lower acidity and toxic metal concentrations, raise pH and often lower sulphate concentrations and salinity (Taylor et al., 2005). The active treatment methods involve regular reagent inputs for continued operation, thus are labour intensive. The AMD treatment systems can be used for both operational mine sites and occasionally post-closure scenarios (Taylor et al., 2005). Within the two categories of AMD treatment technologies, the methods can also be classified as (1) traditional (or conventional) treatment methods (i.e., methods that follow the pattern of an ordinary wastewater treatment plant) and (2) methods taking a new and original approach (innovative treatment methods, e.g., anoxic limestone drains [ALDs], constructed wetlands, etc.). However, it is noted that most of the traditional methods tend to follow the active treatment route, whereas the other category tends to involve a more passive treatment approach.

7.2.1 Classification of Conventional Active Treatment Methods

There are a number of methods that are considered as active treatment processes. The most common technique involves oxidation of AMD through aeration especially for solutions containing iron, dosing with an alkali to raise the pH for neutralisation followed by sedimentation for solid liquid separation. This process is similar to that of traditional wastewater treatment plants (Younger et al., 2002). Other traditional or active treatments common to wastewater treatment plants include sulphurisation, adsorption, ion exchange and membrane processes like filtration and reverse osmosis (RO) (Younger et al., 2002). Some of these processes will be considered in the subsequent sections.

7.2.1.1 Aeration

Since the principal contaminant is often dissolved ferrous iron, a key aspect of treating AMD is aeration. Iron is generally present in mine water as Fe^{2+} although Fe^{3+} can also be detected. Since ferrous iron (Fe^{2+}) is highly soluble over a larger pH range compared to ferric iron (Fe^{3+}), a larger amount of neutralizing agent is generally required to reduce the amount of ferrous iron in solution through precipitation as a hydroxide as compared to that needed to precipitate the iron present in the ferric state. Even at lower Fe^{2+} concentrations, aeration increases the level of dissolved oxygen and promotes oxidation of iron and manganese, increases chemical treatment efficiency and decreases remediation costs (INAP, 2012). As a result, most processes involving neutralisation as a form of mine water treatment generally tend to incorporate a pre-treatment step involving ferrous iron oxidation to the less soluble ferric form, which can then be precipitated as ferric hydroxide at a much lower pH (Dinardo et al., 1991). For mine water solutions with a high prevalence of iron, it is important to either pre-treat the solutions for iron removal or to initiate and maintain conditions that facilitate the easy removal of iron during the acid mine treatment process. In most cases, however, aeration is commonly applied simultaneously with addition of lime and flocculant to increase pH, oxidise metals species and precipitate metal hydroxides that are then treated through settlement, filtering or other processes (INAP, 2009). The iron precipitate generated is suitable for disposal or for application in a number of other uses. Research into possible uses and/ or means of disposal indicates the potential for the use of by-products as coagulants, mine backfill, or in the production of concrete, bricks, pigments or magnetite (MEND, 1994) as discussed in Chapter 9.

7.2.1.2 Treatment of Acid Mine Drainage Using the Neutralisation Process

The use of alkali neutralisation to remove dissolved metals from industrial effluent, mine water and groundwater is a well-established technique. In the process, sufficient alkalinity is added to raise the pH of AMD so that insoluble metal hydroxides precipitate and settle out of the AMD. The conventional approach to the neutralisation process involves three main steps such as (1) neutralisation of acidity, (2) aeration which is a key component for iron treatment, and (3) a flocculation and clarification stage (Taylor et al., 2005). The treatment process relies on the application of neutralizing agents such as lime, limestone or a combination of both to raise the solution pH, lower the solubility of dissolved metal ions and subsequently remove them as hydroxides (Skousen, 1988; Coulton et al., 2003). Flocculation agents are added to facilitate the formation of stable solids that can be easily filtered out and removed from the effluent.

For the actual treatment of AMD, the conventional neutralisation treatment process involves contacting an AMD solution with a controlled dosage of lime in a mixing tank to attain a desired pH set point (Aubé, 2004). The slurry is then contacted with a flocculant and fed to a clarifier for solid/ liquid separation. The sludge is collected from the bottom of the clarifier and can either be pumped to a storage area or pressure filtered to increase its density prior to transportation. The sludge produced by a conventional plant tends to settle to between 2% and 5% and can be mechanically dewatered to between 25% and 40% solids (Coulton, 2003). However, the low-density sludge produced requires significant power for pumping and a large storage area which presents high operational and disposal costs (MEND, 1994). This remarkable disadvantage has led to a slight modification of the conventional process flowsheet to include a sludge recycle stream resulting in the production of sludge with higher density. The high-density sludge (HDS) process substantially reduces the sludge volume and is a now a standard procedure in the AMD treatment industry.

The HDS process is a proven, well-established and understood process that has been in operation for a number of years and has been applied widely in a number of acid mine water treatment plants. The process is an improvement of the conventional neutralisation processes which tend to generate a high volume of precipitation sludge as a final product. It has been widely adopted by the mining industry where the relatively high metal concentrations and large flows make sludge volume minimisation imperative.

The key difference associated with HDS plants from the conventional plants is that the neutralisation phase is undertaken in two stages and a proportion of alkaline treatment sludge from the thickener underflow is recycled back through the plant to the first phase of the neutralisation process (Aubé, 2004). Basically, the recycled sludge is mixed with the lime slurry in a sludge/lime mixing tank and the resulting solution is mixed with acidic drainage in the chemical oxidation reactor (neutralisation tank). The recirculated sludge acts as seed for further metals precipitation, thus allowing for the generation of much higher density sludge than achieved by most conventional alkali neutralisation processes.

Figure 7.1 shows the typical flowsheet for a treatment plant utilizing the HDS method. Precipitation reactions come into completion in the lime reactor tank in which air is also added in order to help oxidise ferrous iron to ferric iron (Kuyucak, 2006). A typical HDS plant will produce a sludge that settles to between 35% and 50% solids and can be mechanically dewatered to between 50% and 70% solids (Coulton et al., 2003). The HDS process results in a substantial reduction in sludge volume leading to a reduction in sludge disposal costs. There is also an increase in sludge stability, both chemically and physically and a much higher quality effluent is produced (Aubé, 2004; Taylor et al., 2005). The HDS process is, therefore, particularly advantageous where large amounts of sludge are generated or where sludge disposal costs are significant. When this process is implemented, the toxicity

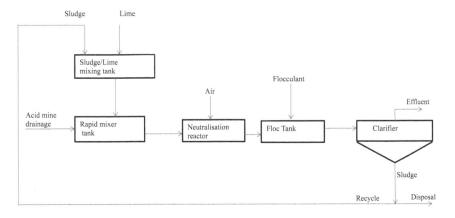

FIGURE 7.1
The conventional HDS process. (From Aubé, 2004.)

of the AMD solution is reduced to the extent where the treated water is, in most cases, regarded as non-toxic to animals and many organisms (Corbett, 2001).

The process has been implemented with a few variations made on the original flowsheet in order to meet the different needs of various plants, such as the differences in quality and quantity of the AMD (MEND, 1994; INAP, 2003). The major disadvantage associated with the process, however, is that the lime demand is relatively high, which has a significant impact on the operational costs. In addition, the process produces large volumes of metal hydroxide and gypsum sludge that have no direct value and require safe disposal (INAP, 2003). Lastly, although the quality of the water is much better than that of a conventional neutralisation process, it still does not meet the environmental or potable water quality specifications, and hence, further treatment would be required.

Beside the HDS process, there have also been a number of numerous technologies developed by researchers that focus on the acid-neutralisation step. There is the acid-barium-calcium (ABC) process developed by a team of researchers at the CSIR in South Africa (see Chapter 9 also). The ABC water treatment process is designed to achieve neutralisation as well as metal and sulphate removal (<100 mg/L) from AMD through the optimal and efficient use of readily available and affordable chemicals (De Beer et al., 2012; Maree et al., 2013). The process makes provision for three processing stages, pretreatment with lime and CaS to remove free acid and metals; $BaCO_3$ treatment to form barite; waste processing to recover alkali, barium and calcium in a coal-fired kiln (Merta, 2015). The original process has been further modified to minimise the gypsum generation, thereby reducing the quantity of gypsum crystallised during the water-treatment stage with all the sulphate removed as barite. This design increases the sludge load to the barium sludge processing stage, significantly reducing the gypsum sludge processing whose CAPEX alone is very high and was estimated at about R1 billion in 2004 (Maree et al., 2004). The major disadvantage of the process is the high energy requirement due to the thermal reduction in the barium sludge processing step which is required for $BaCO_3$ recycling (Mottay and Van Staden, 2018).

Another process developed in South Africa is the SAVMIN process which is indicated in Figure 7.2 (see Chapter 9 also). This process was developed by researchers at Mintek, South Africa and focuses on the removal of sulphate from the AMD solutions through the precipitation of Ettringite (Smit, 1999). The process has a number of sequential steps for the selective precipitation of insoluble complexes and the recycling of some of the reagents that are used in the process. The main treatment steps involve the following (Smit, 1999): (1) heavy metal precipitation, (2) gypsum crystallisation, (3) selective sulphate removal by ettringite precipitation using aluminium hydroxide, (4) aluminium recovery and (5) softening and pH adjustment by re-carbonation.

The end products are potable water and a number of potentially saleable by-products such as metal rich gypsum sludge, relatively pure gypsum sludge and calcium carbonate sludge. A major disadvantage of the process

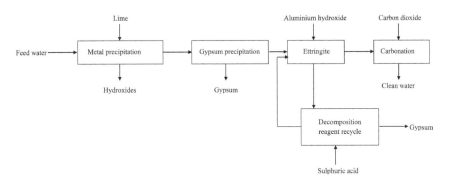

FIGURE 7.2
The SAVMIN process flow diagram. (From INAP, 2003.)

is the amount of the sludge produced which adds to the disposal costs. A mine water treatment demonstration plant, employing the SAVMIN technology, was established in partnership with gold mining company Sibanye-Stillwater at Randfontein, west of Johannesburg in 2018 (Mining Review, 2018).

7.2.1.3 Treatment of Acid Mine Drainage Using Sulphide Precipitation

Sulphide precipitation has been demonstrated to be an effective alternative to neutralisation using alkaline reagents for removing various heavy metals from industrial wastewater (Bhattacharyya et al., 1979; Peters et al., 1984; Abdulkadir, 2009). However, these processes have not been applied significantly on a commercial scale in the past due to a number of disadvantages. The costs of sulphide reagents are relatively higher than those used in hydroxide precipitation. In addition, the sulphide reagents used tend to produce hydrogen sulphide fumes when contacted with acidic wastes. However, this can be prevented by maintaining the pH of the solution between 8.0 and 9.5 and may require ventilation of the treatment tanks (MEND, 1994). The other disadvantage is that excess sulphide ions must also be present to drive the precipitation reaction to completion. Since the sulphide ion itself is toxic, sulphide addition must be carefully controlled to maximise heavy metals precipitation with a minimum of excess sulphide to avoid the necessity of post-treatment (MEND, 1994). Lastly, the sulphide sludge generated is more prone to oxidation resulting in resolubilisation of the metals as sulphates and as a result the sludge should, therefore, be stored carefully or be recycled for metal recovery (Peters et al., 1984). However, recent advances in technology and process development have seen the minimisation of some of the aforementioned challenges leading to a significant shift and popularity of the sulphide precipitation processes. This is largely because the solubilities of the metal sulphide precipitates are dramatically lower than those of hydroxide precipitates and, thus considerably lower metal concentrations in the effluent can be achieved. Furthermore, the metal sulphide precipitates exhibit better thickening and dewatering characteristics (Peters et al., 1984). The valuable metals can also be recovered selectively as a compact metal sulphide precipitate, which can be reprocessed at an appropriate stage in the flowsheet of a smelter.

The basic mechanism of the sulphidisation process involves the addition of a soluble metal sulphide to form an insoluble metal sulphide. The sulphide ion used in the precipitation processes can be supplied through a chemical or a biological source. Chemicals such as calcium sulphide (CaS), sodium sulphide (Na_2S), sodium hydrogen sulphide (NaHS) and iron sulphide (FeS) are commonly used in the process and these dissociate in solution to provide the sulphide ion necessary for metal sulphide precipitation (Kim, 1981; Peters and Ku, 1984; Lewis, 2010). The NaHS in particular is commonly used in waste water treatment following conventional lime treatment to reduce concentrations of residual metals, particularly cadmium.

The sulphide precipitation process can be represented by Equations 7.1 and 7.2, in which FeS, the precipitating reagent, dissociates to generate the sulphide ion that subsequently reacts with the metal ions (M^{2+}) in solution to precipitate a metal sulphide, MS

$$FeS = Fe^{2+} + S^{2-} \tag{7.1}$$

$$M^{2+} + S^{2-} = MS \tag{7.2}$$

The BQE Water firm (formerly BioteQ) in Canada has developed a sulphide precipitation technology that uses chemical (ChemSulphide® process) sources of sulphide to selectively precipitate dissolved metals from wastewater. Successful commercial operations have shown that high quality effluents that comply with stringent discharge limits can be produced and that metals can be recovered selectively into saleable high-grade concentrates from AMD or leach solutions (Nodwell and Kratochvil, 2012; Kratochvil et al., 2015).

The ChemSulphide process uses chemical sulphides such as NaHS for the removal of metals from contaminated wastewaters (Stedman, 2010). In the process, the chemical sulphide is added to a contactor tank where it mixes with the water to be treated under controlled conditions to selectively precipitate metals as a metal sulphides. The precipitated metals and treated water are then pumped to a clarifier tank where the clean water is separated from the metal solids and is either discharged to the local environment or recycled (Nodwell and Kratochvil, 2012). Thereafter, the metal solids are filtered to remove excess water, thus producing a high-grade metal product in the form of for example, copper sulphide that is suitable for refining. The rate of recovery for the metals is greater than 99%, and the recovered metal products are of a sufficiently high grade to be suitable for refining (Stedman, 2010). This technology has been applied since 2001 at multiple sites (Canada,

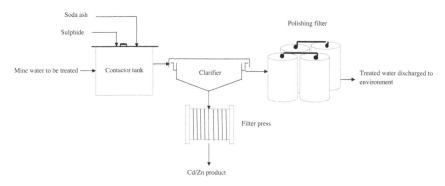

FIGURE 7.3
Wellington Oro ChemSulphide process flow diagram. (From Kratochvil et al., 2015.)

the United States, Australia and China) with different site conditions and requirements. The ChemSulphide process can be integrated with the HDS process, for the recovery of valuable metals, control of iron and sulphate and production of value-added construction materials from waste sludge, thus leading to a subsequent reduction or elimination of waste sludge (Lopez et al., 2009; Kratochvil et al., 2015).

The Wellington Oro Mine in the United States (Figure 7.3) is one example where the ChemSulphide technology has been successfully applied for the treatment of acid mine solutions (Lopez et al., 2009; Kratochvil et al., 2015). The mine, which is non-operational and treats post-closure solutions, utilises the ChemSulphide process to cost effectively remove dissolved metal ions of zinc and cadmium from the mine solution stream in order to meet strict effluent discharge targets (Nodwell and Kratochvil, 2012; Kratochvil et al., 2015). The dosing of NaHS is carefully controlled so that zinc and cadmium are removed to meet discharge limits and also to ensure that excess hydrogen sulphide gas is not produced (Smit, 1999). A small amount of soda ash is also added to the process to maintain the pH at the optimal range for sulphide precipitation. The precipitated metal sulphides can be sold to smelters for further processing thus helping to offset some of the process costs.

Another example is the Raglan Mine, an active nickel mine in Northern Quebec, which is owned and operated by Xstrata Nickel and was commissioned in 2004 with the first full year of operation in 2005 (Jones et al., 2006). The water treatment plant at Raglan employs the ChemSulphide process to selectively recover nickel from low-grade contaminated mine water. The treated water quality produced in the plant can be released directly into the local environment whilst the nickel sulphide precipitate is dewatered and shipped periodically to the Raglan concentrator where it is added to the flotation concentrate for shipment to the smelters (Lawrence and Fleming, 2007). Another mining operation that applies the ChemSulphide process is Jiangxi Copper in China. The company utilises this technology for a water treatment plant at its Dexing site in southeast China. Through this technology, the company is able to remove copper from mine wastewater to produce a high-grade marketable copper product, along with water that is clean enough to be safely discharged to the environment or recycled into the mining process.

Industrial effluents can also be treated using biogenic sulphide in the form of H_2S generated through the reduction of elemental sulphur in a bioreactor (Kuyucak, 2001, Lawrence et al., 2007). In fact, in recent years, the development of biological technologies for process-related applications in the mining and metallurgical industries has made great progress. For these applications, the focus is on minimizing the use of chemicals used for precipitation of metals and caustic used for removing sulphur dioxide from dilute off-gas streams (Dijkman et al., 1999). This subsequently translates to large-scale savings of costs. In addition, with regard to AMD treatment, an environmental problem can be solved while the metals recovered as sulphides generate a revenue stream.

In a biological process, the biogenic sulphide gas is transferred to a gas/liquid anaerobic agitated contactor in order to selectively precipitate the metals to be recovered as sulphides (Huisman et al., 2006; Lawrence and Fleming, 2007; Adams et al., 2008). In this set-up, there is no direct contact between the bacteria and the liquid stream treated, so issues with regard to possible toxic compounds in the liquid stream or temperature fluctuations are of no concern. Figure 7.4 shows a generic flowsheet for the metal recovery using biogenic H_2S gas (Boonstra and Buisman, 2003).

The advantage of the biogenic sulphide approach is that it can be very profitable in very low metal concentration solutions as the costs are relatively lower compared to NaHS or Na_2S (EPA, 2003). Compared to the addition of a NaHS or Na_2S solution, another obvious advantage of the use of biogenic H_2S is that no sodium ions are introduced to the precipitation circuit (Dijkman et al., 1999). In addition to the significant cost savings, biogenic sulphide produced on demand at site provides the additional benefit of improved safety due to the elimination of the transportation, handling and storage required for chemical sulphide reagents (Lawrence and Fleming, 2007).

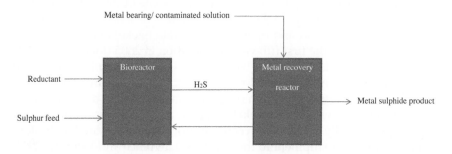

FIGURE 7.4
Process diagram for metal recovery using biogenic H_2S. (From Boonstra and Buisman, 2003.)

Sulphide can also be generated by the biological reduction of sulphate, although the generation of sulfide from this source is limited to passive water treatment applications in which low flows containing low concentrations of metals can be treated to precipitate metals in natural wetlands and sediments (Lawrence and Fleming, 2007). One example of the process that uses biological reduction of sulphate is the Thiopaq process. The Thiopaq process is a well-established process which uses sulphate reduction reactions to generate biogenic sulphide that can be used in the treatment of metal sulfate solutions such as AMD. The Thiopaq system utilises two distinct microbiological populations and processes: (1) conversion of sulphate to sulphide by sulphate-reducing bacteria (SRB) and precipitation of metal sulphides, and (2) conversion of any excess hydrogen sulphide produced to elemental sulphur, using sulphide-oxidizing bacteria (Johnson and Hallberg, 2005).

The bacterial reduction of sulphate to sulphide is an electrochemical process and a reductant (electron donor) is needed to supply electrons required for the conversion reaction. Examples of typical electron donors used in the Thiopaq process include hydrogen, methanol, ethanol or other organic material as electron donors for the production of biogenic H_2S (Reinsel, 2015). The choice of the type of reductant is not only dependent on the sulphur load to be processed, but also on the geographical location of the installation, reagent availability and cost. In the use of ethanol as the electron donor, this is first converted to acetic acid and hydrogen by bacteria as given in reaction 7.3.

$$CH_3CH_2OH + H_2O \rightarrow CH_3COOH + 2H_2 \tag{7.3}$$

Both hydrogen and acetic acid formed are consumed in the sulphate-reducing reaction taking place in the anaerobic reactor resulting in the formation of biogenic hydrogen sulphide.

$$H_2SO_4 + 4H_2 \rightarrow H_2S + 4H_2O \tag{7.4}$$

The hydrogen sulphide produced can then be employed for the precipitation of metals by contacting it with the solution to be treated. Careful control of pH can allow for selective precipitation of metal sulphides from a solution stream that contains a number of different metal ions.

The BioSulphide® process developed by BQE Water firm is another process that utilises biologically generated hydrogen sulphide gas for the removal of metals from contaminated water. The process is based on the Thiopaq process (Dijkman et al., 1999; Reinsel, 2015) and uses naturally occurring sulphur-reducing bacteria, to produce sulphide which is used to precipitate and recover metals selectively from contaminated industrial water. The BioSulphide process has two components – the biological and chemical stages which are fully integrated, but operate independently. The bioreactor contains a mixture of SRB that reduce the contained sulphate and produce the sulphide used in the chemical stage. Hydrogen or an organic electron donor is supplied to the bioreactor. Raw AMD enters the chemical circuit where it comes into contact with hydrogen sulphide generated in the biological circuit leading to metal sulfide precipitation. The characteristics of the BioSulphide process that make it different from other conventional sulfate reduction processes can be summarised as follows: (1) the biological component of the process is separated from the chemical precipitation/neutralisation stage, (2) only a fraction of the stream volume, as determined by sulphide and/or alkalinity requirements, enters the bioreactors, (3) AMD treatment to discharge quality is achieved entirely with bacterially generated products and (4) metal concentrates, metal sludge and biomass can be produced separately for sale or disposal.

Figure 7.5 shows the flowsheet for the application of the BioSulphide process at Bisbee, Arizona. The plant recovers copper from the drainage of a large low-grade stockpile. Copper had previously been recovered by cementation with iron, but is now precipitated into a high-grade copper sulphide concentrate using H_2S generated in a bioreactor in which elemental sulphur is reduced as shown in the flowsheet.

The BioSulphide process is generally used for high metals loading and higher sulphide demand due to a lower operating cost per tonne of sulphide required (Nodwell and Kratochvil, 2012). The BioSulphide process technology can be integrated with other water treatment technologies to improve overall water treatment. For example, it can be introduced upstream of an existing lime treatment plant to recover metals contained in the contaminated water. In such a case, lime plant economics are improved with lower lime consumption and the volume and toxicity of the sludge is reduced significantly.

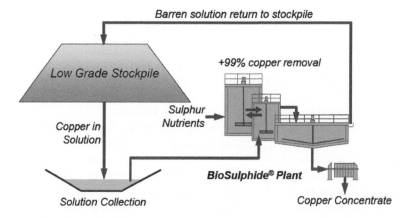

FIGURE 7.5
Simplified flowsheet showing the use of biogenic sulphide for copper recovery at Bisbee. (From Lawrence and Fleming, 2007.)

Another process that utilises biologically produced sulphide is the Rhodes Biological Sulphate Reduction (Biosure) process. This is a cost-effective and proven option for the treatment of AMD so as to mitigate the effect of AMD on water quality. The Biosure process utilises SRB that reduce sulphate to sulphide under anaerobic conditions. The bacteria require an organic source of readily biodegradable carbon such as primary sewage sludge and organic waste from the dairy and abattoir industries. The sulphide produced can be reacted with the metals in the AMD to form metal sulphides, which then precipitate and can be removed from the water in a clarifier. However, the metal sulphide sludge needs further treatment to prevent pollution of the environment at the disposal site; otherwise it will revert back to acid and sulphate on exposure to moisture and oxygen in the atmosphere, which is the cause of AMD in the first instance (DWA, 2013).

The Biosure process has been tested on a large scale at a mine shaft in Grootvlei, South Africa by the East Rand Water Care Company (ERWAT) (Godongwana et al., 2015). The process has been shown to be able to potentially produce water that complies with the general standard of waste water. The major disadvantage of the Biosure process is its limitation with regard to the carbon source. Large volumes of primary sewage sludge and/or external readily biodegradable carbon sources are required to meet the sulphate reduction demand of the AMD (DWA, 2013). The availability of the carbon source, i.e., sewage sludge, determines the placement of the treatment works and, in most cases, there is a lack of adequate sewage facilities at the AMD point source. Therefore, the application of the process at commercial scale becomes feasible only in the presence of an adequate source of readily available and biodegradable carbon to augment the carbon from sewage or alternatively to form the main carbon source.

7.2.1.4 Treatment of Acid Mine Drainage Using Reverse Osmosis

Membrane filtration by RO is a well-established strategy for heavy metal removal as it is capable of achieving strict metal discharge criteria whilst providing high efficiency, easy operation and saving space (Fu and Wang, 2011). It is an increasingly popular wastewater treatment option in chemical and environmental engineering and accounts for more than 20% of the world's desalination capacity (Shahalam et al., 2002). Large-scale applications of this process exist for AMD treatment and South Africa in particular have developed multistage RO concentration and gypsum precipitation process that have been used on a commercial scale in mining communities (DWA, 2013).

In the RO process, cellophane-like membranes separate purified water from contaminated water. A pressure is applied to the concentrated side of the membrane forcing purified water into the dilute side and the rejected impurities from the concentrated side are washed away in the reject water. The RO can be used to recover trace metals from AMD, remove 90% to 99% of the total dissolved solids to produce water of high quality enough to be reused as process water (Slater et al., 1983; Mortazavi, 2008) (see Chapter 9 also). The advantages of RO include low cost, high selectivity and flux, low capital and operating cost, chemical resistance and resistance to fouling, but the disadvantages include the feed to the membrane being very specific.

As a result of the high scaling tendency of the membranes in the presence of metal precipitates, this desalination process requires that the feed to the membranes be very specific with the removal of metals present in solution undertaken prior to the RO process. This usually involves the application of an upstream treatment plant such as the HDS process to ensure that metals are removed to the standards required by the suppliers of RO membranes. Some of the proven processes involving the commercial application of the RO techniques include the high pressure reverse osmosis (HiPRO) process. Modified processes including seeded reverse osmosis (SPARRO) that uses a suspension of salt crystals to promote precipitation have also been developed to improve the efficiency of the RO applications (Pulles et al., 1992).

The HiPRO™ technology developed by Aveng Water makes use of multiple stages of ultrafiltration and RO membrane systems which produce supersaturated brine streams (less than 3% of the total feed), from which sparingly soluble salts may be released in precipitation reactors. However, depending on the feed water quality, a zero liquid discharge (zero brine) solution is possible, thus, eliminating the need for high-cost evaporation and crystallizer plants. Solid waste generation can be eliminated through the production of useful by-products such as calcium sulphate of saleable grade as well as metal sulphate products. This process has been successfully applied at a number of mining operations in South Africa such as the eMalahleni water reclamation plant (Gunther and Mey, 2008; Karakatsanis and Cogho, 2010). A modular mine water treatment plant has also been installed at the Anglo American Thermal Coal's Kromdraai opencast mine.

The advantages of the HiPRO process include the following: the process has been proven at large-scale commercial operations, produces highest quality drinking water which meets the South African National Standard (SANS), has a very high water recovery that is usually in excess of 98% and generates potentially useful by-products (Karakatsanis and Cogho, 2010).

7.2.1.5 Treatment of Acid Mine Drainage Using Ion Exchange Technology

Ion-exchange processes have been used for the extraction of several metals from AMD and for the conversion of AMD to potable water (Feng et al., 2000; INAP, 2003). The ion-exchange process involves an exchange of ions or molecules between solid and liquid with no substantial change to the solid structure. It is a reversible chemical reaction where an ion from the solution is exchanged for similarly charged ion (typically hydrogen or hydroxyl) attached to an immobile solid particle (resin) thus rendering the target ion immobile (Bowell, 2000). Various resin forms are available to remove either cations or anions and a range of silica-based and polymeric resins can be used for metal recovery or removal (Dinardo et al., 1991). Synthetic organic resins are the predominant type since their characteristics can be tailored to specific applications. In the treatment of sulphate containing water, a two-stage process that uses cationic and anionic resins to remove calcium or magnesium and sulphate, respectively, is usually employed (Robertson et al., 1993). The resins are then regenerated, which typically requires acidic and basic reagents.

The appropriate ion-exchange resin is selected based on the acid mine water chemistry and the specific parameters that need to be removed from solution. Large flows generally require a full-scale treatment plant, but for small to intermediate flows, standard tank sizes that allow systems to be set up quickly can be used (ITRC, 2015). Many ion-exchange technologies appear to be technically effective at achieving water quality targets, but few have proven to be commercially viable or are in widespread use for the treatment of acidic mine water. The GYP-CIX process (Figure 7.6) is a fluidised bed ion-exchange process developed in South Africa to remove sulphate from water that is close to gypsum saturation and, therefore, it can be used as a polishing step after lime precipitation (Reinsel, 2015). The GYP-CIX process can tolerate relatively high concentrations of calcium, however, the TDSs need to be less than 4000 mg/L (Mottay and Van Staden, 2018). The process requires pre-treatment to remove metals, which may interfere and decrease the efficiency of resins in the downstream ion-exchange process. The resins

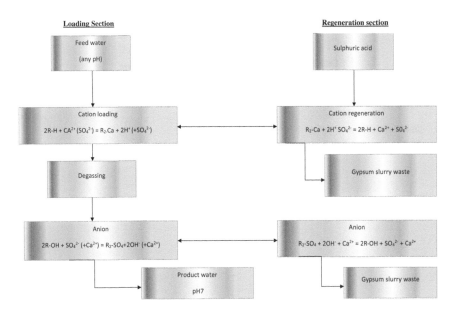

FIGURE 7.6
Simplified process flow diagram of the GYP-CIX process for treatment of mine water. (From INAP, 2003.)

are designed so as to target calcium and sulphate in order to reduce gypsum levels in the effluent thereby reducing the total dissolved solids and corrosion potential (Smit, 1999; Bowell, 2000).

The cation resin exchanges Ca^{2+}, Mg^{2+} and other cations (i.e., metal ions) by the following reaction:

$$2R-H \; + Ca^{2+} \rightarrow R_2 - Ca + 2H^+ \qquad (7.5)$$

The anion resin exchanges SO_4^{2-}, Cl^- and other anions by the following reaction:

$$2R-OH + SO_4^{2-} \rightarrow R_2SO_4 + 2OH^- \qquad (7.6)$$

Unlike conventional ion-exchange technology, the GYP-CIX process uses $Ca(OH)_2$ and H_2SO_4 (lowest cost industrial reagents) to regenerate the ion-exchange resins. The cationic and anionic exchange processes result in the generation of a large amount of gypsum that can be sold commercially thus offsetting treatment costs whilst the treated water meets standards for reuse. The continuous precipitation of gypsum in the regeneration of ion-exchange resins also allows the reuse of the regeneration solutions (INAP, 2003).

Another process that uses ion-exchange chemistry which is also largely based on the GYP-CIX process is the Sulf-IX™ process (Figure 7.7), developed by BQE Water (formerly BioteQ) (Lawrence, 2007; Lopez et al., 2009). It overcomes difficulties of the GYP-CIX process associated with limited process

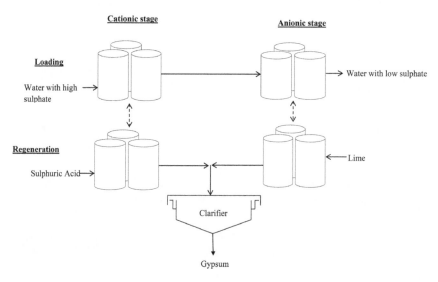

FIGURE 7.7
The Sulf-IX™ process. (From Lawrence, 2007.)

flexibility for varying feed chemistry, mechanical entrainment of gypsum in the regeneration stage and limitations on sulphate removal when magnesium is present in significant concentration in the feed water (Lopez et al., 2009). The process uses low-cost, off-the-shelf resins to remove calcium and sulphate ions from water in various concentrations. Removal of sulphate to meet new regulations worldwide is more efficient when using the Sulf-IX™ process compared with biological sulphate reduction and membrane technology. The process requires no pre-treatment and leaves no residual waste for special disposal (Nodwell and Kratochvil, 2012). The resins are regenerated using the low-cost reagents, sulphuric acid and lime, so that the only products of the process are clean water (water recoveries to up to 97%) that can be discharged or reused and solid gypsum product that can be used for the production of fertiliser and building materials (Nodwell and Kratochvil, 2012). The Sulf-IX process has been extensively piloted to remove sulphate from water at sites in Canada, the United States and Chile (Nodwell and Kratochvil, 2012).

Table 7.1 gives a summary of the active processes discussed in the sections above, highlighting the advantages and disadvantages of each technology.

7.2.2 Classification of Innovative Passive Treatment Methods

Passive treatment systems have been developed that do not require continuous chemical inputs and they take advantage of natural chemical and biological processes to cleanse contaminated mines (Skousen, 2002). Passive treatment systems are commonly, but not exclusively aggregate-carbonate based, with or without the inclusion of organic matter (Taylor et al., 2005). A variety of passive treatment systems have become the most predominate innovative technologies applied in the treatment of AMD solutions apart from traditional choices. These treatment technologies include constructed wetlands, anoxic limestone dams, permeable reactive barriers (PRBs), etc, and are mostly based on the same principle. The PRBs and constructed wetland technologies can all utilise alkaline agents and SRB to treat mine drainage. In low-flow and low-acidity situations, passive systems can be reliably implemented as a single permanent solution for many AMD problems. These passive treatment systems can provide a potentially long-term solution to the AMD problem although their success has been limited in cases where excessive volumes, high iron loadings and excessively low pH values are encountered (INAP, 2003). In addition, relative to chemical treatment, passive systems require longer retention times and greater space, provide less certain treatment efficiency and are subject to failure in the long term (Skousen, 1998).

The mechanisms of metal removal and retention in passive treatment systems are varied and include oxidation, precipitation as hydroxides and carbonates under aerobic conditions, precipitation as sulphides and hydroxysulphate under anaerobic conditions, complexation and adsorption onto organic matter, ion exchange with organic matter and uptake by plants

TABLE 7.1

Summary of the AMD Active Treatment Methods

Process	Process Technology	Advantages	Disadvantages	Possible Improvements
HDS process	Chemical neutralisation	• Inexpensive • Low maintenance • Removes trace metals • Produces water quality suitable for irrigation or reuse in the mine • Can be used as a cost-effective pre-treatment method to other processes • HDS process proven technology	• Limited sulphate removal • High amount of sludge produced • Costs associated with handling and safe disposal of potential unstable sludge	• Reduction in the production of waste or recycling of sludge
MINTEK SAVMIN	Chemical neutralisation	• Reduce sulphate to very low levels <200 mg/L • High quality water generated even at fluctuating feed sulphate levels • It regenerates the required $Al_2(OH)_3$ reagent for reuse, which results in a significant cost reduction • The waste products from the process can be disposed of either as a stable waste, or, in certain instances, constitute a usable by-product	• High amount of sludge produced • Success depends on high level of gypsum crystallisation. • Process can be complicated to control	• Reduction in the production of waste or recycling of sludge
BQE ChemSulphide	Sulphide precipitation	• Treated effluent meets regulatory compliance • Saleable products that can generate revenue, reduce and eliminate waste	• The cost of sulphide precipitants is high • Requires controlled addition of sulphide precipitant to maximise heavy metals precipitation with a minimum of excess sulphide to avoid the necessity of post-treatment	• Versatility in the case where AMD may contain significant alkalinity

Process	Method	Benefits	Limitations	Research needs
Paques Thiopaq	Microbial sulphate reduction	• Can effectively treat solutions with low metal concentrations	• Sensitivity of SRB to high metal concentrations and low pH • High process costs due to use of ethanol or butanol as carbon source and electron donor	• Explore microbial communities that can work at optimal process conditions, e.g., use of acid tolerant bacteria
BQE BioSulphide	Microbial sulphate reduction	• Can tolerate high metals loading and higher sulphide demand • Produces saleable metal products and clean water • Can be integrated with other water-treatment technologies	• Hydrogen sulphide (H_2S) produced as metabolic end • Product requires proper management to avoid pollution and health risk to personnel	• Explore use of other microbial communities • Explore alternative microbial substrates
Rhodes BioSure	Microbial sulphate reduction	• Effective in reducing sulphate in AMD • Produces water that complies with the general standard of waste water. • Generates stable bio solids	• Limited to geographical locations with sludge/waste facilities next to AMD point source • Hydrogen sulphide (H_2S) produced as metabolic end product and released into the external environment can cause air pollution and be a health risk	• Explore easily available carbon sources • Robust hydrogen sulphide capturing systems
HiPRO	Reverse osmosis	• Produces high quality water at variable water recoveries • Applied commercially for water desalination	• Membrane lifetime affected by fouling • Mine water needs to be pre-treated. • Relatively expensive	• Not suitable for scaling water-membrane extension
GYP-CIX	Ion exchange	• Can treat most waste water including scaling mine waters • Low cost reagent used to regenerate the resin • Produces very good quality water	• Volume of gypsum sludge produced in the ion-exchange-resin regeneration process	• Reduction in the production of waste or recycling of sludge • Reduction in the frequency of ion resin regeneration
BQE SULF-IX™ Process	Chemical precipitation + Ion exchange	• Low cost reagent used to regenerate the resin • Produces very good quality water • Saleable products that can generate revenue	• Requires controlled addition of sulphide precipitant to maximise heavy metals precipitation with a minimum of excess sulphide to avoid the necessity of post-treatment	• Reduction in the frequency of ion resin regeneration

Sources: Smit, 1999; INAP, 2003; Stedman, 2010.

(phyto-remediation) (INAP, 2003). However, the environmental conditions in the different passive treatment systems dictate the dominant metals removal mechanisms. Some of the passive treatment systems that have been employed for the treatment of AMD solutions are discussed in the subsequent sections.

7.2.2.1 Constructed Wetlands

Constructed treatment wetlands are engineered systems, designed and constructed to utilise the natural functions of wetland vegetation, soils and their microbial populations to treat contaminants in surface water, groundwater or waste stream (ITRC, 2003). They can be considered as treatment systems that use natural processes to stabilise, sequester, accumulate, degrade, metabolise and/or mineralise contaminants and have the ability to remove organic and inorganic compounds and suspended solids (ITRC, 2003). During the past few decades, wetlands have been established as systems that have high potential for meeting wastewater treatment and water quality objectives in a controlled manner. Constructed treatment wetlands can be used alone or in conjunction with other technologies to extend the operational lifespan of the systems or enhance the removal performance of specific constituents during the treatment of AMD (Brodie et al., 1991; Hedin et al., 1994; Sheoran and Sheoran, 2006). This flexibility further makes the technology applicable to many types of contaminants in many types of situations (ITRC, 2015).

A wetland is usually composed of two distinctive zones, oxidative zone which is vegetated with aquatic plants and reducing zone which is the sedimentation zone rich in SRB (Kuyucak, 2006). The water treatment mechanisms are biological, chemical and physical, which include physical filtration and sedimentation, biological uptake, transformation of nutrients by anaerobic and aerobic bacteria, plant roots and metabolism, metal exchange reactions as well as chemical processes (precipitation, absorption and decomposition) that purify and treat the wastewater (WWG, n.d.; Hedin et al., 1994). Other beneficial reactions in wetlands include generation of alkalinity due to microbial mineralisation of dead organic matter, microbial dissimilatory reduction of Fe oxyhydroxides and SO_4 and dissolution of carbonates (Skousen, 2002).

Wetlands can be classified as either aerobic or anaerobic. The main difference in these systems is the biological and chemical processes promoted and the design of water flow direction (ITRC, 2015). Aerobic wetlands are essentially shallow ponds designed to precipitate metals from water under aerobic conditions usually in a horizontal flow system. The aerobic wetlands are designed with depths no more than 30 cm that lower suspended solids and provide a substrate and increased water retention times (due to reduced flow rates) for the reaction between influent alkalinity and acidity that is generated from AMD (ITRC, 2015). The reaction is via metal oxidation and hydrolysis, and oxygen infiltration is encouraged thereby causing precipitation and physical retention of metals as oxyhdroxides, hydroxides and carbonates within the wetland (Skousen, 2002; Taylor et al., 2005). Aerobic wetlands are often used for net alkaline waters and predominantly for just aeration and precipitation of metals (Skousen and Ziemkiewicz, 2005).

Anaerobic systems primarily rely on chemical and microbial reduction reactions to precipitate metals and neutralise acidity. They generate alkalinity through bacterial activity and the use of Fe^{3+} as a terminal electron acceptor (Fripp et al., 2000) and are most effective in the treatment of small flow acidic water. The water infiltrates through thick, permeable organic material that becomes anaerobic due to high biological oxygen demand (Skousen, 2002). Since anaerobic wetlands produce alkalinity, their use can be extended to poor quality, net acidic, low pH, high Fe and high dissolved oxygen AMD (INAP, 2003).

The attractiveness of constructed wetland treatment lies in its ability to produce a near-neutral water product which can be readily discharged (Johnson and Hallberg, 2005). They also have significantly lower total lifetime costs and often lower capital costs than conventional treatment systems (ITRC, 2003); once constructed they can operate for long periods of time with minimal operations and maintenance. However, constructed wetlands are limited by the metal loads they deal with; hence adequate pre-treatment is required especially when dealing with high volumes and/or highly acidic water. Furthermore, some of the metal precipitates retained in sediments are unstable when exposed to oxygen; hence, it is crucial that the wetland sediments remain largely or permanently submerged (Johnson and Hallberg, 2005).

7.2.2.2 Anoxic Limestone Drains

Anoxic lime drains (ALDs) are well-known passive treatment systems that can be an effective and established technology for the treatment of acid mine water. The ALDs are buried cells or trenches of limestone engineered to intercept anoxic, acidic mine water and add alkalinity through dissolution of the limestone (Watzlaf et al., 2000). The water is constrained to flow through a bed of limestone gravel held within a drain that is impermeable to both air and water (Johnson and Hallberg, 2005). This creates an environment which is high in carbon dioxide and low in oxygen (INAP, 2003) and increases the dissolution of limestone while preventing the precipitation of iron hydroxide. Minimal iron precipitation is essential since iron hydroxides generated tend to inhibit limestone dissolution and clog the drain (Skousen, 1998). The ALDs can be used to treat AMD flows of various rates, alone or in combination with other treatment systems, and can be installed in a wide variety of locations with the use of commonly available construction equipment. They have been applied on a large scale for the treatment of acidic mine water from a number of abandoned mine plants in the United States (ITRC, 2015).

Although ALDs produce alkalinity at a lower cost than constructed wetlands, they are not suitable for treating all AMD waters. They are suitable to treat AMD that has low concentrations of ferric iron, dissolved oxygen and aluminium. In situations where the AMD contains significant concentrations of ferric iron or aluminium, the long-term performance of ALD is not good. When any of the three parameters (i.e., concentrations of ferric iron, dissolved oxygen and aluminium) are elevated, armoring of limestone can occur resulting in slow dissolution rate of limestone (Skousen, 1998; EPA, 2014). When the dissolution rate slows, there is a higher buildup of ferric iron and aluminium on the limestone, which eventually clogs the open pore spaces, resulting in abnormal flow paths that can reduce both the retention time of AMD within the ALD and the reactive surface area of the limestone (EPA, 2014). This may, in turn, cause failure of the drain within 6 months of construction (Watzlaf et al., 2000; Johnson and Hallberg, 2005). Moreover, problems also occur where ALDs are used to treat aerated mine waters due to iron oxidation. Hence, passage of AMD through an anoxic pond prior to the ALD may be necessary to lower dissolved oxygen concentrations to prevent this oxidation (Skousen, 1998). It is also important that the solution to be treated remains in an anoxic state prior to entering the ALD to prevent metals from precipitating out of the mine drainage and causing premature failure of the ALD (Cravotta and Trahan, 1999). This can be achieved by constructing the ALD directly on top of the discharge, allowing the acidic water to flow through the limestone, adding calcium carbonate to the water and increasing the alkalinity and pH while maintaining anoxic conditions (Skousen, 1996).

7.2.2.3 Permeable Reactive Barriers

The PRBs are barriers that are placed in the path of groundwater flow allowing the water to flow through easily and the barriers react with specific chemicals of concern (Blowes et al., 2000). The PRBs are often designed to provide a source management remedy or as an on-site containment remedy (ITRC, 1999). The treatment zone may be created directly using reactive materials such as iron or indirectly using materials designed to stimulate secondary processes, such as by adding carbon substrate and nutrients to enhance microbial activity. In this way, contaminant treatment may occur through physical, chemical or biological processes (ITRC, 2015). Other reactive media, such as limestone, compost, zeolites, granular activated carbon, apatite and others, have also been employed in recent years and offer treatment options for controlling pH, metals and radionuclides (ITRC, 2005). In PRBs that are designed to treat AMD, the barriers are generally composed of solid organic matter, like municipal compost, leaf compost and wood chips/sawdust (Blowes et al., 2000). Organic matter encourages the proliferation of SRB that reduce sulphate to biogenic sulphide resulting in the subsequent formation of metal sulphides.

The PRB technology has been applied at a good number of sites worldwide, including some full-scale installations to treat chlorinated solvent compounds (ITRC, 2005). The advantages of using a PRB include a relatively low cost of operation and monitoring and the absence of above-ground structures. Although barriers often have very long theoretical treatment lifetimes when only the material and the contaminants of concern are considered, actual lifetimes can become considerably shorter if there are other reactive substances present in the environment (Blowes et al., 2000). Depending on several site-specific conditions, PRBs are expected to last 10–30 years before reactivity or hydraulic issues result in the need for maintenance. Disposal issues could also develop in the PRB treatment media after the contaminants are concentrated within the barrier system. This design is the most important in PRB systems that retain the contaminants, as opposed to PRB systems which degrade the contaminants as they flow through the system (ITRC, 1999).

7.3 Concluding Remarks

The active and passive techniques have been developed to treat AMD. It is noted that the most suitable treatment depends on the overall treatment performance compared to other technologies (Barakat, 2011), technical factors such as fitting into the life cycle of the mine, operational factors such as utility requirements and maintenance, environmental impact such as waste disposal as well as economics parameter such as the capital investment and operational costs. Most of these treatment methods suffer from lack of economic sustainability with the biggest challenge being the high operational costs associated with low revenue generation, if any. An alternative, therefore, in the treatment of AMD is to consider it as a valuable resource and look at the recovery of water that would satisfy the needs of a variety of mining and non-mining users and other valuable and saleable by-products such as metal sulphides and hydroxides that could be used to offset some of the operational costs (Simate and Ndlovu, 2014). This approach which is widely discussed in Chapter 9 has the potential to open a range of flowsheet options that treat mine water as a resource rather than a pollution problem (Warkentin et al., 2010). Indeed, the production of industrially valuable products may address the problem of AMD in a holistic and sustainable manner. As already stated, this approach is considered in Chapter 9 which falls under Part III of this book.

References

Abdulkadir, M. I. (2009). Optimisation and application of plant-based waste materials for the remediation of selected trace metals (Cd, Pb and Mn) and oxyhalides (BrO_3^-, $ClO3^-$ and IO_3^-) in aqueous system. MSc dissertation, University of South Africa.

Adams, M., Lawrence, R. and Bratty, M. (2008). Biogenic sulphide for cyanide recycle and copper recovery in gold-copper ore processing. Minerals Engineering 21(6): 509–517.

Aubé, B. (2004). The science of treating acid mine drainage and smelter effluents. Available at https://gaftp.epa.gov/gkm/SITE_FILE_MATERIALS/3.6.17/R08-1126068.PDF [Accessed 27 July 2020].

Barakat, M.A. (2011). New trends in removing heavy metals from industrial wastewater. Arabian Journal of Chemistry 4: 361–377.

Bhattacharyya, D.A., Jumawan, A.B. and Grieves, R.B. (1979). Separation of toxic heavy metals by sulfide precipitation. Separation Science and Technology 14: 441–452.

Blowes, D.W., Ptacek, C.J., Benner, S.G., McRae, C.B., Timothy, A. and Puls, R.W. (2000). Treatment of inorganic contaminants using permeable reactive barriers. Journal of Contaminant Hydrology 45(1–2): 123–137.

Boonstra, J. and Buisman, C.J.N. (2003). Biotechnology for sustainable hydrometallurgy. Available at https://d3pcsg2wjq9izr.cloudfront.net/files/587/articles/5497/paques8.pdf [Accessed 28 July 2020].

Bowell, R.J. (2000). Sulphate and salt minerals: The problem of treating mine waste. Available at http://www.srk.co.za/files/File/UK%20PDFs/pubart_sulphat1.pdf [Accessed 12 September 2015].

Brodie, G.A., Britt, C.R. and Taylor, H.N. (1991). Use of passive anoxic limestone drains to enhance performance of acid drainage treatment wetlands. Available at http://www.asmr.us/Publications/Conference%20Proceedings/1991%20Meeting%20Vol%201/Brodie%20211-228.pdf [Accessed 23 September 2015].

Corbett, C.J. (2001). The Rhodes BioSURE process in the treatment of acid mine drainage wastewaters. Submitted in fulfilment of the requirements for the degree of Master of Science of Rhodes University, Grahamstown, 2001.

Coulton, R., Bullen, C.J., Dolan, J., Hallett, C., Wright, J. and Marsden, C. (2003). Wheal Jane mine water active treatment plant – Design, construction and operation. Land Contamination and Reclamation 11(2): 245–252.

Cravotta, C.A. and Trahan, M.K. (1999). Limestone drains to increase pH and remove dissolved metals from acidic mine drainage. Applied Geochemistry 14(5): 581–606

De Beer, M., Maree, J.P., Wilsenach, J., Motaung, S., Bologo, L. and Radebe, V. (2012). Acid mine water reclamation using the ABC process. CSIR: Natural Resources and the Environment, Pretoria.

Department of Water Affairs (DWA). (2013). Feasibility study for a long-term solution to address the acid mine drainage associated with the East, Central and West Rand underground mining basins in the Gauteng Province. Available at http://www.environment.co.za/acid-mine-drainage-amd/feasibility-study-long-term-solution-to-acid-mine-drainage-amd-east-central-and-west-rand-underground-mining-basins-in-the-gauteng-province.html [Accessed 23 September 2015].

Dijkman, H., Buisman, C.J.N. and Bayer, H.G. (1999). Biotechnology in the mining and metallurgical industries: Cost savings through selective precipitation of metal sulfides. Available at file:///C:/Users/a0009328/Downloads/0f3175368e5ca21c20000000.pdf [Accessed 23 September 2015].

Dinardo, O., Kondos, P.D., MacKinnon, D.J., McCready, R.G.L., Riveros, P.A. and Skaff, M. (1991). Study on metals recovery/recycling from acid mine drainage.

Available at http://mend-nedem.org/wp-content/uploads/2013/01/3.21.1a.pdf [Accessed 23 September 2015].

EPA. (2003). Control and treatment technology for the metal finishing, industry: Sulphide precipitation. Available at http://cfpub.epa.gov/si/si_public_record_Report.cfm?dirEntryId= 44445&CFID=38307209&CFTOKEN=61664089 [Accessed 10 September 2015].

EPA. (2014). Reference guide to treatment technologies for mining-influenced water. Available at https://clu-in.org/download/issues/mining/Reference_Guide_to_Treatment_Technologies_for_MIW.pdf. [Accessed 2 November 2020].

Feng, D., Aldrich, C. and Tan, H. (2000). Treatment of acid mine water by use of heavy metal precipitation and ion exchange. Minerals Engineering 13: 623–642.

Fripp, J., Ziemkiewicz, P.F. and Charkavorki, H. (2000). Acid mine drainage treatment. EMRRP-SR-14. Available at http://el.erdc.usace.army.mil/elpubs/pdf/sr14.pdf [Accessed 8 September 2015].

Fu, F. and Wang, Q. (2011). Removal of heavy metal ions from wastewaters: A review. Journal of Environmental Management 92: 407–418.

Godongwana, Z.G., Joubert, J.H.B., Wilken, K. and Pocock, G. (2015). Full-scale operation of The Rhodes Biosure® Process. Available at http://www.ewisa.co.za/literature/files/ID71%20Paper131%20Godongwana%20Z.pdf [Accessed 25 September 2015].

Gunther, P. and Mey, W. (2008). Selection of mine water treatment technologies for the eMalahleni (Witbank) water reclamation project. Available at http://www.ewisa.co.za/literature/files/122%20Gunther.pdf [Accessed 10 September 2015].

Hedin, R.S., Nairn, R.W. and Kleinmann R.L.P. (1994). The passive treatment of coal mine drainage. Available at http://www.osmre.gov/resources/library/pub/ptcmd.pdf [Accessed 12 September 2015].

Huisman, J., Schouten, G. and Schultz, C. (2006). Biologically produced sulphide for purification of process streams, effluent treatment and recovery of metals in the metal and mining industry. Hydrometallurgy 83(1–4): 106–113.

International Network for Acid Prevention (INAP). (2003). Treatment of sulphate in mine effluents. Available at http://www.inap.com.au/public_downloads/Research_Projects/Treatment_of_Sulphate_in_Mine_Effluents_-_Lorax_Report.pdf [Accessed 28 July 2020].

International Network for Acid Prevention (INAP). (2009). Aeration systems for treating CMD. Available at http://mend-nedem.org/wp-content/uploads/2013/01/3.21.1a.pdf [Accessed 23 August 2015].

International Network for Acid Prevention (INAP) (2012). Global acid rock drainage guide (GARD Guide). Available at http://www.gardguide.com [Accessed 23 August 2015].

Interstate Technology and Regulatory Council (ITRC). (1999). Regulatory guidance for permeable reactive barriers designed to remediate inorganics and radionuclide contamination. Available at http://www.itrcweb.org/GuidanceDocuments/PRB-3.pdf [Accessed 17 September 2015].

Interstate Technology and Regulatory Council (ITRC). (2003). Technical and regulatory guidance document for constructed treatment wetlands. Available at http://www.itrcweb.org/GuidanceDocuments/WTLND-1.pdf [Accessed 15 July 2015].

Interstate Technology and Regulatory Council (ITRC). (2005). Permeable reactive barriers: Lessons learned/new directions. Available at http://www.itrcweb.org/Guidance/GetDocument?documentID=68 [Accessed 15 July 2015].

Interstate Technology and Regulatory Council (ITRC). (2015). Mining waste treatment technology selection. Available at http://www.itrcweb.org/miningwaste-guidance/ [Accessed 15 July 2015].

Johnson, D.B. and Hallberg, K.B. (2005). Acid mine drainage remediation options: A review. Science of the Total Environment 338: 3–14.

Jones, L., Bratty, M., Kratochvil, D. and Lawrence, R.W. (2006). Biological sulphide production for process and environmental applications. In: 38th annual Canadian mineral processors conference, 17–19 January 2006, Ottawa, Canada.

Karakatsanis, E. and Cogho, V.E. (2010). Drinking water from mine water using the HiPRO® process – Optimum coal mine water reclamation plant. Available at http://www.imwa.info/docs/imwa_2010/IMWA2010_Karakatsanis_430.pdf [Accessed 20 July 2015].

Kim, B.M. (1981). Treatment of metal containing wastewater with calcium sulfide. AIChE Symposium Series 77(209): 39–48.

Kratochvil, D., Ye, S. and Lopez, O. (2015). Commercial case studies of life cycle cost reduction of ARD treatment with sulfide precipitation. Available at https://www.imwa.info/docs/imwa_2015/IMWA2015_Kratochvi_020.pdf [Accessed 28 July 2020].

Kuyucak, N. (2001). Acid mine drainage – Treatment options for mining effluents. Mining Environmental Management 2001: 14–17.

Kuyucak, N. (2006). Selecting suitable methods for treating mining effluents. Available at http://www.minem.gob.pe/minem/archivos/file/dgaam/publicaciones/curso_cierreminas/02_T%C3%A9cnico/10_Tratamiento%20de%20Aguas/TecTratAgu-L1_AMD%20treatment%20 options.pdf [Accessed 20 August 2015].

Lawrence, R.W. (2007). Commercial water treatment experience in metal and sulphate removal from acidic drainage. Available at http://bc-mlard.ca/files/presentations/2007-8-LAWRENCE-commercial-water-treatment-experience.pdf [Accessed 28 July 2020].

Lawrence, R.W. and Fleming, C.A. (2007). Developments and new applications for biogenic sulphide reagent in hydrometallurgy and mineral processing. Available at https://www.sgs.com/-/media/global/documents/technical-documents/sgs-technical-papers/sgs-min-tp2007-02-biogenic-sulphide-reagent-use-in-hydrometallurgy.pdf [Accessed 28 July 2020].

Lawrence, R.W., Kratochvil, D. and Ramey, D (2007). A new commercial metal recovery technology utilizing on-site biological H_2S production. Available at http://www.hydrocopper.cl/ 2005/esp/resumenes_articulos/13_RICK_LAWRENCE.pdf [Accessed 18 September 2015].

Lewis, A.E. (2010). Review of metal sulphide precipitation. Hydrometallurgy 104: 222–234.

Lopez, O., Sanguinetti, D., Bratty, M. and Kratochvil, D. (2009). Green technologies for sulphate and metal removal in mining and metallurgical effluents. Available at http://www.bioteq.ca/wp-content/uploads/2014/11/BioteQ-2009-Enviromine.pdf [Accessed 12 June 2015].

Maree, J.P., Hlabela P., Nengovhela R., Geldenhuys A.J., Mbhele N., Nevhulaudzi T. and Waanders, F.B. (2004). Mine Water Environment 23(4):196–204.

Maree, J.P., Mujur, M., Bologo, V., Daniels, N. and Mpholoane, D. (2013). Neutralisation treatment of AMD at affordable cost. Water SA 39(2).

MEND (1994). Acid mine drainage – status of chemical treatment and sludge management practices. Available at http://mend-nedem.org/wp-content/uploads/3321.pdf [Accessed 5 September 2015].

Merta, E. (2015). Precipitation with barium salts. Available at http://wiki.gtk.fi/web/mine closedure/wiki/wiki/Wiki/Precipitation+with+barium+salts/pop_up;jsessionid=ffb6a2ab8d80f56fcd17380647af [Accessed 17 September 2017].

Mining Review (2018). Mintek demonstrates mine effluent treatment capability. Available at https://www.miningreview.com/top-stories/mintek-mine-effluent-treatment-technology/ [Accessed 5 May 2020].

Mortazavi, S. (2008). Application of membrane separation technology to mitigation of mine effluent and acidic drainage. Available at http://mend-nedem.org/wp-content/uploads/2013/01/3.15.1.pdf [Accessed 25 August 2015].

Mottay, R. and Van Staden, P. (2018). Positioning the SAVMIN, Mintek Internal Report No.42844, 16 March 2018.

Nodwell, M. and Kratochvil, D. (2012). Sulphide precipitation and ion exchange technologies to treat acid mine drainage. Available at http://www.bioteq.ca/wp-content/uploads/2014/11/BioteQ-2012-ICARD.pdf [Accessed 25 August 2015].

Peters, R.W. and Ku, Y. (1984). Batch precipitation studies f or heavy metal removal by sulfide precipitation. AIChE Symposium Series 81(243): 9–27.

Peters, R.W., Ku, Y. and Bhattacharyya, D. (1984). The effect of chelating agents on the removal of heavy metals by sulphide precipitation. In: LaGrega, M.D. and Longs, D.A. (Eds.), Toxic and hazardous wastes, proceedings of the sixteenth mid-Atlantic industrial waste conference, pp. 289–317.

Pulles, W., Juby, G.J.G. and Busby, R.W. (1992). Development of the slurry precipitation and recycle reverse osmosis (SPARRO) technology for desalinating mine waters. Water Science and Technology 25: 177–192.

Reinsel, M. (2015). Sulfate removal technologies: A review. Available at https://www.wateronline.com/doc/sulfate-removal-technologies-a-review-0001 [Accessed 28 July 2020].

Robertson, A.M., Everett, A.J. and du Plessis, J. (1993). Sulfate removal by the GYP-CIX process following lime treatment. Available at file:///C:/Users/a0009328/Downloads/export_2015-09-26%2015-30-27.pdf [Accessed 15 January 2016].

Shahalam, A.M., Al-Harthy, A. and Al-Zawhry, A. (2002). Feed water pretreatment in RO systems in the Middle East. Desalination 150: 235–245.

Sheoran, A.S. and Sheoran, V. (2006). Heavy metal removal mechanism of acid mine drainage in wetlands: a critical review. Minerals Engineering 19: 105–116.

Simate, G.S. and Ndlovu, S. (2014). Acid mine drainage: Challenges and opportunities. Journal of Environmental Chemical Engineering 2: 1785–1803.

Skousen, J. (1996). Anoxic limestone drains for acid mine drainage treatment. In: Skousen, J. G. and Ziemkiewicz, P. F.(Eds.), Mine drainage: control and treatment, pp. 261–266. West Virginia University and the National Mine Land Reclamation Center, Morgantown, WV.

Skousen, J. (1998). Overview of passive systems for treating acid mine drainage. Available at http://anr.ext.wvu.edu/resources/295/1256049359.pdf [Accessed 20 March 2016].

Skousen, J. and Ziemkiewicz, P.F. (2005). Performance of 116 passive treatment systems for acid mine drainage. Available at http://citeseerx.ist.psu.edu/viewdoc/download?doi=10.1.1.518.9980&rep=rep1&type=pdf [Accessed 2 November 2020].

Skousen, J.G. (1988). Chemicals for treating acid mine drainage. Green Lands 18: 36–40.

Skousen, J.G. (2002). A brief overview of control and treatment technologies for acid mine drainage. Available at http://citeseerx.ist.psu.edu/viewdoc/download?doi=10.1.1.183.132&rep=rep1&type=pdf [Accessed 18 February 2016].

Slater, C.S., Ahlert, R.C. and Uchrin, C.G. (1983). Applications of reverse osmosis to complex industrial wastewater treatment. Desalination 48(2): 171–187.

Smit, J.P. (1999). The treatment of polluted mine water. Available at http://www.imwa.info/docs/imwa_1999/IMWA1999_Smit_467.pdf [Accessed 28 July 2020].

Stedman, L. (2010). Minewater treatment at a profit. Available at https://www.bqewater.com/wp-content/uploads/2012/06/2010-June-Water-21-Minewater-treatment-at-a-profit.pdf [Accessed 4 May 2020].

Taylor, J., Pape, S. and Murphy, N. (2005). A summary of passive and active treatment technologies for acid and metalliferous drainage (AMD). Available at http://www.earthsystems.com.au/wp-content/uploads/2012/02/AMD_Treatment_Technologies_06.pdf [Accessed 15 February 2016].

Warkentin, D., Chow, M. and Nacu, A. (2010). Expanding sulphide use for metal recovery from mine water. Available at http://www.imwa.info/docs/imwa_2010/IMWA2010_Warkentin _480.pdf [Accessed 10 February 2016].

Watzlaf, G. R., Schroeder, K.T. and Kairies, C.L. (2000). Long-term performance of anoxic limestone drains. Available at http://www.imwa.info/bibliographie/19_2_098-110.pdf [Accessed 15 November 2016].

WWG. (n.d.). Constructed wetlands to treat wastewater. Available at http://www.wastewatergardens.com/pdf/WWG_AboutConstructedWetlands.pdf [Accessed 28 September 2015].

Younger, P.L., Banwart, S.A. and Hedin, R.S. (2002). Mine water: Hydrology, pollution, remediation. Kluwer Academic Press, The Netherlands.

8

Life-Cycle Assessment of Acid Mine Drainage Treatment Processes

Kevin Harding

CONTENTS

8.1 Introduction

There is a general trend to promote products, processes, and services by claiming that these are environmentally friendly. To support these claims, several techniques and labelling mechanisms exist. These include life-cycle assessment (LCA), carbon footprinting and water footprinting (the impact on global warming and water, respectively) as well as various eco-labelling schemes, including Energy Star, the Forestry Stewardship Council (FSC) (Bratt et al., 2011) and many others. While many of the techniques have genuine claims at being 'better' for the environment, one of the most robust and popular quantitative tools for environmental comparison is LCA. This section gives details on the basic concepts of LCA.

The LCA is a method used to assess the full environmental impacts of a system. It is a quantitative tool for evaluating the full effects that a system

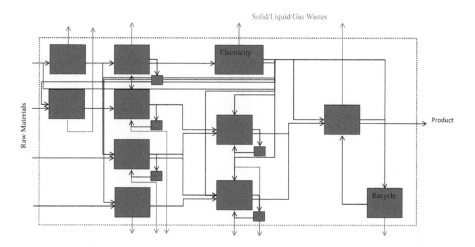

FIGURE 8.1
'Cradle-to-grate' representation of an LCA, including reuse/recycling. When considering the final disposal of the product, this would be a 'cradle-to-grave' flowsheet.

has on the environment, including extraction, processing, manufacturing, transportation, distribution, use, reuse, maintenance, recycling, and final disposal. As a result, the LCA typically includes the 'cradle-to-grave' impacts of the system in question, but it may also include 'cradle-to-gate' (Figure 8.1) or 'gate-to-gate' impacts.

One of the advantages of LCA is that it can focus on a product, process, or service. While LCA is known to look at the entire system, it also has the advantage of being able to look at individual parts within the system, e.g., transportation, packaging, or manufacturing. In addition, it can consider broader geographic aspects (such as continent, country, or smaller geographic aspect, depending on available data) while avoiding social or political arguments. Since LCA is quantitative, the results should be reproducible, thereby enhancing its credibility.

The LCA by its nature is data intensive. Obtaining the required data may be difficult due to the quantity of data needed, the difficulty in measurement of the data, and the sensitivity of certain data (particularly if the data is not from one's own company). As a result of high volumes of data required, LCA practitioners often manipulate data using specialised software packages. This allows easier and faster calculations and standardisation (which also allows for data transfer across databases). To standardise studies, practitioners perform LCAs according to the framework defined by ISO (ISO 14040:2006; ISO 14044:2006), thus allowing for mutual understanding amongst practitioners.

8.2 Performing a Life-Cycle Assessment

8.2.1 Framework

According to the ISO standards governing LCA (ISO 14040:2006 and ISO 14044:2006), the LCA comprises four stages: (1) goal and scope, (2) inventory analysis, (3) impact assessment, and (4) interpretation (Figure 8.2). Completing each stage is not always linear, but it is possible to return to earlier stages, thus forming the iterative nature of LCA.

8.2.2 Goal and Scope

The goal and scope of the LCA define the investigation. For example, the following questions may arise during the stage: Are the goals of the study to perform a strength and weakness analysis, product comparison, product improvement, or eco-labelling? Or, is the question simply to determine the environmental impact of something?

The first step in the goal and scope stage is to define a functional unit (FU). This is a numerical definition of the product/process/service under investigation, e.g., the production of 1 ton of paper. The FU is often extremely

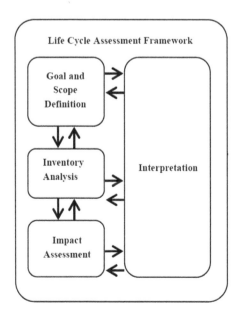

FIGURE 8.2
Phases of the LCA. (Modified from ISO 14040: 2006.)

specific as this defines most of what is to follow in the study. For example, instead of saying: "driving 100 km in a passenger car", it may be better to state it as: "driving 100 km in an Acme Model X passenger car at a constant speed of 60 km/hr, without passengers". However, the FU may depend on the level of detail required in the study.

Another aspect of the goal and scope is to define the system boundaries, and what parts of the process to include. The LCA could be a cradle-to-gate (everything) assessment, or, if there is a valid reason not to include aspects of the process, cradle-to-gate (excluding recycling), or gate-to-gate (excluding certain inputs and recycling). In determining the overall system boundary under investigation, and the FU chosen, the included aspects are determined. Typically, the boundaries include as much of the process as possible (raw material processing, manufacturing, transportation, electricity production, warehousing, use, and recycling), but this may not be possible, or the information on a certain aspect may simply be missing.

When looking at life cycles of large processes, multiple products may result, e.g., in a petroleum refinery. If the primary product of interest (and the FU) is unclear or difficult to split across the entire process (e.g., petrol, diesel, or LPG which all come from the same production facility), it may be necessary to apply allocation or division of multiple products. This is made possible by mass, economic, or any other basis. This means that instead of placing a quantitative burden on a single FU, the study splits the quantitative burden proportionally across the various products. For example, if a petroleum refinery produces 40% liquid products versus 60% gaseous products, then if one wants the LCA of the liquid products, the boundary may include the entire facility. However, the LCA will only attribute 40% of the result to the liquid products.

In the real world, processes are complex, such that adding recycling and disposal steps can complicate the allocation of the original FU and system boundary decisions further. For example, should products, sub-products, and materials recycled back to the same process, or used as raw materials in another process (where they may be replacing a less environmentally friendly process), be included in the boundaries or not (Figure 8.3)?

8.2.3 Life-Cycle Inventory Analysis

The second step of the LCA is the life-cycle inventory analysis (LCI). In engineering terms, this is equivalent to a mass and energy balance. The LCA practitioner needs to collect all inputs and outputs within the defined system, including the economic (services and goods) and environmental (resources and emissions) values. This requires that the full flowsheet, including processes required in the manufacture, use, and final disposal are known. The flowsheet starts with raw materials and ends with final use or disposal and includes all emissions (air, water, solid) along the chain, as defined in the goal and scope of the LCA. Data required to perform this stage of the LCA is

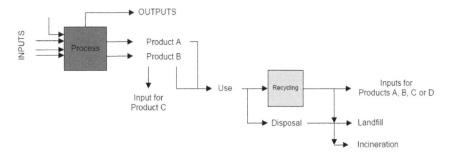

FIGURE 8.3
Simple schematic diagram showing how system boundaries and allocation of the FU are complicated when recycling and disposal are included.

available from literature, LCA databases, industry data, government records, or may be obtained through physical measurements. This is often the most difficult and time-consuming step in the LCA. Data obtained from systems that were calculated or measured for the specific study is known as the foreground process, while data collected from secondary sources, e.g., databases, is known as the background process (Figure 8.4).

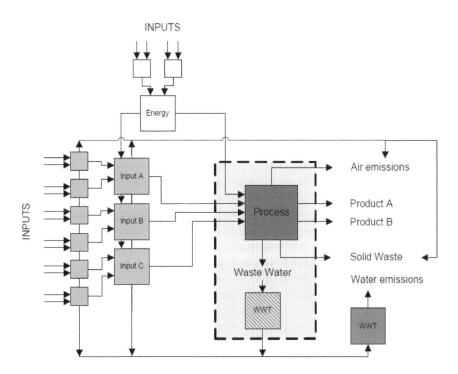

FIGURE 8.4
Schematic diagram of foreground (shaded block) versus background processes (everything else).

The ultimate step in the inventory analysis is to construct an inventory table. This involves converting all the material and energy balance numbers, which are typically in the thousands, across the flowsheet into values relative to the FU. For example, one may be looking at the LCA of coal production with a FU of 1 kg. Through investigations, however, one may have the diesel requirements for the entire facility in the entire year. Therefore, it is part of the process to convert an annual diesel value to the equivalent quantity in terms of per kilogram of coal.

The numbers are typically all mass values. In this step, energy values would have become mass values since a kilowatt of energy would require a certain mass of coal or other fuel as may be applicable. While this is a complicated step of the LCA, there are LCA software that may be helpful in analysing and/or dealing with this step.

It is important to note that some issues may not be picked up in numbers, e.g., soil erosion, noise, rainfall/evaporation, or the exact site location, e.g., prime real estates versus more slum-like areas. This should not be forgotten when analysing and reporting results since these issues could play a big part in decision making.

8.2.4 Life-Cycle Impact Assessment

The next step is life-cycle impact assessment (LCIA). In this step, the material numbers are converted from the inventory analysis into a manageable number of representative values. The LCA practitioner converts thousands of numbers from the LCI to a meaningful representation in three steps as follows: classification, characterisation/normalisation, and valuation.

8.2.4.1 Classification

Each of the thousands of inventory table data points is classified according to an environmental problem. These could include environmental issues such as resource depletion, energy depletion, global warming, photochemical oxidation, acidification, toxicity, ozone depletion, eutrophication, and many others (midpoint categories). Material from the inventory table may contribute to more than one category. It should be noted, in this context, that carbon- and water footprinting are essentially subsets of LCA, each with its own potential set of rules around setting boundaries and classifications. Various organisations promote their own set of classification categories including the methods such as ReCiPe, CML, Eco-indicator, TRACI, GLAM, and many others (Guinée, 2015). Each of the categories also has a weighting factor for characterisation as explained in the subsequent section.

8.2.4.2 Characterisation/Normalisation

Once all the materials have been classified into a category, each material is given a weighting factor depending on their relative impact to the category in question. Weighting factors are defined in terms of some other chemicals, e.g., global warming is given in terms of CO_2 equivalents; 1 kg of CH_4 equals 25 kg $CO_{2\,eq}$ (IPCC GWP100a method). Weighting factors multiplied by the amount of material give the contribution of such a material to the final LCA scores. Summing all the contributions from each material in a category gives a final LCA score for that category.

It must be noted that values from characterisation can seem meaningless in isolation. However, they are more meaningful when used as a comparison to a similar product/process (or during an improvement assessment), or when the assessment is broken down to show the contributions of individual parts of the whole system, e.g., transport, manufacture, use, or disposal. In other words, the LCA result of a plastic bottle from company A (or production line X) is more understandable when compared to the product of company B (or production line Y).

Another way to make sense of the numbers from characterisation is when many LCAs have been performed and a comparison is made from the studies. This is termed normalisation. This involves comparing the size of a contribution against the total contribution to the problem over one year. For example, if a process is emitting one tonne of pollutant and the global number is one million tonnes per year, then the normalised value for the process is one-millionth of a year (10^{-6} years) or about 32 seconds.

The readers must take note that the information presented in this chapter has been overly simplified. In advanced studies, characterisation itself can be broken further into midpoint or endpoint categories as may be seen from an environmental or human health perspective, and areas of protection (human health, ecosystem quality, and natural resources). Impact assessment, in general, is the area of LCA that defines characterisation, and this is where it is constantly under methodological scrutiny. For more in-depth analysis of the topic, it is suggested that the readers find a specialised textbook on the topic such as Guinée (2015) or visit up-to-date websites that have more information, e.g., the UN Environment Programme's LCInitiative.

8.2.4.3 Valuation

A last step in the quantification procedure in the LCA is the valuation step. This is the conversion of characterisation scores into a final single value. Each characterisation value gets a weighting, and all weighted values are summed to give a single score. While valuation and single scores are sometimes available, they are not common. Therefore, often, LCA practitioners leave results at a midpoint characterisation value (Figure 8.5).

To perform the steps in the LCIA phase, specialist LCA software are available, i.e., both open-source and proprietary software options are available. The important aspect of the software is the availability of databases on which to build new LCAs. Such databases are available from several sources, with some of them giving users tens of thousands of datasets.

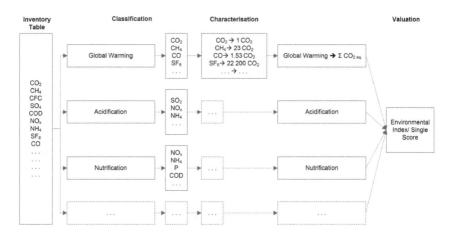

FIGURE 8.5
Representative steps of the LCA from obtaining an inventory table, through classification, characterisation, and valuation.

8.2.5 Interpretation

The design of LCAs are such that they are to be questioned at each stage of the process; this is the role of the interpretation phase. It is particularly easy in the earlier stages to have numbers without much knowledge of what they mean. Due to an iterative nature of LCA, improving parts that are problematic during the interpretation phase (after each other stage) strengthens the confidence of the LCA results. The interpretation phase is also purposefully designed such that the practitioner is reminded to look back at what the exact scope of the study was and ensure the results meet the expectations of what should have been investigated.

8.3 Typical Studies of Life-Cycle Assessments

Researchers have presented LCA studies for a wide range of processes. These include bio-diesels (Harding et al., 2008), chemicals (Fantke and Ernstoff, 2018), food (Cucurachi et al., 2019), plastics (Harding et al., 2007), and many more. It is also common to find studies that look only at certain aspects of the LCA, e.g., global warming or carbon footprinting, as well as water footprinting (Pfister et al., 2017). Consultants have also undertaken many more studies for their corporate clients which they keep for internal use only.

In the mining and mineral processing industries, there is a similar interest in LCA studies. These include various minerals and processes, from full LCA studies to impacts of the use of water (Haggard et al., 2015; Northey et al., 2013; Northey et al., 2016; Osman et al., 2017; Ranchod et al., 2015).

There have been few published studies that have used LCA methods to determine the environmental impacts of different acid mine drainage (AMD) treatment technologies. Such studies include both active and passive remediation techniques in different countries which are valuable examples of what is possible with AMD (Table 8.1).

The first case study presented in this chapter is for the active versus passive AMD treatment technologies in New Zealand (Hengen et al., 2014). As discussed in some sections of this chapter, using LCA, it is possible to look at different scenarios and what-if options. In a study by Hengen et al. (2014), eight scenarios were investigated to determine the impacts of various processing options. This is common in LCA where different transport options or suppliers may be of interest. Where comparative studies are involved, and the aim is to compare different options, it is common to draw the system boundary to a 'gate'. This is useful when the excluded portions of the scenarios are identical, and there is more interest in the differences (the actual comparative LCAs investigated) than the absolute final impact of the entire process. In the Hengen et al. (2014) study, it was shown that passive treatment systems had lower environmental impacts than active treatment technologies. Furthermore, the study found that reducing transport distances and using recycled materials could improve environmental burdens.

The LCA is also an effective way to look at different processes, whose final products have the same use (FU). For example, Tuazon and Corder (2008) determined the life-cycle impacts of lime and reused red mud as neutralizing agents for AMD. While these are two distinct products, which would have originated through completely different processing routes, their use (remediating AMD) would be the same and are thus comparable.

However, LCA does not only compare processes. A useful aspect may be to know the hotspots (areas of concern) in a full process. While LCA software gives life-cycle impact values for the entire process, it is also possible to break them down so as to determine which aspect of the process contributed what proportion. This is something that Masindi et al. (2018) studied when looking at distinct aspects of a single process for AMD treatment. Even though there is no other process to compare results to, it can be useful to know that one part of the process might be contributing proportionally higher to LCA impacts than another part. In their study, Masindi et al. (2018) found that electricity was a large contributor to LCA impacts. Reduction of energy needs or replacement of electricity use with renewable energy sources would thus reduce the entire LCA score substantially.

As mentioned previously, setting the goal and scope can be particularly important. This is because not understanding the actual problem in the first place can make deciphering results confusing. Another aspect could be the time under consideration. Depending on the year the data was obtained, or the length of time under consideration, results can vary. Reid et al. (2009) is one such example, where it was found that results could vary depending on what temporal boundaries were set. In this study, if a mine closure period

TABLE 8.1

Summary of Case Studies on AMD Treatment Scenarios

Title	Location	Aim	Description	LCA Details	Significant Findings	Recommendations	References
1 LCA of active and passive AMD treatment technologies at a coal mine	Stockton Coal Mine, New Zealand	To compare the environmental impacts of different implemented and optional active and passive AMD treatment methods	Eight scenarios investigated, including limestone and the use of mussel shells	Cradle to gate analysis, including transport and construction.	Lime slaking had the greatest LCA impacts, and passive treatment had fewer impacts. Reduced transport significantly reduced scores	Design considerations should include utilizing materials with reduced processing, sourcing local materials and minimizing pumping energy	Hengen et al. (2014)
2 LCA of seawater neutralised red mud for AMD treatment	Mount Morgan, Queensland, Australia	To compare the environmental impacts of different neutralants, lime and seawater neutralised red mud, in controlling acidic water discharges	An investigation into scenarios of using lime (one scenario) versus reused red mud (three scenarios)	Equivalent neutralisation ability used as a basis. Transport included	Red mud has the potential for reuse, but due to the requirement of approximately 12 times more mud by mass, insufficient conclusions were drawn	Further studies involving more detailed analysis are needed, including management issues associated with the physical and chemical stability of red mud	Tuazon and Corder (2008)

3	Assessing the sustainability of AMD treatment	Mpumalanga Province, South Africa	To assess the environmental sustainability of a typical AMD treatment process, identify environmental hotspots and avenues to improve its environmental sustainability	Investigating AMD from an integrated coal mine treatment system using magnesite, lime, soda ash and CO_2	LCA of a single integrated system, including process contribution analysis. Economic and social aspects also investigated	The system had a high environmental impact, due to electricity requirements, particularly due to South African coal-fired electricity generation	Use of renewable energy and sourcing gaseous CO_2 from other production processes, e.g., flue gas, can reduce the environmental impact by up to 81%. Mineral recovery from AMD can also mitigate the environmental footprint. More research is needed	Masindi et al. (2018)
4	LCA of mine tailings management	Quebec, Canada	To quantify and compare the environmental burdens of different mine tailings management methods, specifically relating to the contribution of the land-use category	To develop the inventory of six tailings management scenarios to limit the formation of AMD	Life-cycle stages included development, operation, and closure	The best scenario was shown, but it was seen that different temporal boundaries could affect results	Future results should be applied with caution since mineral ore grade, topography of the site, and soil characteristics could significantly influence the environmental impacts	Reid et al. (2009)

(Continued)

TABLE 8.1

Summary of Case Studies on AMD Treatment Scenarios (*Continued*)

Title	Location	Aim	Description	LCA Details	Significant Findings	Recommendations	References
5 LCA of a passive remediation system	Mina concepcion sulphide mine, Southwest Spain (Iberian Pyrite Belt)	To perform an LCA for dispersed alkaline substrate (DAS) technology, effective for metal-rich and acid waters	LCA to quantify the environmental impacts associated with an AMD passive treatment plant	LCA included construction, operation, waste handling, and water treatment	Construction creates initial environmental impacts, but upstream manufacturing impacts are most significant	The results could be used to support decision making of a restoration plan for the Odiel River basin. Results may also contribute to more environmentally friendly mining by supplying insights into the environmental impacts related to AMD treatment	Martínez et al. (2019)
6 LCA of flotation process to prevent acid rock drainage	Not location specific	To evaluate the environmental consequences of a desulfurisation flotation unit for the prevention of acid rock drainage	LCA for the treatment of a tailings slurry generated from a base-metal sulfide ore	One scenario using conventional dewatering and further processing, with a second scenario using desulfurisation flotation and further processing	Desulfurisation flotation resulted in a decrease in some LCAI scores, but increases in others	While holistic and systemic, LCA tools can be deficient in reliably and comprehensively accessing all environmental impacts for solid mineral waste systems. Future studies will need to be more detailed	Broadhurst et al. (2015)

of under ten years was used, versus greater than ten years, the best to worst ecosystem quality scenario results were inverted. It was also shown that the longer the period under consideration, the greater the differences in these environmental impacts.

A further consideration to the time aspect is how to include the construction phase of the process, which should be included if the LCA is truly 'cradle-to-grave'. In the first year of operation, a substantial part of the environmental impact would be due to the construction phase, while as time passes, the relative average yearly impact of the construction phase becomes smaller due to less (or no) construction in those years. At some point, it may become insignificant enough not to be of concern, as is the case in a study by Martínez et al. (2019), where it was determined that after 4.5 years, the impacts of construction were insignificant. Actually, it is a common assumption that the construction phase is considered negligible and ignored. This is particularly true for projects that have processing facilities with projected lifespans of a few decades or more, or because this information is no longer available, or difficult to find.

The final case study presented looked at a desulfurisation flotation process versus a conventional system to prevent acid rock drainage formation at a base metals facility (Broadhurst et al. 2015). The study showed that while the LCA method was able to determine improvements in some LCIA category results, there could be conflicting results in other categories. It also showed that although the LCA method is holistic and systemic, there could be cases where the impact of categories are not appropriate for the system being investigated.

8.4 Concluding Remarks

Without rigorous scientific analysis, it is impossible to determine objectively if a process is environmentally superior to any alternatives. The LCA is one tool that allows for such an analysis to be carried out. Using quantitative analysis, LCA can give repeatable and dependable results to determine the environmental impacts of a process, product, or service. A good example where LCA can be beneficial is when comparing active or passive AMD treatment systems.

References

Bratt, C., Hallstedt, S., Robèrt, K.H., Broman, G. and Oldmark, J. (2011). Assessment of eco-labelling criteria development from a strategic sustainability perspective. Journal of Cleaner Production 19(14), 1631–1638.

Broadhurst, J.L., Kunene, M.C., von Blottnitz, H., Franzidis, J.-P. (2015). Life cycle assessment of the desulfurization flotation process to prevent acid rock drainage: A base metal case study. Minerals Engineering 76, 126–134.

Cucurachi, S., Scherer, L., Guinée, J. and Tukker, A. (2019). Life cycle assessment of food Systems. One Earth 1(3), 292–297.

Fantke, P. and Ernstoff, A. (2018). LCA of chemicals and chemical products. In: Hauschild, M., Rosenbaum, R. and Olsen, S. (Editor), Life Cycle Assessment. Springer International Publishing, New York, pp. 783–815.

Guinée, J.B. (2015). Selection of impact categories and classification of LCI results to impact categories. In: Hauschild, M. and Huijbregts, M.A.J. (Editors), Life Cycle Impact Assessment, LCA Compendium – The Complete World of Life Cycle Assessment. Springer, Dordrecht, pp. 17–37.

Haggard, E.L., Sheridan, C.M. and Harding, K.G. (2015). Quantification of water usage at a South African platinum processing plant. Water SA 41(2), 279–286.

Harding, K.G., Dennis, J.S., Von Blottnitz, H. and Harrison, S.T.L. (2007). Environmental analysis of plastic production processes: Comparing petroleum-based polypropylene and polyethylene with biologically-based poly-beta-hydroxybutyric acid using life cycle analysis. Journal of Biotechnology 130(1), 57–66.

Harding, K.G., Dennis, J.S., von Blottnitz, H. and Harrison, S.T.L. (2008). A life-cycle comparison between inorganic and biological catalysis for the production of biodiesel. Journal of Cleaner Production 16(13), 1368–1378.

Hengen, T.J., Squillace, M.K., O'sullivan, A.D. and Stone, J.J. (2014). Life cycle assessment analysis of active and passive acid mine drainage treatment technologies. Resources, Conservation and Recycling 86, 160–167.

ISO 14040, 2006. Environmental management – Life cycle assessment – Principles and framework. Available at https://www.iso.org/standard/37456.html [Accessed 14 May 2020].

ISO 14044:2006. ISO 14044:2006 – Environmental management – Life cycle assessment – Requirements and guidelines. Available at https://www.iso.org/standard/38498.html [Accessed 14 May 2020].

Martínez, N.M., Basallote, M.D., Meyer, A., Cánovas, C.R., Macías, F. and Schneider, P. (2019). Life cycle assessment of a passive remediation system for acid mine drainage: Towards more sustainable mining activity. Journal of Cleaner Production 211, 1100–1111.

Masindi, V., Chatzisymeon, E., Kortidis, I. and Foteinis, S. (2018). Assessing the sustainability of acid mine drainage (AMD) treatment in South Africa. Science of the Total Environment 635, 793–802.

Northey, S., Haque, N. and Mudd, G. (2013). Using sustainability reporting to assess the environmental footprint of copper mining. Journal of Cleaner Production 40, 118–128.

Northey, S.A., Mudd, G.M., Saarivuori, E., Wessman-Jaakelainen, H. and Haque, N. (2016). Water footprinting and mining: Where are the limitations and opportunities? Journal of Cleaner Production 135, 1098–1116.

Osman, A., Crundwell, F., Harding, K.G. and Sheridan, C.M. (2017). Application of the water footprinting method and water accounting framework to a base metal refining process. Water SA 43(4), 722.

Pfister, S., Boulay, A.-M., Berger, M., Hadjikakou, M., Motoshita, M., Hess, T., Ridoutt, B., Weinzettel, J., Scherer, L., Döll, P., Manzardo, A., Núñez, M., Verones, F., Humbert, S., Buxmann, K., Harding, K., Benini, L., Oki, T., Finkbeiner, M. and

Henderson, A. (2017). Understanding the LCA and ISO water footprint: A response to Hoekstra (2016) "A critique on the water-scarcity weighted water footprint in LCA." Ecological Indicators 72, 352–359.

Ranchod, N., Sheridan, C.M.C., Pint, N., Slatter, K. and Harding, K.G. (2015). Assessing the blue-water footprint of an opencast platinum mine in South Africa. Water SA 41(2), 287.

Reid, C., Bécaert, V., Aubertin, M., Rosenbaum, R.K. and Deschênes, L. (2009). Life cycle assessment of mine tailings management in Canada. Journal of Cleaner Production 17(4), 471–479.

Tuazon, D. and Corder, G.D.. (2008). Life cycle assessment of seawater neutralised red mud for treatment of acid mine drainage. Resources, Conservation and Recycling 52(11), 1307–1314.

Part III

Reuse, Recycle and Recovery Processes of Valuable Materials from Acid Mine Drainage

The problems of AMD, which include contamination of groundwater and surface water, are growing exponentially. Whilst a wide range of technologies are available for preventing AMD generation and/or treating AMD before discharge as illustrated in Part II of the book, most of these technologies consider AMD as a nuisance that needs to be quickly disposed after the minimum required treatment. However, in the recent past, there has been an emerging worldwide paradigm towards environmental responsibility and sustainable development. In fact, the reuse, recycle and recovery of industrially useful materials and products are the emerging pragmatic approaches to mitigating the challenges associated with AMD. Therefore, the main focus of this part of the book is to bring together a number of studies in which novel methods have been developed for the reuse, recycle and recovery of industrially useful materials from AMD.

Part III

Reuse, Recycle and Recovery Processes of Valuable Materials from Acid Mine Drainage

9

Recovery Processes and Utilisation of Valuable Materials from Acid Mine Drainage

Geoffrey S. Simate

CONTENTS

9.1 Introduction

The title of this book is embedded in this chapter. On the other hand, the previous chapters narrated how worthless AMD is, and the need for it to be either prevented or treated in order to avert a serious global environmental challenge encountered during mining operations whether in operating mines or closed mines. As discussed in several chapters of this book, AMD is characterised by low pH and contains elevated concentrations of toxic metal ions (Fe, Zn, Cd, Al, Cu, Pb), dissolved anions (sulphates, nitrates, chlorides, arsenates, etc.), hardness, and suspended solids (Sheoran and Sheoran, 2006; Dhir, 2018) all of varying compositions depending on the original mineral deposit types (Sheoran and Sheoran, 2006). For decades, the need for sustainable AMD prevention and treatment approaches has led to research that has focused on resource recovery of useful products from AMD (García et al., 2013; Simate and Ndlovu, 2014; Naidu et al., 2019). For example, Simate and Ndlovu (2014) explored the generation of a wide range of industrially useful materials from AMD. In fact, recent research on AMD treatment techniques has shown that most of the AMD constituents can be considered as valuable resources (García et al., 2013; Kefeni et al., 2017). Furthermore, Rodríguez-Galán et al. (2019) argue that the possibility of recovery of valuable products from AMD waste is a potential option since not only environmental impact is reduced, but also economic advantages can be achieved. In view of the aforementioned discussion, overall, this chapter provides insights into previous, current, and future holistic approaches developed towards the recovery and utilisation of resources from AMD for the purpose of sustainable development. More specifically, the chapter deals with the 3Rs – Reuse, Recycle, Recovery – as the most important strategies for dealing with AMD. Reuse in the context of this chapter refers to the use or application of a significant component of the AMD such as water for a specific purpose after undergoing some treatment. Recycling is defined as the conversion of the entire waste such as AMD sludge into a new valuable product or application. Recovery on the other hand refers to the extraction of valuable resources or ingredients from the AMD with the aid of processing and/or reprocessing techniques.

9.2 Recovery of Metals

A metal is basically any chemical element that is considered as a good conductor of both electricity and heat (Geiger and Cooper, 2010). From the chemistry point of view, metals readily lose electrons to form cations and

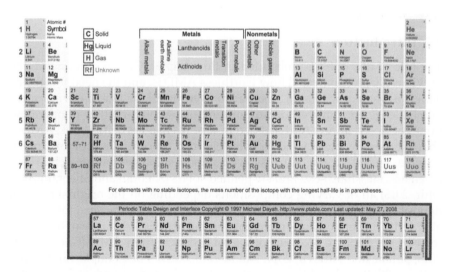

FIGURE 9.1
Periodic table showing metal, metalloid, and non-metal elements.

ionic bonds with non-metals. Metals comprise the greater part of the periodic table of elements, while non-metallic elements are only found on the right-hand side of the periodic table (see Figure 9.1). A diagonal line, drawn from boron (B) to astatine (At), separates the metals from the non-metals (Geiger and Cooper, 2010). Most elements on the diagonal line are metalloids, which are called semiconductors sometimes because they display electrical characteristics that are common to both conductors and insulators (Geiger and Cooper, 2010). Amongst the 88 elements present in the periodic table, about 30 elements may be considered as toxic heavy metals, and it is generally accepted that all the heavy metals have a specific gravity (density) greater than 5.0 g/cm^3 (Parker, 1989; Lozet and Mathieu, 1991; Morris, 1992; Duffus, 2002).

Generally, metals and their alloys are divided into two classes – ferrous and non-ferrous (Hall and Giglio, 2010). Ferrous metals consist of iron, steel, and alloys related to them that are magnetic in nature (Dunaway, 1984). Non-ferrous metals are those that contain either none, or very small amounts of ferrous metals, and they are generally divided into the aluminium, copper, magnesium, lead, and similar groups (Dunaway, 1984).

It is indisputable that metals are an integral part of human life and world development (Hall and Giglio, 2010). According to Norgate and Rankin (2002), metals are actually well suited for sustainable development goals as they have unlimited life span and the potential for unlimited recyclability. However, metal demand and usage around the globe are rising coupled with the depletion of high-grade reserves (Sethurajan, 2015). Norgate and Rankin (2002) also argue that the source of primary metals is finite. As a result, it has become increasingly important to consider other sources of metals. For example, AMD which is considered as an environmental contaminant is now viewed as a potential source of valuable metals (Naidu et al., 2019). Moreover, according to García et al. (2013), AMD contains several metals and metalloids of particular interest such as copper, iron, manganese, aluminium, and zinc. Kwon et al. (2016) also argue that though most of the current remediation technologies consider AMD as a waste material, in actual fact, AMD contains several species of metals including valuable metals. In addition, the increases in prices of many metal commodities have revived questions as to whether metals can be economically recovered from AMD (Smith et al., 2013). Nordstrom et al. (2017) acknowledge that metal recovery from mine waters and effluents is not a new approach, but one that has occurred largely opportunistically over several millennia. The focus of this section of the chapter is to discuss and evaluate different techniques that facilitate the recovery of metals from AMD. The techniques are divided into four categories: settling and sedimentation, selective precipitation, selective adsorption, ion-exchange, and electrochemical treatment.

9.2.1 Settling and Sedimentation

9.2.1.1 Concepts of Settling and Sedimentation

Many approaches can be used for the treatment of mine waters, with the treatment solution dependent on the mine water contamination. Particle settling and sedimentation represent the simplest processes employed for treating mine impacted water (McCauley, 2011). The two processes have been credited for efficiently removing heavy metals associated with particulate matter in AMD particularly in natural and constructed wetlands (Hammer, 1997; Sheoran and Sheoran, 2006). However, despite the effectiveness of settling and sedimentation, the processes rely on a range of other processes like precipitation, co-precipitation, and adsorption that have to occur first in order to aggregate heavy metals into particles large enough to sink (Walker and Hurl, 2002; Sheoran and Sheoran, 2006). For example, chemicals may be added before the AMD flows into sedimentation ponds so that metals in the AMD can settle out or precipitate (Skousen, 2014). In fact, the processes can be accomplished by either active or passive techniques (Skousen et al., 2000; Waters et al., 2003), and see Chapter 7 for details of the two techniques. In view of the two techniques, Trumm (2010) has developed a selection criterion for the two techniques. However, in both passive and active treatment systems, neutralisation (or pH control) is the most commonly used approach (Taylor et al., 2005). By increasing the pH to create alkaline conditions (e.g., pH ≥ 9.5), the solubility of most metals can be significantly decreased by precipitation (Taylor et al., 2005). Ideally, dissolved metals will precipitate from AMD as loose, open-structured mass of tiny grains called flocs or sludge (Skousen, 2014).

This technique, conventional precipitation, has many disadvantages including being expensive, labour intensive, and ineffective when metal concentrations are below ppm levels (Brown, 1996). This conventional method of treating AMD by chemical neutralisation using lime or limestone also results in huge amount of sludge (Kefeni et al., 2015). In addition, it requires being pumped, dewatered, trucked, and processed or even buried, and subsequently monitored for any leachates (Brown, 1996). In view of these disadvantages, Simate and Ndlovu (2014) argue that a suitable method should be based on recovery and reuse of the heavy metals. In other words, to use AMD as a resource, the separation of metals as well as their removal from AMD is required (Bessho et al., 2017). Kefeni et al. (2015) also argue that in order to achieve sound environmental protection and sustainable remedy, it is imperative that the means of recovery of the valuable minerals and reuse of the resources from AMD should be developed. Therefore, there is no doubt that it is desirable to have an AMD remediation technique that would also lead to recycling of major and/or minor metal elements. Such methods are discussed in Sections 9.2.2–9.2.5.

9.2.1.2 Typical Studies of Settling and Sedimentation

Several chemicals are used in AMD settling and sedimentation processes. According to Skousen et al. (2000), each chemical has characteristics that make it more or less appropriate for a specific condition. In addition, Skousen et al. (2000) state that the best choice among alternatives depends on both technical and economic factors. The technical factors include acidity levels, flow, and the types and concentrations of metals in the AMD. The economic factors include prices of reagents, labour, machinery and equipment, the number of years that treatment will be needed, and the interest rate.

INAP (2009) and ITRC (2010a) state that raising the pH of AMD solution using alkaline agents causes certain dissolved metals (e.g., cadmium, copper, iron, lead, manganese, and zinc) to precipitate as hydroxides. A coagulant and/or flocculant may also be added to enhance flocculation, and the solution may be transferred to a clarifier in order to settle the solids and thus separate them from the cleaned overflow effluent. The resultant metal-hydroxide sludge extracted from the bottom of the clarifier usually contains a large percentage of bound water, limiting the potential for reuse, and is disposed of as a solid waste. The amount of sludge generated can be reduced by employing a high-density sludge (HDS) treatment technique. In HDS processes, the precipitated hydroxide sludge is recycled to a conditioning tank, where it is mixed with the alkali reagent. The sludge/alkali slurry is then added into the AMD to raise the pH and cause additional metal precipitation. The reconditioning of the sludge provides for precipitation sites for the dissolved metals to bond, thus increasing the overall density of the sludge in the clarifier underflow.

A two-step neutralisation process, leading to the formation of ferrites, was developed by Herrera et al. (2007) for the treatment of AMD. The process applied in the study used MgO and NaOH as the first and second neutralisers, respectively. In the first neutralisation step, MgO was used to raise the pH of AMD to around 4.5 so as to eliminate aluminium and to reduce the silica concentration. In the second neutralisation step NaOH was used to complete the neutralisation and to co-precipitate ferrous and ferric hydroxides, from which ferrite could be formed. The results of the study showed that the two-step neutralisation process is a very promising option in AMD remediation because it can reduce a sludge volume in the order of 20–80% compared to the conventional lime neutralisation process. Additionally, an industrial use of the ferrite sludge produced would reduce the amount of sludge to be disposed of by about 90%. The results of batch and continuous flow tests showed that MgO performed better than NaOH because of its higher neutralizing capacity per unit weight and because the use of MgO avoids the massive precipitation of Ca species in the second neutralisation step. Large amounts of Ca species would hinder the ferritisation of the precipitates.

A study by Kaur et al. (2018) compared the performance of Bayer liquor and Bayer precipitates with commercially available alkali commonly used in the treatment of AMD water in view of material requirements and discharge water quality. The research questions addressed were: (1) can the waste alkali materials raise the pH to the required levels to meet water discharge limits; (2) is it possible to reduce dissolved metal concentrations to satisfy regulations; and (3) what is scientific explanation for differences in performance for the various alkalis. In a study by Kaur et al. (2018), the treatment of mine pit water involved the addition of known amounts of lime, Bayer hydrotalcite, Bayer liquor, sodium carbonate, and sodium hydroxide to 25 mL of AMD water at ambient temperature. The resultant mixture was then agitated for 24 h before being centrifuged. With respect to the research questions, firstly, all investigated alkaline materials successfully raised the pH of mine pit water so as to meet discharge limits, i.e. pH 6.5–8.5. Secondly, the study found that lime and Bayer precipitates were more effective in removing the metals present in mine pit water than either sodium hydroxide, sodium carbonate, or Bayer liquor. Thirdly, the ability of the precipitates to encapsulate heavy metals was determined to be more important than surface area. Mechanistically, larger precipitates were found to positively influence the removal of heavy metals, with lime and Bayer precipitates forming the largest precipitates. Sludge produced after treatment with Bayer precipitates was more stable and showed minimum metal leaching as compared to sludge produced after treatment with other alkali. The mass of material required for attaining the desired pH was higher for Bayer precipitates compared to lime, but the capital cost for a system using lime was considered high due to its hydrophobic nature and the resultant extensive mixing required.

Olds et al. (2013) evaluated the effect of surface area of precipitates formed by neutralisation of AMD using three alkalinity reagents (NaOH, $Ca(OH)_2$, and $CaCO_3$) on the sorption of Ni and Zn, and their subsequent removal. Jar stirrers were used for neutralisation of the AMD in 1-L beakers, with different mixing regimes for different alkaline reagents. The NaOH neutralised samples were dosed with NaOH followed by a 1-min rapid mixing phase (100 rpm) and a 25-min flocculation phase (20 rpm), as neutralisation by NaOH was instantaneous. By contrast, the $Ca(OH)_2$ and $CaCO_3$ neutralised samples were dosed with slurry of reagents followed by a 60-min rapid mixing phase (100 rpm). At the end of the flocculation/mixing period, the beakers of neutralised AMD were allowed to settle for 2 h. The results of the study indicated that the removal of Ni and Zn by sorption onto AMD precipitates is influenced by the surface area of the floc formed during neutralisation. Neutralisation of AMD by NaOH and $Ca(OH)_2$ produces large fluffy floc, with surface areas, an order of magnitude greater than floc formed by $CaCO_3$ neutralisation. As a result, significantly more Ni and Zn were sorbed and co-precipitated on the NaOH and $Ca(OH)_2$ floc.

Bologo et al. (2009) proposed a process whereby $Mg(OH)_2$ was used for the neutralisation of free acid and subsequent raising of the mine wastewater pH to above 7 to facilitate rapid iron (II) oxidation and precipitation as $Fe(OH)_3$. In other words, the investigation aimed to demonstrate that $Mg(OH)_2$ treatment, in combination with lime treatment, offers an attractive solution for the treatment of acid mine water that is rich in Fe(II) and other metals. In this approach, Fe(II) with a concentration of 900 mg/L was completely oxidised within 10-min reaction time and precipitated as $Fe(OH)_3$ together with other metal hydroxides. The precipitated $Fe(OH)_3$ together with other metals hydroxides was separated from the water. In the next stage, magnesium was precipitated with lime as $Mg(OH)_2$ and was separated from the water together with gypsum. Bologo et al. (2009) suggested that $Mg(OH)_2$ can be separated from the gypsum by treating it with CO_2 to form $Mg(HCO_3)_2$ or with H_2SO_4 to form $MgSO_4$. This study demonstrated that the integrated $Mg(OH)_2$ and lime process can be applied for treating acid mine water effectively. In addition, by using $Mg(OH)_2$ instead of $Ca(OH)_2$ or $CaCO_3$, gypsum precipitation can be avoided and metal hydroxides can be precipitated separately from gypsum.

9.2.1.3 Recovery of Metals from the Sedimentation Sludge

The metals within the sludge that consist of particulate matter and other impurities can be economically extracted using a number of technologies (Sethurajan, 2015). For example, different pyrometallurgical and hydrometallurgical processes have been developed for the extraction of metals from metallurgical sediments and/or sludge (Sethurajan, 2015). Pyrometallurgy employs high temperatures to carry out smelting and refining operations to extract metals, whereas hydrometallurgy uses aqueous solutions to separate the desired metals (Nassaralla, 2001). Hydrometallurgy has a number of advantages including low initial capital investment, low energy requirements, and high-purity grade of the metal produced compared to pyrometallurgy (Nassaralla, 2001); and hence it is mainly adopted.

Sethurajan (2015) proposed different approaches for the extraction of heavy metals from the metallurgical sludges and/or sediments. These processes include (1) thermal treatment coupled with high concentrated acid leaching, (2) acid leaching, (3) combination of pyrometallurgical (roasting) and hydrometallurgical processes (sulphuric acid, water and NaCl), (4) hydrometallurgical process including leaching, cementation and refining, and (4) bioleaching with iron oxidizing bacteria (acidithiobacillus ferrooxidans, acidithiobacillus caldus), and archaea (sulpholobus metallicus), and many other techniques. After the metals have been extracted from the sludge, other processes discussed in Sections 9.2.3–9.2.5 may be used to recover them.

9.2.1.4 Summary

There are many techniques that can be used for the treatment of AMD, with the treatment option dependent on the degree of contamination. The simplest of all the techniques is generally termed settling and sedimentation (or conventional precipitation method). The traditional solution to treat AMD and the first of the settling and sedimentation methodology involves collecting and chemically treating acidified effluents in a centralised treatment plant. Alternatively, the second method involves routing effluents through natural or constructed wetlands within which microbial communities perform the same function. Both methods use pH control (or neutralisation) as a means of precipitating the metals in the AMD, which results in the formation of sludge. Thereafter, the metals can be economically extracted from the sludge using a number of pyrometallurgical and hydrometallurgical technologies.

9.2.2 Selective Precipitation

9.2.2.1 Concepts of Selective Precipitation

Selective precipitation is based on solubility differences among the metal compounds (Rodríguez-Galán et al., 2019). A lot of studies have shown that it is one of the most promising ways to overcome the problems of the conventional precipitation technique discussed in Section 9.2.1 (Wei et al., 2005; Simate and Ndlovu, 2014; Oh et al., 2016; Rodríguez-Galán et al., 2019). The main advantages of this method are the reduced volume of the produced sludge and the valorisation of metals (Oh et al., 2016; Rodríguez-Galán et al., 2019).

The most common reactive agents for metal precipitation are hydroxides or sulphides (Rodríguez-Galán et al., 2019). Amongst the two chemicals, sulphides have various advantages compared to using hydroxides (Lewis, 2010; Oh et al., 2016). Firstly, there is a high possibility of metal separation using sulphides because of the more distinct solubility of various metal sulphides (Sampaio et al., 2009; Oh et al., 2016). Secondly, metal sulphides also have other advantages including faster reaction rates, low solubility over a wide range of pHs, better settling properties, and higher potential for reuse by smelting (Gharabaghi et al., 2012; Oh et al., 2016; Uça, 2017). Furthermore, the use of sulphide not only allows producing effluents with metal concentrations in the order of magnitude of ppm and ppb, but also gives the possibility of precipitation at low pH and selective precipitation for metal reuse (Sampaio et al., 2009). Sulphide precipitation can be effected using either solid (FeS, CaS), aqueous (Na_2S, NaHS, NH_4S), or gaseous (H_2S) sulphide sources (Lewis, 2010; Patil et al., 2016). There is also the possibility of using the degeneration reaction of sodium thiosulphate ($Na_2S_2O_3$) as a source of sulphide for metal precipitation (Lewis, 2010).

9.2.2.2 Typical Studies of Selective Precipitation

This section discusses metal recovery from AMD and/or industrial wastewater using a selective precipitation process based on solubility characteristics of the major and minor metals in the wastewater. The examples discussed involve the use of various neutralisation reagents.

The experimental study performed by Wei et al. (2005) involved the treatment of AMD from a bond-forfeited coal mine site (Upper Freeport seam) that is located in north central West Virginia. The objective of the study was to recover iron and aluminium from AMD by selective precipitation. The untreated AMD water samples were collected and stored in closed high-density polyethylene bottles and kept at 4°C. Thereafter, the AMD water was removed from the fridge and bubbled with compressed air for at least 24 h to ensure the complete oxidation of Fe^{2+}. The water was then filtered to remove debris and suspended solids. Thereafter, it was referred to as "raw" AMD and was used as a feed solution for metal solubility and recovery experiments. A two-step process was used for metal recovery tests: iron precipitation followed by aluminium precipitation. For iron precipitation, raw AMD water samples of 500 mL were neutralised with 10-N caustic soda (NaOH) solution to pH end points between 3.0 and 4.5, at 0.5 standard unit intervals to assess iron recovery at different pHs. After iron recovery (at pH 3.5), the filtered AMD water was used as a feed solution for aluminium precipitation. Samples of 500 mL each were then neutralised with 10-N NaOH to pH end points from 4.5 to 8.0 to determine the aluminium recovery performance over a range of pHs. In addition to NaOH, similar pH adjustment tests were conducted using soda ash (Na_2CO_3), ammonia (NH_4OH), quick lime (CaO), and hydrated lime ($Ca(OH)_2$) for iron recovery at pH of 3.5 and aluminium recovery at pH of 6.5 so as to assess the performances of different neutralisation reagents. The NH_4OH solution of 25–30%, as acquired, and 1-M Na_2CO_3 solution were added for pH adjustment. The CaO and $Ca(OH)_2$ were applied as fine powders.

The results of the study by Wei et al. (2005) showed that separate iron and aluminium hydroxide products with relatively high purity were successfully recovered via iron precipitation at pH 3.5–4.0 followed by aluminium precipitation at pH 6.0–7.0, while simultaneously meeting the National Pollutant Discharge Elimination System (NPDES) effluent discharge standards by the United States. In this study, iron precipitate recovery of >98.6% with a purity of >93.4% was achieved, while aluminium precipitate purity reached >92.1% at a recovery of >97.2%. In addition, the study found that during each metal recovery operation, other metals remained in solution, which ensured the relatively high purity of precipitate products. All of the five neutralisation reagents used in the study that are commonly used in AMD treatment were found to be suitable for iron and aluminium recovery. However, Wei et al. (2005) recommended the application of ammonia and caustic soda for metal recovery in the full-scale processes because unreacted hydrated lime might pose a threat to the purity of precipitate products if lime or hydrated lime was added. It was also suggested that oxidants should be added in the full-scale systems in order to enhance the oxidation of ferrous iron.

The aim of a study by Luptáková et al. (2010) was to precipitate heavy metals from AMD using bacterially produced hydrogen sulphide combined with intermediate steps of metals precipitation by sodium hydroxide at various pH values as shown in Figure 9.2. The experiments were conducted with AMD obtained from the abandoned and flooded deposit of Smolník (Slovak Republic).

Briefly, the study comprised several process steps, which could be grouped into three main stages: (1) biological hydrogen sulphide production using sulphates reducing bacteria, (2) selective heavy metals precipitation by the bacterially produced hydrogen sulphide, and (3) three intermediate steps of metal hydroxide precipitation by 1-M NaOH at various pH values. The solids produced in each step were separated from the remaining solution by filtration. The results of the study indicated that the process by Luptáková et al. (2010) is able to sequentially precipitate Cu^{2+}, Zn^{2+}, and Fe^{3+} in the form of sulphides, Al^{3+}, Fe^{2+}, and Mn^{2+} in the form of hydroxides. Table 9.1 shows the results obtained by Mačingová and Luptáková (2012) in a similar study of selective sequential precipitation process. As can be seen from the table, selective recovery of various metal precipitates was achieved.

Mulopo (2015) studied the sulphidation behaviour of Fe(II) using CaS derived from waste gypsum as a sulphidation agent, together with the possibility of selective precipitation of Pb, Zn, Ni, Co, and Fe(II) from various AMD solutions. The study is an appropriate case illustrating a simple strategy for integrated recycling of two mining waste streams (AMD and gypsum) and highlighted the need for the mining industry to break away from the traditional "linear" *cul-de-sac* disposal of wastes and think of new sustainable ways of waste management. A tubular muffle furnace consisting of a 750 mm long, 24 mm diameter mullite tube mounted horizontally and equipped with a temperature controller was used for the thermal reduction of waste gypsum to calcium sulphide. Basically, CaS was produced by carbothermal reduction of waste gypsum at a temperature of 1025–1030°C

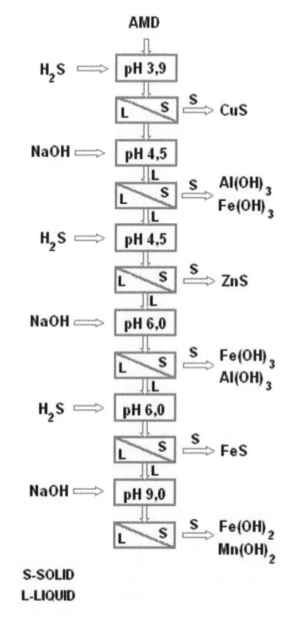

FIGURE 9.2
Schematic illustration of the six-step precipitation process for recovery of metals from acid mine drainage. (From Luptakova et al., 2010.)

TABLE 9.1

Conditions and Results of Selective Sequential Precipitation Process

Step	1	2	3	4	5	6
pH	2.8	3.7	3.7	5.0	5.0	9.5
Reagent	H_2O_2	NaOH	H_2S	NaOH	H_2S	NaOH
Removed metals	Fe	Fe	Cu	Al, Zn	Zn	Mn
Proportion (%)	99.99	99.99	99.99	98.94:1.06	99.99	99.99

Source: Mačingová and Luptáková, 2012.

for 45 min using a C/CaSO$_4$ molar ratio of 2. The CaS yield obtained was about 78%. In this study, the effect of sulphide addition to the AMD system was investigated using sulphide/total metal mole ratios of 0, 0.5, 0.75, 1.0, 1.5, 2.0, and 2.5. Appropriate amounts of CaS were added to the AMD to give a total of 1-L mixture and batch experiments were carried out in plastic beakers equipped with overhead stirrers fitted with radial turbine impellers. The experiments were run at appropriate pH values using a pH cascading approach. The metal removal in the batch experiments were carried out for at least 5 min at a particular pH or until a steady pH was attained. The results showed that sulphidation was dependent on the pH, sulphide dosage, and metal concentration. The selective sulphidation of metals also showed significant dependence on the respective metal sulphide solubility order as a function of pH. It was found that Pb, Zn, Ni, and Co could be removed as metal sulphides at lower pH values while Fe(II) remained in solution, thus enabling ferrous iron to be separated from the other metals, which is a great advantage for metal recovery. The results of the study clearly demonstrated that selective metal removal and recovery as metal sulphides may be achieved conveniently using CaS as the sulphidation medium. However, the purity of CaS obtained by the thermal reduction of waste gypsum and mass transfer limitations associated with the AMD-CaS system was found to be critical for the process development. In addition, the settling characteristics of the precipitates were poor, but this could probably be improved by the use of an anionic polymer at low pH.

Uça (2017) successfully precipitated metals separately in a pH-controlled system by using sulphide produced in an ethanol fed anaerobic baffled reactor. For simplicity, equation 9.1 shows the reaction of the sulphate reduction in the reactor using an organic source such as formaldehyde in the presence of sulphate reducing bacteria. Basically, sulphidogenic bacteria generate hydrogen sulphide primarily by using either sulphate or elemental sulphur as an electron acceptor, and an organic (e.g., ethanol) or inorganic (e.g., hydrogen) electron donor (Ňancucheo and Johnson, 2012).

$$SO_4^{2-} + 2CH_2O \rightarrow H_2S + 2HCO_3^- \tag{9.1}$$

Simulated AMD (pH 2.5) contained approximately 120 ± 2.4 mg/L Cu^{2+} and 124 ± 3.1 mg/L Fe^{2+}. In this study, the sulphide and alkalinity in the anaerobic baffled reactor effluent was used for the selective metal recovery experiments. Nitrogen gas was bubbled through the anaerobic baffled reactor effluent in order to transport the sulphide gas from the reactor to the metal mixture bottle. This method allowed the transportation of H_2S only and leaving the alkalinity in the sulphide bottle. As a result, the pH in the bottle containing a mixture of metals was not increased, thus iron was not precipitated. Since the pH was low, only copper ions reacted with the hydrogen sulphide gas according to the following reaction:

$$Cu^{2+} + H_2S \rightarrow CuS + H^+ \tag{9.2}$$

Over 99% of the Cu^{2+} was removed in first 2 min. A complete removal of 120 ± 2.4 mg/L Cu^{2+} was achieved in 20 min. After filtration of the CuS precipitate, the metal mixture bottle was mixed with anaerobic baffled reactor effluent, which had high concentrations of sulphide and HCO_3^-. As a result, Fe^{2+} was precipitated as FeS at elevated pHs. The results showed that the supernatant Fe^{2+} concentration 5 min after mixing was only 0.3 ± 0.02 mg/L, which implies that most of the Fe^{2+} ion had precipitated.

In a study by Sampaio et al. (2009), Cu was continuously and selectively precipitated from Zn using Na_2S. Selective precipitation was based on the control of pS ($= -\log [S^{2-}]$) and pH. Here, having the solubility product defined as $K_{SP} = (Me^{2+})(S^{2-})$, it means that different sulphide concentrations (S^{2-} potentials) are required to precipitate different metals. Therefore, the addition of sulphide to selectively precipitate heavy metals can be controlled using an ion selective electrode for sulphide (S^{2-}), a so-called pS electrode (Sampaio et al., 2009). In this study, selective precipitation of copper from zinc was achieved at pS and pH of 25 and 3, respectively.

Another example of a technique that uses biologically generated precipitating hydrogen sulphide gas was described by Huisman et al. (2006). In this process (Thioteq technology), just like other similar processes that biologically generate hydrogen sulphide, gaseous or dissolved H_2S is produced on-site and on-demand in an engineered high rate bioreactor. According to Huisman et al. (2006), the Thioteq process consists of two stages: a biological and a chemical. The water to be treated only passes through the chemical stage. Sulphide is produced in the biological stage and transported to the chemical (precipitation) stage with a carrier gas. The properties of the sulphide gas from the bioreactor and the contactor design result in metal sulphides with good settleability and filterability. The chemical stage consists of a gas–liquid contactor. As already stated, the sulphide is transported to the contactor with the help of a carrier gas (e.g., a mixture of CO_2 and N_2) and the metal-loaded water (e.g., AMD) is fed to the contactor. Metals like copper precipitate as sulphides according to reaction 9.2. In the Thioteq process, copper can be precipitated as a sulphide usually without pH adjustment and without significant precipitation of other heavy metals present in the water. The result is a product with a high copper sulphide content usually greater than 90%. Other metals such as zinc and nickel can be recovered as separate high-grade sulphide products when the number of precipitation stages are increased. In addition, a pH control using an alkali source might be required to meet the optimum precipitation conditions. The precipitated metal concentrates are recovered in a clarifier and then dewatered using a filter press. These metal sulphides can be transported to smelters as high quality concentrates.

Several precipitating agents have also been developed in the last few years for chemically and selectively precipitating divalent and univalent heavy metals from water and effluents (Blais et al., 2008). Selective metal precipitating agents include dithiocarbamate (Matlock et al., 2002a), Thio-Red (Matlock et al., 2002a), and dipropyl dithiophosphate (Ying and Fang, 2006). The principal advantages of using metal precipitating agents are (Blais et al., 2008): (1) the formation of metal compounds which have a very low solubility, and (2) the lesser production of metallic residue in comparison to the production of metallic sludge using common chemicals, like sodium hydroxyde or lime. However, the high cost of the metal precipitating reagents inhibits their use for different industrial applications (Meunier et al. 2002; Blais et al., 2008).

A good example of an alternative metal precipitating reagent is 1,3-benzenediamidoethanethiol dianion (BDET, known commercially as MetX). Research has shown that BDET can reduce the concentrations of a wide variety of divalent metals in water and sediments to below wastewater discharge limits. In fact the ligand has been found to selectively and irreversibly bind soft heavy metals from aqueous solutions (Matlock et al., 2002b). Furthermore, it has been demonstrated that the metal-BDET precipitates are insoluble in aqueous solution and in common organic solvents and are stable over pH ranges of 0.0–14.0 (Matlock et al., 2001). In order to explore the utilisation of BDET for iron and metal binding under AMD conditions, an abandoned coal mine was selected for study in Pikeville, Kentucky by Matlock et al. (2002b). The study by Matlock et al. (2002b) involved treating water within the coal mine as well as water being discharged from the mine using BDET. As a result of a variety of metals present in the AMD waters, multiple BDET-metal compounds were expected to be produced, and thus each of the metal-BDET precipitates was identified using NMR, IR, and XRD. The study found that BDET-Fe precipitates were predominant. Therefore, in addition to analysing the multiple BDET-metal precipitates, pure samples of BDET-Fe were prepared and analysed for stability using NMR, IR, Raman, XRD, and elemental analyses. The study found that BDET was able to remove > 90% of several toxic or problematic metals from AMD samples. For example, the concentrations of metals such as iron were reduced at pH of 4.5 from 194 ppm to below 0.009 ppm. In the leaching experiments conducted for stability tests at pH 0.0, 4.0, and 6.5, maximum leaching was seen for pH 0.0 solutions on day 7 (16.7 mg of Fe leached from 1000 mg BDET-Fe). During the 30-day leaching period, no additional leaching was seen after 7 days for samples tested at pH 0.0, 4.0, and 6.5.

Another developed method for heavy metal removal based on chelating precipitants is termed CH collector method. This is simply a solid material which binds heavy metals to its surface (Turhanen and Vepsäläinen, 2013). Typically, such chelating precipitants contain groups with replaceable hydrogen atoms such as carboxyl (–COOH), hydroxyl (–OH), mercapto (–SH), or sulphonic (–SO$_3$H) groups, together with functional groups of basic character, such as amino (–NH$_2$), amino (cyclic)(–NH–), carbonyl or thio keto with which the reacting metal is coordinated to form a four-, five-, or six-member ring. This invention is unique in that ion channels are formed inside the material in which metal ions are collected from the solution (Turhanen and Vepsäläinen, 2013). The collection of the heavy metal ions does not require a separate precipitation step or any adjustments to the solution's pH. Unlike traditional methods, the CH collector method also allows the recovery of metals to occur from very small concentrations. The new method enabled the complete removal of uranium from the water collected from a Finnish mine (Turhanen and Vepsäläinen, 2013). The study showed that there was no need to pre-treat the water even though it contained high concentrations of other metals. The efficiency of the method was also tested on scandium and it was removed from wastewater with 98% recovery (Turhanen and Vepsäläinen, 2013).

9.2.2.3 Summary

The large volumes of sludge that are produced through the active treatment of AMD place a huge burden on the mining industry because the sludge requires further processing and/or final disposal. Moreover, the AMD sludge contains a mixture of various metal oxides/hydroxides that are of little to no practical value. However, the recovery of individual metals has potential commercial value. This section of the chapter has shown that based on the solubility of the metals dissolved in AMD, selective precipitation process can be developed to recover high-purity metals. A variety of precipitation reagents are available. However, the most common reactive agents for metal precipitation are hydroxides and sulphides; and amongst the two chemicals, sulphides have various advantages compared to using hydroxides.

9.2.3 Selective Adsorption

9.2.3.1 Concepts of Selective Adsorption

In the past number of years, adsorption has been one of the most widely used techniques amongst the different methods developed for removal of toxic metals from polluted natural water or industrial wastewater mainly due to its low cost and environmental friendliness (Larsson et al., 2018). Patil et al.

(2016) define adsorption as the accumulation of one substance on the surface of another; and that the mechanism of adsorption can be one or a combination of several phenomena, including chemical complex formation at the surface of the adsorbent, electrical attraction (a phenomenon involved in almost all chemical mechanisms, including complex formation), and exclusion of the adsorbate from the bulk solution. Al-Rashdi et al. (2011) consider adsorption as a mass transfer process in which a substance is transferred from a liquid phase to the surface of a solid and becomes bound by physical and/ or chemical interactions.

Bessho et al. (2019) reiterated that adsorption is one of the effective techniques for recovery of metal ions from water, and that separation and recovery of metals by adsorption can achieve both purification of the AMD and metal recovery. Rodríguez-Galán et al. (2019) also state that adsorption is the most employed technique commercially since it can recover 99% of the metals.

9.2.3.2 Selected Typical Studies of Selective Adsorption

The adsorption process has evolved over the years to become an important method for AMD treatment and metal recovery technique (Bessho et al., 2017; Bessho et al., 2019; Rodríguez-Galán et al., 2019). This chapter deals with only a few examples of materials used in the adsorption process for recovery and/ or removal of metals from AMD. Most of the adsorbents, particularly, the newly developed sustainable adsorbents for industrial wastewater including AMD treatment, are discussed elsewhere in other references.

The possibility of using different types of cross-linked gelatin hydrogels for recovery and removal of metals from acidic wastewater by adsorption was investigated by Bessho et al. (2017). Gelatins isolated from a porcine skin by an acid process (Type A) and a bovine skin by an alkaline process (Type B) were used in the study; and glutaraldehyde was used for crosslinking of Type A and Type B gelatin hydrogels. Copper was used as a model substance in the study because it is a common AMD metal. Metal recovery by gelatin hydrogels mainly consists of adsorption to gelatin molecules and absorption into hydrogels. The results of the study showed that the crosslinked Type B gelatin hydrogels recovered more Cu than Type A gelatin which appears to indicate that Type B gelatin is a more suitable adsorbent material to recover cationic metal ions. It must be noted that the liberation of a proton from the carboxyl group is required for adsorption of metals to gelatin (Bessho et al., 2017). Therefore, in the case of Type B gelatin, it appears that the increased Cu recovery was mainly due to electrostatic adsorption of Cu^{2+} to the carboxyl group without a proton. In other words, the Cu recovery using Type B gelatin hydrogels was mainly affected by carboxyl groups on side chains of gelatin hydrogels at higher pH. On the other hand, Type A rarely had carboxyl groups as side chains. In fact, amongst the two metal recovery mechanisms, adsorption was rarely detected with Type A gelatin. It is possible that Type B gelatin hydrogels also recovered Cu by absorption, but at higher pH, adsorption surely contributed to Cu recovery. Doubtless, in this study, the copper recovery by Type B gelatin hydrogels was dependent on pH. In order to develop gelatin hydrogels having a high-performance adsorption capacity for metal recovery from AMD, preparation of "mixed" gelatin hydrogels blended with other natural organic compound was suggested by Bessho et al. (2019). However, the study by Bessho et al. (2019) only investigated the metal recovery efficiency of chitosan with a view of preparing, in future research, "mixed" gelatin hydrogels including some organic solids with a high metal adsorption capacity. Thus, metal recovery efficiency of chitosan was investigated in a study by Bessho et al. (2019) using 2 mM of three kinds of simulated metal solutions (i.e., copper, zinc, and manganese). The results showed that at pH 5, 0.1 g of chitosan recovered approximately 90% of Cu from 50 mL of the simulated solution. At relatively higher pH (>3.0), over 90% of Cu recovery was achieved. However, Cu recovery was not detected at pH 2.0. Thus, it was considered that adjustment of solution pH allowed the recovery of Cu from acidic wastewater. Chitosan mainly has lots of amino and hydroxyl groups. This implies that Cu recovery using chitosan was mainly induced by formation of chelate compounds. Ideally, the Cu recovery using chitosan was affected by pH. In contrast, lower Zn was recovered in comparison to Cu. The Mn recovery was hardly detected within the pH range of 2–5. From these results, and in particular, because chitosan had a high performance for Cu recovery, Bessho et al. (2019) suggested that "mixed" gelatin hydrogel blended with chitosan had a potential for the high-performance adsorbent for Cu recovery.

In 1997, Tavlarides and Doerkar developed a new class of adsorbents for selective separation and recovery of metal ions from dilute aqueous solutions (Tavlarides and Doerkar, 1997a, b, c, d). According to Doerkar and Tavlarides (1998), selectivity of these materials is a function of immobilised ligands on the ceramic supports and the condition under which the metal ion solutions are treated. These adsorbents differ from polymeric resins/ion-exchange resins and have the following advantages: (1) selectivity for separation of desired metal ions, (2) reduced or no interference from the accompanying cations (Na^+, Ca^{2+}, Mg^{2+}), anions (NO_3^-, SO_4^{2-}, Cl^-), and complexing agents, (3) easy and selective regeneration of the adsorption bed which yields concentrated solutions for recovery, (4) high mechanical strength for fixed bed applications, (5) higher metal ion uptake due to open pore structure, higher porosity, and non-swelling characteristics, and (6) no irreversible adsorption of organics. Following the development of the adsorbents, Doerkar and Tavlarides (1998) undertook a study to (1) evaluate the performance of adsorbents for separation and recovery of iron, copper, zinc, cadmium, and lead from Berkeley Pit simulated waters, and (2) propose an integrated process with fixed bed set-ups and specify operating conditions. The adsorbents studied include ICAA-A, ICAA-B with a substituted quinoline group (Tavlarides and Doerkar, 1997d), ICAA-C with a substituted oxime, ICAA-D

TABLE 9.2

Target Metals and Other Constituents in Berkeley Pit Water

Ions	Concentration (mg/L)	Other Constituents
Aluminium	2.60	pH = 2.85
Cadmium	2.14	$[Fe^{3+}]/[Fe^{2+}] = 0.15$
Calcium	456	
Copper	172	
Iron	1068	
Lead	0.031	
Magnesium	409	
Manganese	185	
Sodium	76.5	
Zinc	550	
Nitrate	<0.1 (as N)	
Sulphate	7600	

Source: Doerkar and Tavlarides, 1998.

with the thio/amine group (Tavlarides and Doerkar, 1997a), ICAA-E with the phosphoric acid group (Tavlarides and Doerkar, 1997a), and ICAA-F with a thiophosphinic group (Tavlarides and Doerkar, 1997a). The batch shake-out tests and breakthrough curve studies were executed using simulated Berkeley Pit water. Simulated Berkeley Pit water containing targeted metal ions (Fe, Cu, Zn, Cd, Pb), sulphate ions, and a pH of 2.8 was prepared according to the compositions shown in Table 9.2.

The results showed that the prepared adsorbents ICAA-A, ICAA-B, or ICAA-C and ICAA-D have the potential to remove and recover Fe^{3+}, Cu^{2+}, Zn^{2+}, Cd^{2+} and Pb^{2+} from Berkeley Pit waters. Given the selectivity of the adsorbents and the potential to synthesise an adsorbent for Fe^{2+}, an integrated adsorption process was devised. In one scheme the first bed comprised of ICAA-A for Fe^{3+} removal, the second bed comprised of ICAA-C for Cu^{2+} removal, and the third bed comprised of ICAA-D for removal of Zn^{2+}, Cd^{2+}, Pb^{2+}, and some Fe^{2+}. In a second scheme, another adsorbent can be synthesised to remove Fe^{2+} and a bed of this material can be used between the second and third bed. The study showed that the effluent from the last bed of either scheme was not acidic and can be discharged in an environmentally safe manner if other toxic metal ions are removed. Furthermore, the processes do not require adjustment of the pH of the feed stream. A number of the adsorbents were found to be able to retain a stable adsorption capacity after 20 cycles of adsorption and stripping.

The main objective of a study by Mohan and Chander (2006) was to remove and recover metal ions (Fe^{2+}, Fe^{3+}, Mn^{2+}, and Ca^{2+} ions) from AMD using lignite as a low-cost sorbent in single and multiple column set-ups operating in downward flow modes. The results showed that the lignite usage rate was higher in a single column (0.981 g/L) compared to multiple columns (0.085 g/L), three columns in this study. In this study lignite usage was defined as follows:

$$\text{Lignite usage rate (g/L)} = \frac{\text{Weight of lignite in column (g)}}{\text{Volume at breakthrough (L)}} \quad (9.3)$$

Some studies have shown that using multi-stage treatments of heavy metal solutions with lignite could reduce the pollutants to acceptable discharge limits at a lower cost than using conventional heavy metal treatment processes (Simate et al., 2016). Mohan and Chander (2006) used a batch mode to examine the effect of pH. The sorption of Fe^{2+}, Mn^{2+}, and Fe^{3+} on lignite was found to increase with an increase in pH of the test solution. However, it must be noted that for sorption studies, the pH of solution must be less than the pH for precipitation of respective metal ions (Mohan and Chander, 2006). In their study, Mohan and Chander (2006) observed that the sorption of Fe^{2+} was very low at $pH_{in} \leq 2$, but it increased from 6% to 84% at pH of 4.0. The insignificant removal of metal ions at low pH is attributed to the competition between the protons and the metal ions for the same binding sites (Schiewer and Volesky, 1995). Furthermore, the increase in the positive surface charges at low pH results in a higher electrostatic repulsion between the surface and metal ions (Reddad et al., 2002; Wang et al., 2006). However, at $pH_{in} > 4.0$, the removal of Fe^{2+} was considered to have taken place by sorption as well as precipitation, i.e., the OH^- ions from the solution formed some complexes with Fe^{2+} (e.g., $Fe^{2+} + 2OH^- \rightleftarrows Fe(OH)_2$) (Kuhr et al., 1997; Arpa et al., 2000; Karabulut et al., 2000; Butler et al., 2007; Simate et al., 2016). Similarly, the removal of Mn^{2+} was also negligible at $pH_{in} \leq 2$, but increased with an increase in pH_{in} though precipitation only occurred at $pH_{in} \geq 8$. These results show that solution pH is a significant factor that determines the degree of metal adsorption. Furthermore, the equilibrium solution pH is a major parameter governing the extent of metal adsorption (Simate et al., 2016). Coals that generated higher solution pHs were found to exhibit the largest metal adsorption.

Stanković et al. (2009) presented the results of the batch and column adsorption for copper and some associated ions by utilizing sawdust of deciduous trees (i.e., linden and poplar) as a low-cost adsorbent. The AMD from an abandoned copper mine, as well as synthetic solutions of the ions (Cu^{2+}, Zn^{2+}, Mn^{2+}, and Fe^{2+}) that are the main constituents of the AMD were both used as a model-system in the study. The adsorption of heavy metal ions strongly depended on the process time, the initial pH of the aqueous phase and the kind of ions adsorbed. The adsorption process was fast and after ten to twenty minutes of contact time, the system reached equilibrium. More specifically, the adsorption capacity of the studied sawdust was significantly affected by the initial pH of the solution and the kind of metal ions adsorbed. At lower pH of solutions the adsorption percentage decreased leading to a zero adsorption percentage at pH < 1.1. Maximum adsorption percentage was achieved at 3.5 < pH < 5. It was found that both poplar and linden sawdust have almost equal adsorption capacities against copper ions. The highest adsorption percentage ($\approx$80%) was achieved for Cu^{2+}, while for Fe^{2+} it was slightly above 10%. The other considered ions (Zn^{2+} and Mn^{2+}) were within this interval. The following gives the ranking of the ability of the considered ions to be adsorbed on sawdust: $Cu^{2+} > Zn^{2+} > Mn^{2+} > Fe^{2+}$. The used sawdust had shown certain selectivity in the adsorption of heavy metal ions. Calculated selectivity coefficients over ferrous ions were: $\beta_{Cu^{2+}-Fe^{2+}} \approx 22$; $\beta_{Zn^{2+}-Fe^{2+}} \approx 11$; $\beta_{Mn^{2+}-Fe^{2+}} \approx 2$. The results obtained in the batch mode were validated in the column experiments using the real mine water originating from an AMD of a copper mine. Very high degree of copper ions adsorption was achieved in the column adsorption (>99.7%) experiments before the breakthrough point. Both kinds of sawdust used had an equal ability to adsorb copper ions. The pH of solution increased slightly during the adsorption which implied that there was a co-adsorption of protons contained in a treated solution. After completing the adsorption, instead of desorption, the loaded sawdust was drained, dried, and burned; the copper bearing ash was then leached with a controlled volume of sulphuric acid solution to concentrate copper therein. The obtained leach solution had the concentration of copper higher than 15 g dm^{-3} and the amount of H_2SO_4 was high enough to serve as a supporting electrolyte suitable to be treated by the electrowinning technique for recovery of copper.

With the advent of nanotechnology, various types of nanomaterials with large surface area and small diffusion resistance have been developed and are now receiving considerable attention in water treatment (Zhang, 2003; Hu et al., 2006; Simate, 2012; Simate et al., 2012). For example, nanoscale iron particles have been established as effective reductants and catalysts for a variety of contaminants including heavy metals (Zhang, 2003). Iron nanoparticles possess the advantages of large surface area, high number of surface active sites, and high magnetic properties, which lead to high adsorption efficiency, high removal rate of contaminants, and easy and rapid separation of adsorbent from solution via magnetic field (Hu et al., 2006). In addition, it is possible that after magnetic separation by the external magnetic field, the harmful components can be removed from the magnetic particles, which can then be reused (Ponder et al., 2000; Oliveira et al., 2003; Hu et al., 2006).

The adsorption studies by Hu et al. (2006) showed that the nanoscale maghemite (γ-Fe$_2$O$_3$) synthesised using a sol-gel method was very effective for selective removal of Cr(VI), Cu(II), and Ni(II) from wastewaters. The removal efficiency was highly pH dependent, which also governed the selective adsorption of metals from the solution. The optimal pH for the selective removal of Cr, Cu, and Ni was found to be 2.5, 6.5, and 8.5, respectively. Regeneration and readsorption studies demonstrated that the maghemite nanoparticles could be recovered efficiently for the readsorption of the metal ions, and metals could be highly concentrated for recycling.

Several studies have also been carried out to assess the technical feasibility of various kinds of raw and surface oxidised carbon nanotubes (CNTs) for sorption of various metals from aqueous solutions (Rao et al., 2007). The CNTs are carbon nanomaterials that were re-discovered by Iijima (Iijima, 1991). These materials have shown exceptional adsorption capabilities and high adsorption efficiencies for various organic pollutants (Lu et al., 2005), inorganic pollutants (Li et al., 2003a), and heavy metals (Li et al., 2003b; Li et al., 2006; Li et al., 2007). The CNTs are particularly attractive as sorbents because, on the basis of mass, they have larger surface areas than bulk particles, and that they can be functionalised with various chemical groups to increase their affinity towards target compounds (Savage and Diallo, 2005; Simate et al., 2012; Simate, 2012). The CNTs also have small size and are hollow with layered structures (Wu, 2007), which are important attributes for adsorption.

The studies have shown that the sorption capacities of metal ions by raw CNTs are very low, but significantly increased after oxidisation by HNO_3, NaOCl, and KMnO4 solutions (Rao et al., 2007). In fact, this is another advantage of CNTs in that they can be functionalised (or oxidised) with various kinds of chemical agents depending on the adsorption objective. The removal efficiency was also highly pH dependent, thus controlling the selective adsorption of metals from the solution. The sorption/desorption studies showed that CNTs could be regenerated and reused consecutively several times without significant loss in adsorbent capacity which signify the appropriateness of CNTs for commercial applications. Therefore, the superior sorption capacity and effective desorption of heavy metal ions suggest that the CNTs are promising sorbents for environmental protection applications (Rao et al., 2007).

The aim of a study by Ríos et al. (2008) was to evaluate the use of low-cost sorbents like coal fly ash, natural clinker, and synthetic zeolites to clean up AMD generated at the Parys Mountain copper–lead–zinc deposit, Anglesey (North Wales), and to remove heavy metals and ammonium from AMD. Coal fly ash is a by-product of coal combustion that has been regarded as a problematic solid waste (Skousen et al., 2013; Ram and Masto, 2014), mainly due to the presence of potentially toxic trace elements (e.g., Cd, Cr, Ni, Pb) and organic compounds (e.g., polychlorinated biphenys, polycyclic aromatic hydrocarbons) (Shaheen et al., 2014). Natural clinker is a product of coal-bed fires ignited by natural processes (Rios et al., 2008). Zeolites are naturally occurring alumino-silicates with a three-dimensional framework structure bearing AlO_4 and SiO_4 tetrahedra (Motsi et al., 2009). These are linked to each other by sharing all of the oxygen to form interconnected cages and channels (Englert and Rubio, 2005; Motsi et al., 2009) where exchangeable cations are present which counter-balance the negative charge on the zeolite surface generated from isomorphous substitution (Barrer, 1978; Dyer, 1988; Motsi et al., 2009). The manufacture of synthetic zeolite is also possible (Simate et al., 2016). In this study synthetic zeolites were prepared by two methods, namely, (1) classic hydrothermal synthesis using natural clinker, and (2) alkaline fusion prior to hydrothermal synthesis using both coal fly ash and natural clinker. The sorption of Cu, Pb, Zn, Ni, Cr, Fe, As, and ammonium onto coal fly ash, natural clinker, and synthetic zeolites was studied in laboratory-batch experiments, which were carried out at room temperature to investigate the efficiency of the sorbents for removing heavy metals and ammonium from AMD.

With a rise in pH values as the sorbent dosage was increased, the results in a study by Ríos et al. (2008) suggest that pH is strongly affected by the sorbent material rather than the AMD composition and particularly a higher sorbent dosage. It was noted from the study that there is a possibility of two competing reactions, namely, (1) release of alkalinity from sorbents, and (2) the removal of acidity from AMD components. The study found that at higher sorbent dosage the acidity from AMD components is overrun and pH is bound to increase whereas with the lower sorbent dosage the alkalinity from the sorbent is exceeded by the acidity from the AMD components and the pH remains low. In other words, the pH played a very important role in the sorption/removal of the contaminants and a higher adsorbent ratio in the treatment of AMD promoted an increase of the pH and vice versa. The results also revealed that the heavy metal removal was depended on the sorbent material and the applied dosage. For example, coal fly ash and natural clinker did not show good efficiency as sorbents to neutralise the AMD, but their synthetic products (e.g., coal fly ash-based faujasite; natural clinker-based faujasite; natural clinker-based Na-phillipsite) were effective as ion exchangers in removing acidity, Fe, Zn, and Cu from AMD. Amongst the two variants of the zeolite (i.e., faujasite and Na-phillipsite), faujasite was more effective. In fact, the results of the adsorption experiments suggest that faujasite can be applied in wastewater treatment as an immobiliser of pollutants, and the selectivity of faujasite for metal removal was as follows in decreasing order: Fe > As > Pb > Zn > Cu > Ni > Cr. In general, the results of the study showed that different sorbents contain considerable amounts of accessory phases that partly dissolve during the batch reaction, which may explain the sudden increase or decrease in metal concentration and, therefore, the release rate of the metal elements is controlled by the dissolution of the sorbent.

In addition to cation-exchange reactions, precipitation of hydroxide species (mainly of Fe) also played an important role in the sorption and co-precipitation, and thus led to the immobilisation of metals in the batch experiments. The efficiency in the removal of ammonium by coal fly ash and natural clinker was poor. However, the reaction between synthetic zeolites (coal fly ash-based faujasite; natural clinker-based faujasite; natural clinker-based Na-phillipsite) and AMD after 24 h of contact time produced lower ammonium concentrations in the solution. In fact, the study indicated that natural clinker-based faujasite produced a complete removal of ammonium after 24 h of contact time when a dosage of 1 g was used.

The adsorption behaviour of natural zeolite (clinoptilolite) was studied by Motsi et al. (2009) in order to determine its applicability in treating AMD containing 400, 20, 20, and 120 mg L^{-1} of Fe^{3+}, Cu^{2+}, Mn^{2+}, and Zn^{2+}, respectively. Batch experimental tests for single and multi-component solutions were performed to ascertain both the rate of adsorption and the uptake at equilibrium. The optimum conditions for the treatment process were investigated by observing the influence of pH levels, the presence of competing ions, varying the mass of zeolite and thermal modification of the natural zeolite (calcination and microwaves). The adsorption studies showed rapid uptake, in general, for the first 40 min, and after the initial rapid period, the rate of adsorption decreased. The study found that about 80%, 95%, 90%, and 99% of Fe^{3+}, Mn^{2+}, Zn^{2+}, and Cu^{2+}, respectively, were adsorbed from single component solutions in the first stage.

For multi-component solutions in a study by Motsi et al. (2009), only the adsorption of Fe^{3+} was significantly unaffected by the presence of competing ions. This may be because the main mechanism responsible for Fe^{3+} removal from solution is believed to be precipitation. The adsorption of the other three cations was affected significantly. For example, the amount adsorbed from multi-component solutions of concentration of 40 mg L^{-1} decreased by 33%, 41%, and 39% for Cu^{2+}, Zn^{2+}, and Mn^{2+}, respectively, compared to their respective single component solutions. When the solution concentration was increased from 40 mg L^{-1} to 120 mg L^{-1}, the relative decrease in the amount adsorbed between the multi-component and single component cases increased further.

The study by Motsi et al. (2009) also indicated that the removal of the heavy metal ions was not only due to ion exchange, but also due to precipitation of metal hydroxides from the solution. This observation is similar to other previous studies (Khur et al., 1997; Arpa et al., 2000; Karabulut et al., 2000; Butler et al., 2007; Ríos et al., 2008; Simate et al., 2016). It is noted that natural zeolites are generally weakly acidic in nature and that sodium form exchangers are selective for hydrogen (R–Na + H_2O ↔ RH + Na^+ + OH^-), which leads to high pH values when the exchanger is equilibrated with relatively dilute electrolyte solutions (Erdem et al., 2004; Motsi et al., 2009) making metal hydroxide precipitation feasible. In other words, as reaction proceeds the solution pH increases which promotes metal precipitation. The rate of adsorption was also found to be directly proportional to the pH value of the solution. Adsorption decreased in more acidic solutions, due to hydrogen ion competition. The rate of adsorption and capacity also depended on the mass of the adsorbent and heat treatment by either microwaves or heating in a furnace. For zeolite exposed to microwave radiation, the adsorption rate increased with exposure time, but only up to a certain limit. The adsorption rate of the zeolite exposed to microwave radiation began to decrease as exposure time approached 30 min. The rate of adsorption by calcined zeolite was also found to be faster compared to untreated zeolite, but the efficiency decreased for zeolite exposed to very high temperatures (e.g., 800°C). The increase in rate of adsorption capacity as a result of thermal treatment may be attributed to the removal of water from the internal channels of natural zeolite which leaves the channels vacant and hence increases the adsorption capacity of the zeolite (Turner et al., 2000; Ohgushi and Nagae, 2003; Motsi et al., 2009). The removal of water also resulted in a change in the surface area of the samples after thermal treatment (Motsi et al., 2009). The samples that were exposed to extreme thermal conditions had lower surface areas due to thermal runaway, whereby the zeolite structure collapses (Ohgushi and Nagae, 2005; Akdeniz and Ulku, 2007; Motsi et al., 2009). When the structure collapses the porosity of natural zeolite decreases and thus the adsorption capacity is reduced (Motsi et al., 2009). The study also found that efficiency of metal removal from solution by natural zeolite is inversely proportional to the initial solution concentration. According to the equilibrium studies, the selectivity sequence of metals by natural zeolite can be given as $Fe^{3+} > Zn^{2+} > Cu^{2+} > Mn^{2+}$, with good fits being obtained using Langmuir and Freundlich adsorption isotherms. The significance of this study is that preliminary tests using AMD samples from Wheal Jane Mine, UK, showed that natural zeolite has great potential as an alternative low-cost adsorbent in the treatment of AMD.

Petriláková and Bálintová (2011) utilised five different types of natural and synthetic adsorbents for the removal of Fe, Cu, Al, Mn, and Zn from real AMD (shaft Pech, Smolnik locality, Slovakia) with a pH of 4.2. The following adsorbents were employed in the study: (1) inorganic composite sorbent SLOVAKITE, (2) active carbon (granularity ≤ 1 mm), (3) turf brush PEATSORB, (4) universal crushed sorbent ECO-DRY (REO AMOS Slovakia), and (5) zeolite (granularity 0.5–1 mm, 2.5–5 mm, 4–8 mm). The chemical composition of the AMD was as follows in mg/L: Fe (338), Cu (1.16), Al (44.4), Mn (26), and Zn (5.81). Batch experiments were carried out by mixing various amounts of adsorbents into 100 mL of raw AMD over a period of 24 h. Active carbon was found to be the most efficient for the removal of Fe at 99.98% efficiency. The efficiency of Cu removal from AMD using active carbon and inorganic composite sorbent SLOVAKITE was 98.3% in both cases. Active carbon and inorganic composite sorbent SLOVAKITE adsorbents at 99.98% efficiency were the two most efficient for Al removal. For the Mn, the most effective adsorbent was active carbon (93.08%) and Turf brush PEATSORB (87.69%).

9.2.3.3 Summary

The typical examples given in the use of the adsorption processes to remove metallic pollutants from the AMD clearly demonstrate the feasibility of the adsorption techniques. In other words, the efficiencies of various adsorbents for the removal of heavy metals from AMD were illustrated in a number of studies. Several studies have also evaluated various techniques for recycling of used adsorbents and recovery of the heavy metals from the desorbing agents (Lata et al., 2014).

For regeneration and reuse of adsorbents, various possible regenerating agents such as acids, alkalis, and chelating agents such as ethylene diamine tetraacetic acid (EDTA) have been studied by many researchers. Lata et al. (2014) reviewed and summarised the performance of various desorbing agents for removal of adsorbed metals and regeneration of the saturated adsorbents. A study by Lata et al. (2014) made the following conclusions: (1) the alkalis are efficient desorbing agents for desorption of heavy metal(s) from chemical adsorbents or chemically modified adsorbents, (2) acids are efficient for desorbing bio-adsorbents, and (3) the chelating agent, EDTA, is the most efficient desorbing agent for biomass desorption. The review by Lata et al. (2014) found that many of the adsorbents can be reused effectively after regeneration.

After the solution is separated from the adsorbents, the metal laden solution is often subjected to various processes of purification and concentration before the valuable metals can be recovered either in their metallic state or as chemical compounds. The methodologies may include precipitation, distillation, adsorption, and solvent extraction (Roto, 1998; Parnell, 2019), and the final recovery step may involve precipitation, cementation, or electrometallurgical processing (Parnell, 2019).

9.2.4 Ion Exchange

9.2.4.1 Concepts of Ion Exchange

According to Hubicki and Kołodyńska (2012) ion exchange is the exchange of ions between the substrate and surrounding medium. In addition, Patil et al. (2016) regard ion exchange as a physical treatment technique in which ions dissolved in a liquid or gas interchange with ions on a solid medium. Patil et al. (2016) state further that the ions on the solid medium are associated with functional groups that are attached to the solid medium, which is immersed in the liquid or gas. Typically, ions in dilute concentrations replace ions of like charge that are of lower valence state, but ions in high concentration replace all other ions of like charges (Patil et al., 2016). EPA (2014) define ion exchange as the reversible exchange of contaminant ions with more desirable ions of a similar charge adsorbed onto solid surfaces known as ion-exchange resins. Despite the myriad of definitions for the ion-exchange process, it is also noted that positively charged molecules bind to cation-exchange resins while negatively charged molecules bind to anion-exchange resins (Thermo Scientific, 2007). However, it must be noted that Dinardo et al. (1991) classify ion-exchange resins (i.e., insoluble matrixes or support structures that act as a medium for ion exchange) as anionic, cationic, and chelating. Anionic and cationic resins are used extensively in water purification processes. More specifically, anionic resins are significant in extracting amphoteric elements such as arsenic and metals that form sulphate complexes such as uranium (Dinardo et al., 1991). Cationic resins can have either sulphonic or carboxylic functionality and are essentially non-selective and thus can extract most polyvalent cations including magnesium, iron, calcium, and many others (Dinardo et al., 1991). Chelating resins are relatively new and have high selectivity for some specific metals.

Ion exchange is considered as the most energy efficient and economical technology of all the recovery techniques (Patil et al., 2016). It is the only process that can efficiently treat very dilute solutions in parts per million (ppm) levels on a once-through basis (White and Asfar-Siddique, 1997; Patil et al., 2016). In other words, the technology allows efficient removal of traces of contaminants from solutions (Dąbrowski et al., 2004). Hubicki and Kołodyńska (2012) also emphasise that ion-exchange process is designed to remove traces of ionic impurities from water and process streams and give a product of desired quality. According to Hardwick and Hardwick (2016) the advantage of ion exchange over other methods such as solvent extraction or even precipitation is that the technique can still be viable when feed concentrations have dipped below the economic threshold of the other technologies. Furthermore, according to Dąbrowski et al. (2004), the technique is specifically convenient when there is a need to treat large volumes of diluted solutions. According to EPA (2014), the process is appropriate for desalination and the removal of a number of pollutants including hardness, alkalinity, radioactive waste, ammonia, and metals.

9.2.4.2 Selected Typical Studies of Ion Exchange

Ion exchange has been successfully tested on wastewater from mining operations and, generally, works more effectively for waters in the pH range of 4 to 8 that has low suspended solids and low concentrations of iron and aluminium (EPA, 2014). The more complex the mixture, the harder it is to remove all metals effectively, and the capacity of any resin to remove contaminants is limited by the type of resin, the number of available exchange sites and the chemistry of input water (EPA, 2014).

Hardwick and Hardwick (2016) provided an overview of the potential for the recovery of value from contaminated mine waters and waste streams. The study discussed various metal ions commonly found in AMD and other mine waters and evaluated the levels of concentration the metals would become attractive for recovery using ion-exchange technology. Based on the capital and running costs, Hardwick and Hardwick (2016) determined the capital payback period for various contamination levels for a number of valuable metals that are likely to be found in mine waters. Firstly, the study categorised the constituents of AMD as follows: (1) value containing elements, (2) hazardous elements, (3) low value elements that will co-load, (4) ion-exchange poisons, and (5) elements of little concern with regard to the ion-exchange process. Secondly, the study discussed the viability and performance of the resins and economics of the process based on metal concentrations, presence of impurities (or competing/poisonous ions), flow rate, and price of metals.

Hardwick and Hardwick (2016) suggested that the first influencing factor on the economics to consider is the concentration at which a particular metal is present in a possible feed source. When the concentration of metals in a stream is very low, efficiency of removal by ion-exchange resins is relatively high because at that point it is film diffusion rather than particle diffusion that limits the kinetics. In such a condition, metal leakage is very low, and the operating capacity of the resin increases. However, it is important to note that where the concentrations of the desired metals are high, the rough guideline is that for levels above 1g/L, it may become more economical to investigate another technology, such as solvent extraction for recovery. The study also found that the concentration at which a metal becomes economically viable for recovery is heavily dependent on the sales price of the product produced. If there is a depression in metal pricing, the operation is more likely to become less economical even for relatively high concentration streams.

Undoubtedly, the resin choice is often made on how suitable it is for the recovery of a specific element, but it is possible for the wastewater to contain trace amounts of other metals that have an even higher affinity to the functional groups on the resin, thus taking up active sites and reducing the capacity of the resin for the element of value (Hardwick and Hardwick, 2016). Iron is a very good example. For instance, most common ion-exchange resins have a high selectivity for trivalent iron. The iron may displace the desired element and poison the resins over time. Aminophosphonic resins, for example, may lose useful capacity over a period of time and thus become uneconomical. Therefore, it is imperative to use a resin with a lower selectivity for iron upfront of such resins (e.g., aminophosphonic resins) in order to protect them. Alternatively, iron may be removed by precipitation by raising the pH above 3 and, thereafter, use the filtration process. Radioactive elements when present in the feed solution also may make the recovered metals unsaleable. A very good example is thorium that may be loaded onto resin during the recovery of rare earths (Hubicki and Kołodyńska, 2012). However, thorium may be removed by precipitation or selective elution. Uranium can be recovered separately by a strong base anion resin (Botha et al., 2009), whereas radium may be removed using a strong acid cation resin (Clifford, 2004).

As discussed in Chapter 3, a number of reaction pathways, in the course of AMD generation, lead to the production of aqueous hydrogen cation (H_3O^+) thus resulting in very low pH for AMD (Dold, 2010; Garland, 2011). Studies on recovery of Cu(II) ions have shown that due to low pH values of wastewaters (pH < 2) such as AMD, conventional chelating ion exchangers of functional iminodiacetate and aminophosphonic groups practically do not adsorb Cu(II) ions. Therefore, special chelating ion exchangers characterised by much greater affinity for Cu(II) than for other metal ions have been used instead (Melling and West, 1984; Bolto and Pawłowski, 1987; Dorfner, 1991). For example, Cu(II) ions can be removed from highly acidic solutions (pH > 1.5) using Dowex XFS-4196 chelating ion exchanger which can be

TABLE 9.3

Physical Properties and Specifications of the Ion-Exchange Resins

Parameter	Indion 820 (Strong Base Anion Exchange Resin)	Indion 850 (Weak Base Anion Exchange Resin)
Physical form	Spherical opaque beads	Moist beads
Ionic form as supplied	Chloride	Free base
Moisture holding capacity	49–56%	40%
Particle size	0.3–1.2 mm	0.3–1.2 mm
Uniformity coefficient	1.7 max	1.7 max
Total exchange capacity	1.0 meq mL^{-1}	1.10 meq mL^{-1}
pH range	0–14	0–7

Source: Gaikwad et al., 2009.

easily regenerated by means of sulphuric acid with a concentration of 100 g H_2SO_4 dm^{-3} (Dąbrowski et al., 2004). Amphoteric ion-exchange fibres can also be applied for the removal of Cu(II) ions from acidic wastewaters. Various other types of ion exchangers have also been used for selective removal of Cu(II) ions (Hubicki and Jusiak, 1978; Hubicki and Pawłowski, 1986; Hubicki et al., 1999).

Gaikwad et al. (2009) performed a laboratory scale investigation to remove copper from AMD using ion-exchange resins – Indion 820 and Indion 850. The physical properties and specifications of the resins are given in Table 9.3. The concentrations of copper used in the study ranged from 50 to 250 mg L^{-1} and the dosage of resin from 25 to 700 mg L^{-1}. The effect of the initial concentration of copper ions, dosage of resin and pH on exchange capacities of ion-exchange resins was studied in a batch mode. The mixture of resins and copper solutions was agitated for a predetermined period at room temperature. Thereafter, the resins were separated and the filtrate was analysed by an atomic absorption spectrometer (Chemito, AAS-3000) for the residual concentration of copper.

The ion-exchange process, which is pH dependent, showed maximum removal of copper in the pH range of 2–6 for an initial copper concentration of 50–250 mg L^{-1} and with a resin dosage of 25–700 mg L^{-1}. Five isotherm models (Freundlich, Langmuir, Redlich-Peterson, Temkin, and Dubinin-Radushkevich) were tested and the equilibrium data fitted to all the sorption isotherms very well. The study found that the uptake capacity of Indion 850 was larger than that of Indion 820 due to the intrinsic exchange capacity. Therefore, of the two different ion-exchange resins studied, Indion 850 was considered to be the most efficient in removing copper ions. The uptake of copper by the ion-exchange resins was reversible which showed that the resins have a good potential for the removal/recovery of copper from AMD. In other words, the study showed that ion-exchange resins such as Indion 820 and Indion 850 can be used for an efficient removal of copper from mine wastewater.

In a study by Gaikwad et al. (2010), using the ion-exchange technique, a factorial experimental design method was used to examine the removal of Cu^{2+} ions from AMD wastewater. The strongly acidic cation-exchange resins, Indion 730, were used and the following four factors were varied at three different levels: initial concentration of Cu^{2+} ions (100,150, 200 mg L^{-1}), pH (3, 5, 6), flow rate (5, 10, 15 L h^{-1}), and dosage of the resin or bed height (20, 40, 60 cm). A matrix was established according to the high, middle, and low levels of experimental parameters, coded as +1, 0, and −1, respectively, and, thereafter, 81 experiments with all possible combinations of variables were conducted. The factors were coded so as to simplify the calculations (Simate and Ndlovu, 2008). For each run, 25 L of Cu^{2+} solution was passed through the column (height = 100 cm; diameter = 5 cm) made of glass at 5 L h^{-1} for 100 min. The results of the study clearly showed that ion-exchange process was effective in removing Cu^{2+} ions and can provide a solution for removal of such metals from AMD waste. The results were analysed statistically using Student's t-test, analysis of variance, F-test, and lack of fit so as to establish the most important process variables affecting the removal of Cu^{2+} ions by Indion 730 resins. In this study, pH was found to be the most important variable. It is noted from previous studies that the cation-exchange capacity and selectivity both increase with increasing pH, while the anion-exchange capacity and selectivity decrease with increasing pH (Churms, 1966). The interaction between the resin bed height and initial concentration of Cu^{2+} ions and that between the pH and initial concentration of Cu^{2+} ions were also found to be highly important.

Gordyatskaya (2017) explored the selective removal of toxic metal (Cu, Ni) ions present in AMD with excess ferrous ions using a chelating resin with a di-(2-picolyl)amine functional group (Lewatit TP 220) and the prospect of subsequent recovery of metallic copper with electrowinning. Simulated AMD solution used in the study contained 2 g/L of iron together with manganese, zinc and copper or nickel ions. Metals were added as sulphates and the pH was adjusted to 2 using sulphuric acid. Column dynamic experiments were used to determine the sorption and desorption efficiencies of Cu and Ni ions, separately. The chelating resin TP 220 demonstrated a very high affinity for copper, while manganese, zinc, and iron were not taken up during the experiments. Similar experiments performed related to the removal of nickel from the same AMD matrix, showed that the sorption of nickel was less efficient than that of copper. During the experiments for the removal of copper or nickel, small quantity of iron in the form of hydrated ferric oxide flakes formed a layer on the surface of the column. Therefore, a two-stage desorption process using sulphuric acid in the first step and ammonia solution in the second step was used. Unlike the Dowex XFS 43084 which is not produced anymore, Lewatit TP 220 (analogue of Dowex XFS 4195) could not be regenerated directly with sulphuric acid. In this process, the necessity to strip copper from the loaded resin by a complexation reaction with ammonia solution is the main drawback of the resin. This is because the resulting ammonia solution of copper is not suitable for the electrowinning of Cu. Therefore, a weak base anion exchanger having tetraethylenepentamine (TEPA) functional group on macroporous polyacrylate matrix Purolite A 832 was used as a chelating resin to take up copper from ammonia solution. Thereafter, once copper has been stripped off from TEPA using sulphuric acid, it can easily be recovered by electrolysis.

In view of the high toxicity of lead, its content in water and industrial waste-waters must be reduced to a minimum within the ppb level (Dąbrowski et al., 2004), thus several ion-exchange studies have been carried out to remove Pb(II) ions. Of significant importance is the simultaneous removal of Pb(II) and Cd(II) ions as well as organic ligands using anion exchangers of various types (Dudzińska and Clifford, 1991; Dudzińska and Pawłowski, 1993). Some results showed that the anion exchangers of weakly basic functional groups are characterised by higher affinity for the complexes of Pb(II) and Cd(II) with EDTA than strongly basic anion exchangers and that the anion exchangers of polyacrylic skeleton are characterised by greater affinity for the complexes (i.e., Pb (II) and Cd (II) with EDTA) than the anion exchangers of the same type of polystyrene skeleton (Dąbrowski et al., 2004). Other studies have shown that higher selectivity, great exchangeability as well as reversibility of the sorption-elution process towards Pb (II) ions are characteristics of the sulphine cation exchanger exhibiting-SO_4^{2-} groups (Bogoczek and Kociołek-Balawejder, 1988). On the other hand, chelating ion exchanger with functional iminodiacetate groups, Lewatit TP 207, has been recommended by the Bayer Company for selective removal of metal ions, particularly Pb(II) ions (Bayer, 2000).

The study by Ladeira and Gonçalves (2017) investigated the separation of uranium from the other anions present in the acid water under batch and column mode using the ion-exchange technique. Two strong base resins (Dowex Marathon A and IRA-910U) were compared in the study for their effectiveness in the removal of uranium from high sulphate AMD. The AMD sample was collected nearby the uranium mine in the southeast of Brazil and consisted of acid water generated at waste rock piles. The acid water pH was around 2.7, the uranium concentration was in the range of 6–14 mg L^{-1}, sulphate concentration was nearly 1400 mg L^{-1}, fluoride concentration of 140 mg L^{-1}, and iron concentration of 180 mg L^{-1}. The influence of ions, commonly found in acid waters like sulphate and fluoride, was also assessed in the ion-exchange process. The resins showed a significant capacity for uranium uptake which varied from 66 to 108 mg/g for IRA-910U and 53 to 79 mg/g for Dowex A. This shows that IRA-910U performed significantly better than Dowex A. However, the results showed that both resins performed at only about 40–60% of their theoretical value (1 equiv. g^{-1} for IRA-910U and 1.3 equiv. g^{-1} for Dowex A) probably because of the interference of other anions. The effect of pH on the loading capacity for column experiments was more accentuated than that for batch tests. The results also showed that SO_4^{2-} was the most interfering ion and it had a deleterious effect on the uranium recovery in the pH range studied. Fluoride did not affect uranium removal. Based on the fact that the study was carried out with a real acid water sample it was demonstrated that, although loadings were not considered so high, uranium can be removed efficiently, and elevated recoveries may be achieved.

9.2.4.3 Summary

The ion-exchange methodology has been found to be technologically simpler compared to other techniques and enables efficient removal of even traces of impurities from solutions. In fact, according to Hardwick and Hardwick (2016) when the concentration of impurities in a waste stream is very low, efficiency of removal by ion-exchange resins is relatively high because at that point it is film diffusion rather than particle diffusion that limits the kinetics. By its nature the undesirable ions in waste streams are replaced by the ones on ion-exchange resins, for example, that do not contribute to contamination of the environment (Dąbrowski et al., 2004). In other words, ion exchange enables replacing the undesirable ion by another one which is neutral within the environment. At the moment new types of ion exchangers with specific affinity to specific metal ions or groups of metals are available as an effort to enhance selectivity. To sum it up, the importance of ion exchange with respect to AMD treatment is characterised by two basic approaches according to Dinardo et al. (1991): (1) the selective removal of heavy metals from the AMD solution, and (2) water recovery to produce potable water, which involves the total removal of both anions and cations from AMD.

9.2.5 Electrochemical Treatment

9.2.5.1 Concepts of Electrochemical Treatment

The application of electricity in the treatment of contaminated water, generally referred to as the electrochemical techniques, has been in existence for over a century, and since then the methods are still highly reliable for wastewater treatment (Tran et al., 2017a). In fact, electrochemical technologies have the potential to provide selective and measurable recovery of base metals from dilute mining influenced water (Figueroa and Wolkersdorfer, 2014). However, the technology is not widely used in the treatment of AMD (Figueroa and Wolkersdorfer, 2014).

The nature of the electrochemical process is premised on the usage of electricity to pass a current through an aqueous metal bearing solution, which also contains a cathodic plate and an insoluble anode (Tran et al., 2017b). In other words, an electrochemical system consists of at least two

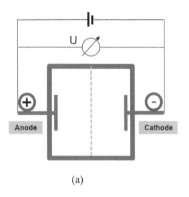

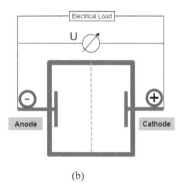

(a)   (b)

FIGURE 9.3
Schematic illustration of the (a) electrolysis process and (b) the galvanic element.

electrodes – an anode and a cathode – and an intermediate space filled with electrolyte (Muddemann et al., 2019). The electrical circuit is closed through electrical wires either with a voltage source (electrolysis cell) or an electrical load (galvanic element). In many applications, a separator (membrane or diaphragm) separates the reactor into anode and cathode compartments. The electrolyte surrounding the anode is named anolyte and the electrolyte on the cathode side is called catholyte. Figure 9.3a shows the general set-up of an electrolysis cell, while the function of a galvanic element is shown in Figure 9.3b.

There are a number of possible mechanisms involved in electrochemical processes including electrocoagulation, electroflotation, and electrodeposition (Chen, 2004; Fu and Wang, 2011; Figueroa and Wolkersdorfer, 2014; Liu et al., 2016; Tran et al., 2017a). Other electrochemical techniques are available including electrochemical oxidation (and reduction), electrochemical precipitation, electrokinetic, and emulsion splitting electrolysis (Chen, 2004; Drogui et al., 2007; Muddemann et al., 2019), but will not be covered in this chapter. Only electrocoagulation, electroflotation, and electrodeposition will be discussed in the chapter.

9.2.5.1.1 Electrocoagulation

The concept of electrocoagulation is based on the use of sacrificial metal anodes to stimulate electrolytic metal precipitation in undivided cells (Bejan and Bunce, 2015). Ideally, electrocoagulation (or sometimes called electroflocculation) involves the generation of coagulants in situ by dissolving either aluminium or iron ions electrically from aluminium or iron electrodes, respectively (Chen, 2004; Fu and Wang, 2011). However, in electrocoagulation, iron is the most widely used electrode followed by aluminium (Singh and Mishra, 2017). Chen (2004) outlined the following reactions as the ones taking place at the anode under alkaline and acidic conditions:

For the aluminium anode:

$$Al - 3e^- \rightarrow Al^{3+} \tag{9.4}$$

$$\text{in alkaline conditions } Al^{3+} + 3OH^- \rightarrow Al(OH)_3 \tag{9.5}$$

$$\text{in acidic conditions: } Al^{3+} + 3H_2O \rightarrow Al(OH)_3 + 3H^+ \tag{9.6}$$

For iron anode:

$$Fe - 2e^- \rightarrow Fe^{2+} \tag{9.7}$$

$$\text{in alkaline conditions: } Fe^{2+} + 3OH^- \rightarrow Fe(OH)_2 \tag{9.8}$$

$$\text{in acidic conditions: } 4Fe^{2+} + O_2 + 2H_2O \rightarrow 4Fe^{3+} + 3OH^- \tag{9.9}$$

In addition, oxygen is also liberated at the anode through the following reaction:

$$2H_2O - 4e^- \rightarrow O_2 + 4H^+ \tag{9.10}$$

During electrocoagulation, water electrolysis at the cathode also takes place and generates hydrogen (Equation 9.11) in form of microbubbles (Muddemann et al., 2019).

$$2H_2O + 2e^- \rightarrow H_2 + 2OH^- \tag{9.11}$$

In addition, the dissolved H^+ ions are also reduced at the cathode as follows:

$$2H^+ + 2e^- \rightarrow H_2 \tag{9.12}$$

From reactions 9.1 to 9.12, it can be summarised that during the electrocoagulation process, the generation of metal ions takes place at the anode and the hydrogen gas is released from the cathode (Chen, 2004).

In the electrocoagulation process, co-precipitation of the sacrificial metal (reactions 9.4 and 9.7) and the metals present in the AMD take place due to pH increase because of the cathodic reduction of H^+ ions (Equation 9.12) (Bejan and Bunce, 2015) or due to the OH^- formed at the cathode through reaction 9.11 (Muddemann et al., 2019). Ideally, the reduction of dissolved hydrogen ion to hydrogen gas as shown in reaction 9.12 results in a pH increase in solution (Jenke and Diebold, 1984). In addition, Muddemann et al. (2019) as well as

Bejan and Bunce (2015) postulate that the OH⁻ ions formed at the cathode enable the coagulation or precipitation of the metal ions present in the AMD and metal ions generated from sacrificial metal anodes as well as flocculation of dissolved or colloidal constituents in water. Mollah et al. (2004) also argue that during electrocoagulation, iron and aluminium almost instantly become polymeric hydroxides, which are excellent coagulating agents. Drogui et al. (2007) propose that in an electrocoagulation process, aluminium or iron ions are produced and they combine with hydroxyl ions (generated by electrolysis of water at the cathode) to form a polymeric coagulant, aluminium hydroxide, or iron hydroxide, which adsorbs colloidal material present in the solution to form insoluble flocs which are then carried away by the hydrogen gas bubbles generated at the cathode to the surface of the liquid. According to Heidmann and Calmano (2008), in the case of aluminium, the metal ions are removed from the solution by several mechanisms including direct reduction at the cathode, hydroxide formation by the hydroxyl ions formed at the cathode, and co-precipitation with the aluminium hydroxides.

In the electrocoagulation process, the hydrogen gas formed at the cathode (reaction 9.12) helps to float the flocculated particles out of the water (Chen, 2004). Muddemann et al. (2019) also re-emphasise that the hydrogen bubbles formed are often used for flotation (i.e., electroflotation) and separation of the formed aggregates simultaneously. Therefore, according to Muddemann et al. (2019) electroflotation is often combined with electrocoagulation. Feng et al. (2016) reiterate that in practice, an electrocoagulation process will be often followed by an electroflotation process and this combined system can be considered as electrocoagulation-flotation process. In other words, the combination of electrocoagulation and electroflotation is called electrocoagulation-flotation process (Azimi et al., 2017). This method achieves better removal percentages.

Chen (2004) suggests that the advantages of electrocoagulation include high particulate removal efficiency, compact treatment facility, relatively low cost, and possibility of complete automation. In addition, Vasudevan et al. (2010) state that some of the advantages of electrocoagulation are its generally low cost, reduced sludge production, and is easy to operate. According to Rodriguez et al. (2007), electrocoagulation may prove to be not only feasible and economically friendly, but also technically and economically superior to conventional technology like chemical precipitation.

9.2.5.1.2 *Electroflotation*

Electroflotation is a separation process in which hydrophobic particles in water or particles generated by other processes (e.g., electrocoagulation) are carried to the aqueous surface by adhering to gas bubbles (Muddemann et al., 2019). In the context of this chapter, electroflotation is considered as a simple solid/liquid separation process that floats pollutants to the surface of a water body by tiny bubbles of hydrogen and oxygen gases generated from water electrolysis (Raju and Khangaonkar, 1984; Fu and Wang, 2011;

Chen, 2014). In this process, hydrogen and oxygen gases are liberated from the electrochemical reactions at the cathode (reaction 9.11) and anode (reaction 9.10), respectively (Chen, 2014). It is also possible to combine electroflotation with electrocoagulation (see the discussion on electrocoagulation) if one of the electrodes is dissolved by electric current during electrolysis. The released metal ions cause coagulation of colloidal molecules, which adhere to the gas bubbles formed by the water electrolysis (Muddemann et al., 2019). According to Kraft (2004) electroflotation is one of the most effective and versatile methods of electrochemical water purification, as micro gas bubbles are produced and the size distribution of the gas bubbles is very narrow. In addition, Srinivasan and Subbaiyan (1989) state that the advantages of electroflotation technique are its high degree of gas saturation and the dimensional uniformity of the generated bubbles which subsequently lead to the highest recovery in a short time span.

9.2.5.1.3 *Electrodeposition*

The recovery of metals using electrochemical techniques has been practiced in the form of electrometallurgy for a very long time (Dubpernel, 1978; Chen, 2004). Figueroa and Wolkersdorfer (2014) also state that electrochemical deposition of metals is widely used in metallurgical processing and treatment of high metal waste streams (e.g. electroplating waste) to recover high-purity metallic forms. In the process of electrochemical deposition of metals, the metal ions are electrochemically reduced and, in contrast to electrocoagulation, removed as metals of valence 0 (Muddemann et al., 2019). The electrochemical mechanism for metal recovery is very simple and is basically the cathodic deposition process (Chen, 2004). In the process, the positively charged metal ions move in the electric field between the electrodes to the negatively polarised cathode, where they are reduced to an element and deposited on the electrode surface according to the following equation (Muddemann et al., 2019):

$$M^{n+} + ne^- \rightarrow M \tag{9.13}$$

In other words, when the dissolved metals migrate towards the cathode, they are reduced and deposited on the cathode (Figueroa and Wolkersdorfer, 2014). In general, the anode composition (effectively inert) is selected to limit its oxidation and thus promote the oxidation of water at the anode (Figueroa and Wolkersdorfer, 2014).

The metals extracted from the wastewater such as AMD using electrodeposition technique can be of a very high purity and thus the recovery of valuable metals is possible (Muddemann et al., 2019). The process is used not only for metal separation from wastewater, but also for large-scale production of metals such as copper and zinc. Indeed, electrodeposition which is commonly referred to as electrowinning is an alternative to conventional smelting technologies (Bejan and Bunce, 2015).

9.2.5.2 Selected Typical Studies of Electrochemical Treatment

This section discusses typical examples of studies performed using electrochemical techniques for the treatment and/or recovery of metals from wastewaters including AMD. Electrochemical treatment of AMD offers possible advantages in terms of operating costs and the opportunity to recover metals (Chartrand and Bunce, 2003; Gaikwad and Gupta, 2008; Figueroa and Wolkersdorfer, 2014; Park et al., 2015). The technique can be operated at ambient temperature and pressure and has a robust performance and capability to adjust to variations in the influent composition and flow rate (Tran et al., 2017b). Furthermore, Tran et al. (2017a,b) including Fu and Wang (2011) emphasised that electrochemical processes are known to be very efficient methods for the treatment of industrial wastewaters, particularly, for the removal of heavy metal ions.

9.2.5.2.1 Electrocoagulation

Nariyan et al. (2017) used electrocoagulation to investigate the removal of copper, silicon, manganese, aluminium, iron, and zinc as well as sulphate from real mine water. Batch experiments with monopolar iron anode and stainless steel cathode as well as monopolar aluminium anode and stainless steel cathode were conducted separately to identify the best electrocoagulation conditions. The removal efficiency in mine water increased with increasing reaction time and increasing current density and the type of electrodes affected the metals and sulphate removal as could be shown by the adsorption isotherms. Based on kinetic modelling, the aluminium electrode was found to be more efficient for metal removal than the iron electrode. The removal behaviour of the metals can be explained by the E_h-pH and the metal's stability diagram. Both the free sorption energy calculation from the Dubinin-Radushkevich isotherm and the k-parameter for the kinetics showed that the removal of copper and silicon were influenced by physical interaction with the two electrodes, while zinc was merely being influenced by physical interaction with the iron electrode in the electrocoagulation process. All of the contaminants, except manganese and sulphate, obeyed a Langmuir isotherm when an iron electrode was applied. However, when an aluminium anode was utilised, the metals presented a different behaviour compared to the iron electrodes. Specifically, silicon, copper, and manganese were obeying a Freundlich isotherm, whereas zinc and sulphate were obeying a Langmuir isotherm. Sulphate was better removed by an aluminium electrode compared to the iron electrodes with a maximum removal rate of 40.5% and 28.9%, respectively.

Mamelkina et al. (2017) studied the possibility of treating mining water using 1 L and 70 L electrocoagulation reactors and pressure filtration. The results indicated that metals were almost completely removed after 1 h of operation. The highest sulphate removal was obtained using aluminium electrodes after 5 h of treatment, while for nitrate after 3 h. Basically, nitrates were also almost completely removed just like metals. No significant effect of current density, reactor configuration and electrode material on metal removal was observed within the variable range investigated. However, sulphate and nitrate removal declined with the increase in treated volume. Cake formed during filtration had high porosity and moisture content.

Venkatasaravanan et al. (2016) investigated the efficiency of electrocoagulation process on the removal of heavy metals from synthetically prepared AMD. The electrocoagulation studies were performed in batch mode using vertically positioned aluminium electrodes (anode and cathode) in a 1 L reactor connected to a DC supply of 0–30 V and 5 A. The reactor was mixed at 200 rpm to avoid the mass transport over the potential of electrocoagulation reactor. As current density and pH were increased from 10 to 25 mA/cm^2 and 5 to 7, respectively, the removal efficiency of Cu and Zn increased drastically from 57.6% to 95.4% and 53.4% to 86.2%, respectively. The energy consumption in the electrocoagulation process varied from 4.7 kWh/m^3 to 29.4 kWh/m^3 for current densities of 10 mA/cm^2 to 25 mA/cm^2. Clearly, this study proved that the electrocoagulation is an efficient process.

A study by Nariyan et al. (2016) investigated the removal of cadmium from real mine water by the electrocoagulation process using iron–stainless steel anode/cathode electrode combinations as well as aluminium–stainless steel anode/cathode electrode combinations. The effects of time, current density, and the type of electrode on the performance of electrocoagulation process were investigated. It was found that the current density had a direct effect on the removal of cadmium. In particular, cadmium was removed better at 70 mA/cm^2 than at 10 mA/cm^2. In addition, the reaction time had a direct effect on cadmium removal. For example, by increasing the time, cadmium was removed at higher removal rates compared to the beginning of the reaction. The type of electrode was also found to have an influence on the removal of cadmium. For example, cadmium was removed much better by an iron–stainless steel anode/cathode combination than by a combination of aluminium–stainless steel anode/cathode electrodes. The removal efficiency of the aluminium–stainless steel anode/cathode combination reached 82%, whereas the cadmium removal efficiency by iron–stainless steel was 100% at 120 min of reaction and 70 mA/cm^2. The best condition in which 100% of cadmium was removed was obtained by using an iron–stainless steel anode/cathode electrodes combination with a current density and reaction time of 70 mA/cm^2 and 120 min, respectively.

The aim of the study by Orescanin and Kollar (2012) was to develop and apply the purification system suitable for the treatment of the AMD accumulated in the "Robule" Lake, which represents the part of the Bor copper mining and smelting complex in Serbia. The study was undertaken in order to minimise adverse effects on the environment caused by the discharge of untreated AMD, which was characterised with low pH value (2.63) and high concentration of heavy metals (up to 610 mg/L) and sulphates (up to 12 000 mg/L). The treatment of the effluent included pre-treatment/pH adjustment with CaO followed by electrocoagulation using iron and aluminium electrode sets. The results showed that the application of the electrochemical method for the treatment of the AMD pre-treated/pH adjusted with CaO resulted in extremely high removal efficiencies of heavy metals from the waste effluent of above 99% in most cases. The removal efficiency increased with increasing initial metal concentration. High degree of the removal of sulphates (over 70%) was also achieved. The concentration of iron in the treated effluent was reduced from 610 to 0.010 mg/L, copper from 82.5 to 0.006 mg/L, manganese from 59 to 0.336 mg/L, and zinc from 41.6 to 0.024 mg/L. The concentrations of other heavy metals in the final effluent were below 0.005 mg/L. The removal efficiencies for the metals and sulphates from the studied AMD are summarised as follows: SO_4^{2-} = 70.83%, Hg = 98.36%, Pb = 97.50%, V = 98.43%, Cr = 99.86%, Mn = 97.96%, Fe = 100.00%, Co = 99.96%, Ni = 99.78%, Cu = 99.99%, and Zn = 99.94%. Since the concentrations of heavy metals in the electrochemically treated AMD (ranging from 0.001 to 0.336 mg/L) are very low, the negative impact of this effluent on the aquatic life and humans is not expected. The waste sludge from the combined treatment process could be reused for the pH adjustment/pre-treatment of the AMD instead of CaO, and afterwards, due to its inertness, it could be used as an overlaying layer of the flotation waste heap during its recovery work. From the presented results, it could be concluded that electrochemical treatment is a suitable approach for the treatment of AMD.

A study by Oncel et al. (2013) compared the removal of heavy metals such as Fe, Al, Ca, Mg, Mn, Zn, Si, Sr, B, Pb, Cr, and As from coal mine drainage wastewater at a laboratory scale using chemical precipitation and electrocoagulation. The chemical precipitation was performed with NaOH, whereas the electrocoagulation process was evaluated via an electrolytic cell using iron plate electrodes. In the chemical precipitation process, the optimum pH for removal of most of the heavy metals from coal mine drainage wastewater was 8 except for Ca, Sr, and B (pH 10 or higher). The removal efficiencies at the optimum pH varied from 28.4% to 99.96%. Influence of current density and operating time in the electrocoagulation process was explored on the removal efficiency and operating cost. Results from the electrocoagulation process showed that the removal of metals present in coal mine drainage wastewater increased with increasing current density and operating time. The electrocoagulation process was able to achieve higher removal efficiencies (>99.9%) at an electrocoagulation time of 40 min, a current density of 500 A/m² and pH of 2.5 as compared to the results obtained with the chemical precipitation at pH 8. The residual metal ion concentrations which varied from 0.00001 to 0.104 mg/L in the electrocoagulation process were below the limiting value for coal mine drainage wastewater discharge. The operating costs at the optimum operating conditions were also determined to be 1.98€/m³ for the electrocoagulation and 4.53€/m³ for the chemical precipitation. There is no doubt from the results that the electrocoagulation process was more effective than the chemical precipitation with respect to the removal efficiency, amount of sludge generated, and operating cost. The study by Oncel et al. (2013) has clearly shown that electrocoagulation has the potential to extensively eliminate disadvantages of the classical treatment techniques in order to achieve a sustainable and economic treatment of polluted wastewater.

Chartrand and Bunce (2003) performed the electrolysis of synthetic AMD solutions containing iron, copper, and nickel both singly and mixtures of the metals using a flow-through cell divided with an ion-exchange membrane. The anode used for all experiments was a dimensionally stabilised anode (DSA) consisting of titanium metal coated with iridium dioxide ($\approx$7.2 cm²), and the cathode was made of platinum. Each set of experiments was carried out at constant current densities at the highest value of which the steady-state pH of the exiting catholyte was about 10 for iron only solutions or 12 for all other solutions. The results showed that iron was successfully removed from a synthetic AMD solution composed of $FeSO_4/H_2SO_4$ via $Fe(OH)_3$ precipitation outside the electrochemical cell after oxygenation of the catholyte. The experiments with copper and nickel were only partly successful due to the fact that metal removal occurred more by hydroxide precipitation than electrocoagulation. The work was extended to an authentic AMD sample containing principally iron and nickel. Electrolysis of authentic AMD was successful in removing iron from solution, but quantitative removal of nickel required re-electrolysis or chemical precipitation. In the study, Chartrand and Bunce (2003) noted that the development of an electrolytic technology for AMD remediation requires more work on the chronology of electrolysis, aeration, and sludge separation, and on cell design so as to optimise mass transfer and permit the in situ separation of the sludges formed when the original AMD contains significant quantities of Fe^{3+}.

Both mine water and industrial effluents are known antecedent of heavy metals (Singh and Mishra, 2017). Therefore, other studies that are important in the use of electrocoagulation technique involved the removal and/or recovery of heavy metals from wastewaters of other industries. For example, a study by Akbal and Camcı (2011) investigated the applicability of an electrocoagulation method in the treatment of metal plating wastewater under various conditions. The results indicated that electrocoagulation can effectively reduce metal ions to very low levels. The metals (Cu, Cr, and Ni) were removed by precipitation as hydroxides by the hydroxyl ions formed at the cathode via water electrolysis (see reaction 9.11) and by co-precipitation with aluminium and iron hydroxides. The results also showed that Cu, Cr, and Ni removal efficiency increased with increasing current density. This was because higher current density increased the production rate of aluminium and iron hydroxide flocs, and that the hydroxide flocs subsequently acted as adsorbents for metal ions and thus removed the metal ions from the wastewater. The highest removal rate for Cu, Cr, and Ni was achieved at a pH of 9.0. The high efficiency of metal removal at higher pH levels was attributed to the precipitation of their hydroxides at the cathode. The Fe–Fe and Fe–Al electrode combinations were more effective for the removal of Cu, Cr, and Ni from the wastewater. The results also indicated that electrocoagulation with Fe–Al electrode pair was very efficient and was able to achieve removal rates of 100% Cu, 100% Cr, and 100% Ni at a current density of 10 mA/cm^2 and pH of 3.0 after an electrocoagulation time of 20 min with energy and electrode consumptions of 10.07 kWh/m^3 and 1.08 kg/m^3, respectively.

A systematic study to ascertain the performance of an electrocoagulation system with aluminium electrodes for removing heavy metal ions (Zn^{2+}, Cu^{2+}, Ni^{2+}, Ag^+, $Cr_2O_7^{2-}$) on laboratory scale was carried out by Heidmann and Calmano (2008). Several parameters – such as initial metal concentration, numbers of metals present, charge loading, and current density – and their influence on the electrocoagulation process were investigated. The study was able to establish the removal mechanisms of Zn, Ni, Cu, Ag, and Cr. The results showed that Zn, Ni, Cu, and Ag are removed by direct reduction at the cathode surface, as hydroxides by the hydroxyl ions formed at the cathode via water electrolysis (see reaction 9.11) and by co-precipitation with aluminium hydroxides. It was proposed that Cr(VI) was reduced first to Cr(III) at the cathode before precipitating as a hydroxide, $Cr(OH)_3$. The results from experiments with five metals indicate a co-precipitation of Cr with the other metals.

Kabdaşlı et al. (2009) experimentally investigated the treatability of a metal plating wastewater (concentration range, 230–280 mg/L) containing complexed metals originating from the nickel and zinc plating process through electrocoagulation technique using stainless steel electrodes. In this study, nickel and zinc were removed by hydroxide precipitation and incorporation in the colloidal material generated by the formation of $Fe(OH)_3$ flocs. The study by Kabdaşlı et al. (2009) demonstrated that the highest TOC abatement (66%) as well as nickel and zinc removals (100%) were achieved with an applied current density of 9 mA/cm^2 to the original electrolyte (chloride) concentration and original pH of the composite sample used.

A study by Shafaei et al. (2015) investigated the removal of Mn^{2+} ions from solutions by an electrocoagulation process with aluminium electrodes. The study found that Mn^{2+} ion was removed by direct reduction at the cathode surface, as hydroxides by the hydroxyl ions formed at the cathode via water electrolysis and by co-precipitation with the aluminium hydroxides. It was found that the optimum initial pH for the removal of Mn^{2+} ions was 7.0. The results from the study also showed that increasing the current density and electrolysis time has a positive effect on the Mn^{2+} removal efficiency. The removal of Mn^{2+} ions was not influenced by the solution conductivity whilst the electrical energy consumption decreased with an increase in the solution conductivity. In addition, the study also found that as the initial concentration of the Mn^{2+} increased, the rate of removal of the contaminant decreased.

A study that investigated electrocoagulation treatment using aluminium sacrificial electrodes for a synthetic water containing Cu, Cr, and Zn heavy metals was conducted by Singh and Mishra (2017). The effects of operational parameters, such as current density, inter-electrode distance, operating time, and pH, were studied and evaluated for maximum efficiency. This study showed that experimental results as well as kinetic modelling data gave high removal rate for all metals and total suspended solids at higher current density except Cu in which the same results were obtained at lower current density. In the case of energy consumption, it was concluded that 0.459 kWh/m^3 was sufficient for the removal of 99% Cu, 59% Cr, and 71% of Zn up to 30 min of treatment time for which 0.450 A current was required. In the context of sludge generated, the study showed that the sludge that accumulated on the top layer of solution was more in comparison to amount of sludge at the bottom of the reactor. In addition, sludge generated was less at bottom at high current density.

9.2.5.2.2 Electroflotation

According to Feng et al. (2016) and many other studies (Azimi et al., 2017; Muddemann et al., 2019), an electrocoagulation process is often followed by an electroflotation process. A combination of electrocoagulation and electroflotation, which actually achieves better removal percentages, is called electrocoagulation-flotation process (Azimi et al., 2017). As already discussed, electroflotation is a simple process where pollutants are floated to the surface of a water body with the aid of tiny bubbles of hydrogen and oxygen gases generated from water electrolysis. Therefore, most of the examples in this section of the chapter pertain to studies where electroflotation was combined with other water treatment systems. In other words, electroflotation is mainly integrated into a process train with other technologies, particularly, electrocoagulation.

The possibility of the removal of metal ions from mining wastewaters through ion flotation was investigated by Alexandrova et al. (1994). Wastewater of an opencast mine containing about 50 mg/L copper ions underwent precipitation with xanthates forming chelate complexes with high hydrophobicity. Electroflotation was used to generate a gaseous phase with sufficient volume and high dispersity so as to effect precipitate flotation. Alexandrova et al. (1994) argue that effective precipitation and adsorbing colloid flotation require the presence of a gaseous phase with sufficiently large area and maximal dispersity due to the sensitivity of precipitated and co-precipitated particles to hydrodynamic conditions. Electroflotation proved appropriate in this study since it combined the electrocoagulation effect and electrolytic gas separation.

Khelifa et al. (2005) used electroflotation to reduce the concentrations of copper and nickel found in real wastewater. Through the electro-generation of gas bubbles (hydrogen and oxygen) at the electrodes and the variation of pH, this technique allowed the precipitation of hydroxides of the polluting metals by alkalisation and subsequent transport by flotation to the surface of the solution (Srinivasan and Subbaiyan, 1989; Alexandrova et al., 1994; Muddemann et al., 2019). The effects of the following parameters were examined: current density, pH, heavy metal concentration, supporting electrolyte concentration, and the nature of the electrodes. By optimizing the operation, heavy metal removal reached 98–99% and maintained final and global concentration to a value lower than the World Health Organisation standard, which is 1 mg/L for nickel and copper. In a later study, Khelifa et al. (2013) demonstrated the feasibility of simultaneous removal of heavy metals and EDTA in an electrolytic undivided cell equipped with Ti/RuO_2 as anode and stainless steel as cathode. In the absence of EDTA, results showed that nickel and copper removal by electroflotation process is pH sensitive; and nickel and copper were substantially removed by electroflotation with removal efficiencies of 99.6% and 97%, respectively. In the presence of EDTA, the metal removal by the electroflotation process was inhibited. The inhibition rate was found to be dependent on EDTA/metal molar ratio. In the study, Khelifa et al. (2013) also used a one-step process, involving the combination of two techniques, i.e., electrochlorination and electroflotation. Active chlorine generated in situ allowed the decomplexation of metal-EDTA. As a result, free metal ions were removed by precipitating and subsequent floating to the surface by rising electro-generated bubbles. The obtained results revealed that, with 0.6 EDTA/metal molar ratio, removal efficiencies were 77% and 78% for nickel and EDTA, respectively, in the case of nickel–EDTA solutions. Removal efficiencies were 89% and 96% for copper and EDTA, respectively, in the case of copper–EDTA solutions. Furthermore, heavy metal removal efficiency by the combined process showed that it was affected by chloride content and current intensity.

The objective of a study by da Mota et al. (2015) was to remove Pb, Ba, and Zn ions from solutions containing 15 mg dm^{-3} of each metal representing a typical concentration of wastewater from washing soil contaminated by drilling fluids from oil wells by electrocoagulation/electroflotation using stainless steel mesh electrodes. The effects of different parameters, including the pH, the electrolysis time, the current density, and the supporting electrolyte dosage were evaluated. The results of the study indicated that it is possible to remove Pb, Ba, and Zn metals by electrocoagulation/ electroflotation and achieving up to 97% removal efficiency with a power consumption of 14 kWh m^{-3}. The optimal conditions of the treatment process were 0.1% sodium dodecyl sulphate (as a foamy) in a molar ratio against the heavy metals of 3:1, current density of approximately 350 Am^{-2}, ionic strength of 3.2×10^{-3} M, pH of 10.0, and 20-min operation time. The results of the study indicated that the proposed electrocoagulation/electroflotation was adequate to simultaneously treat the common heavy metals found in the drilling fluids from oil wells.

9.2.5.2.3 Electrodeposition

Electrowinning, as a form of electrodeposition, is the deposition of a metal from solution due to an applied electrical potential. The dissolved metals migrate towards the cathode where they are reduced and deposit on the cathode. In general, the anode composition (effectively inert) is selected to limit its oxidation and thus promote the oxidation of water at the anode (Figueroa and Wolkersdorfer, 2014). Most importantly, electrochemical recovery processes, particularly electrowinning, have been applied primarily to high metal concentration solutions ($\gg$1000 mg/L) at low pH (<1) in waste streams where competitive metals tend to be at much lower concentrations than the target metal (e.g., electroplating waste) (Figueroa and Wolkersdorfer, 2014). Nordstrom et al. (2017) also reiterate that electrowinning for Cu is generally not done on AMD because concentrations of nearly 1000 mg/L are necessary.

In a study by Gorgievski et al. (2009), the removal of copper from AMDs originating from a closed copper mine "Cerovo" RTB Bor in Serbia containing approximately 1.3 g dm^{-3} of copper and a very small amount of Fe^{2+}/Fe^{3+} ions was successfully performed by direct electrowinning using either a porous copper sheet or carbon felt as the cathode. The cells used in the electrowinning experiments were compared in terms of cell voltage, pH, and copper concentration. A high degree of electrowinning, higher than 92% copper removal rate, a satisfactorily good current efficiency (>60%) and a good, dense metal deposit was obtained with both cathodes. However, the cell with the porous copper cathode had better features than that with the carbon felt cathode in terms of current efficiency and specific energy consumption. Depending on the process time and the applied current, a final copper concentration in the remaining solution of less than 0.1 g dm^{-3} was achieved. The specific energy consumption was approximately 7 kWh kg^{-1} of deposited copper. A dense copper deposit was obtained when a three-dimensional electrode was used. The cell voltage decreased with time due to decreasing pH as a consequence of oxygen evolution as an anode reaction which increases the acid content. When the cells were compared with respect to the achieved cell voltage, it showed that the cell with the porous copper sheet cathode is the most suitable, having the lowest cell voltage. When using the carbon felt cathode, high cell voltage was registered due to the increased potential drop within the cell as well as the higher overvoltage of copper deposition onto carbon. The decrease in pH of the treated mine water with time due to the anodic oxygen evolution causes an increase in conductivity and acidity of the mine water. This fact may have a beneficial effect if the treated water is being recycled in a leaching stage, if present, leading to a decrease in the consumption of sulphuric acid. The opposite effect is an elevated consumption of chemicals needed to neutralise sulphuric acid formed during the electrowinning process prior to its release into a receiving water course.

Stanković et al. (2008) ran experiments with the aim of generating data which would be needed in designing the mine waters' treatment process by direct electrowinning. The mine waters from the open pit mine were characterised with a reasonably high copper concentration ($\approx$1 g dm^{-3}) and with a very low iron ion content allowing the consideration of a direct electrowinning as a method for the copper recovery from the mine waters. Zinc ions also existed in an unexpected amount of about 25 mg dm^{-3}, but do not affect the electrowinning process. The removal of copper from mine waters was performed using two types of electrochemical cells: a cell with inert turbulence promoters and a cell with copper foam both operating in galvanostatic mode at different current densities and at different hydrodynamic conditions. There was no pre-treatment of the mine waters prior to the electrowinning process. Copper concentration was monitored periodically and the cell voltage was recorded thus allowing energy consumption in the process to be determined. The results of the study clearly showed that it was possible to remove copper successfully from the mine waters by direct electrowinning. The process achieved high degree of electrowinning and satisfactory current efficiency. Dense metal deposit was obtained in both electrolytic cells. Cell voltage was considerably high in both cells, but was much higher in the cell with inert turbulent promoters up to 12 to 14 V at the very beginning. However, the voltage fell down over time due to the lowering of pH with time as a consequence of the evolution of oxygen as an anode reaction that increases acid content in the water. Both cells used in the study were compared in a view of their applicability for treatment of mine water. Copper foam allows the application of higher geometrical current densities because of its developed internal surface. This allowed the application of higher operating currents on cells with copper foam and thus achieving lower final copper concentrations in the outlet stream compared to the cell with inert turbulence promoters.

The removal of copper and nickel from aqueous solutions using a simple tank cell, an improved mass transfer inert fluidised bed cell (Chemelec) and a three-dimensional high surface area cell (i.e., graphite packed bed cell) were examined by Campbell et al. (1994). A series of electrolysis experiments were carried out to examine the effect of current density and flow rate on both the efficiency of the process and the quality of the deposit. The packed bed cells were found to be extremely effective in reducing metal concentrations in the remaining solution to below 1 ppm. The most economic concentration range in which each cell should operate was also determined. The study found that the electrochemical recovery of nickel and copper over the concentration range of 20 g/L to less than 1 ppm can be carried out at high current efficiency if the concentration of metal ions is reduced in stages using the three types of electrochemical cells. At concentrations from 20 g/L to 2 g/L a tank cell should be used, 2 g/L to 0.2 g/L a Chemelec cell is required and below 0.2 g/L a graphite packed bed should be used followed by a high surface area cell to reduce the metal level to below 1 ppm. These results show that suitable combinations of cells can be chosen to remove the metals from high concentrations (20 000 ppm) to very low concentrations (<1 ppm).

Ubaldini et al. (2013) demonstrated the technical feasibility of electrowinning as a remediation process for toxic metals removal from AMD. In this study, high recoveries of metals were achieved of about 99% and 93% for Zn and Mn (as MnO_2), respectively, with relatively low power consumptions (i.e., 118 kWh/kg for Zn and 619.05 kWh/kg for MnO_2). The other metals such as Cu and traces of Cd, Ni, Mn (as metallic Mn) co-deposited with the Zn such that the levels of the remaining metals in solution were within the recommended limits suggested from the Peruvian law. The degree of purity of Zn was over 90%. Although all the studied metals deposited on the cathode, manganese deposited on the anode as MnO_2 except traces of metallic Mn that co-deposited with Zn on the cathode. According to Veglio et al. (2003), the following reactions occur at the electrodes during Mn deposition in the electrowinning process:

$$\text{Cathode: } Mn^{2+} + 2e^- \rightarrow Mn; \ E^0 = -1.18V \tag{9.14}$$

$$\text{Anode: } Mn^{2+} + 2H_2O \rightarrow MnO_2 + 4H^+ + 2e^-; \ E^0 = +1.23V \tag{9.15}$$

In a comparative study of manganese removal from pre-treated AMD, Mačingová et al. (2016) used three methods – precipitation using sodium hydroxide alkaline, oxidative precipitation using potassium permanganate, and electrowinning for anodic Mn recovery in MnO_2 form.

The results showed that the three methods are effective and manganese was removed from AMD with levels of the remaining metals in solution complying with environmental requirements. However, when sodium hydroxide was used as a reagent, co-precipitation of manganese and magnesium present in AMD was observed. Therefore, the alkalisation process did not show sufficient selectivity. Oxidative precipitation by potassium permanganate resulted in an enhanced selectivity and purity of the obtained manganese precipitates was achieved. In the process of electrowinning, over 95% of Mn was deposited as MnO_2 with a high-grade degree of purity of about 99% having been attained. With reference to reaction 9.14 and reaction 9.15, in the electrowinning process, manganese was deposited on the anode as MnO_2 as already stated, while only a small amount was deposited on the cathode as Mn.

A number of studies (Marracino et al., 1987; Stanković et al., 2008) have tested the attractiveness of metal foam electrodes for electrolytic treatment of dilute solutions containing metal ions. Panizza et al. (1999) used copper foams as high surface area cathodes in order to remove copper (II) from an industrial effluent that came from the filtration unit in the plant of copper phthalocyanine, which is produced by a discontinuous process from phthalic anhydride, urea, and cupric chloride in Italy. The electrochemical reactor used was an undivided monopolar cell. A series of experiments in a batch recycle mode were performed in order to find the trend of copper removal and current efficiency over time at different flow rate conditions. The influence of initial metal concentration on the removal of copper (II) and current efficiency was also evaluated. The results confirmed that copper foams can be successfully used as cathode materials for copper (II) removal from an industrial effluent. The study found that the best performance was obtained with the higher flow rate which is evident of a process under mass transport control. Working with a flow rate of 1000 L/h, a recovery of copper ions from feed solution greater than 98% was achieved. To obtain such a high copper removal, a very low current efficiency was used due to the great amount of organics present in the wastewater that could lead to the fouling of the electrodes. Furthermore, the study results indicated that the initial copper concentration had no influence on the rate at which metal ions were removed, but the current efficiency increased with initial copper concentration. In a similar and/or follow-up study, Panizza et al. (1999) used a cell that contained a series of 50 cathodes and 51 anodes (48 × 36 cm). The cathodes were vertical polymeric foam (0.6 cm thick) covered with a thin outer layer of copper, and the anodes were O_2-evolving DSA® coated titanium mesh. The cell also had a provision for an air-sparging system to improve mixing of the solution. The cell can be considered as consisting of a cascade of 50 continuous stirred tank reactors (CSTR). The solution flowed through a porous cathode and the electrochemical reactions took place. Afterwards, the solution arrived in the chamber, between two cathodes, where it was mixed by the air sparging, before passing through the following electrode. A cascade with such a high number of CSTR can be well approximated with an ideal plug flow reactor (PFR).

The results of the study by Panizza et al. (1999) have shown that a tank reactor with metal foam cathodes can be approximated successfully to a PFR in order to predict its performance for copper removal from an industrial effluent. High fractional conversion was reached in a single pass, but it was necessary to use moderate flow rate and in this way there was a low cathodic current efficiency (e.g., at 100 L/h, the fractional conversion of copper (X_A) was 86% and the current efficiency (η) was 3.5%). An acceptable compromise was obtained at 500 L/h, having a current efficiency of 16.5% and a fractional conversion of 76.5%. Moreover, the presence of microcrystals of phthalocyanine and sulphates ions in the effluent that was being treated in the study caused the poisoning of the cathode surface, thus increasing the cell voltage and consequently decreasing the current efficiency due to the secondary reactions. In addition, the quality of the copper deposit was poor since it contained copper salts and phthalocyanine that compromised the exploitation of the fully loaded cathodes and anodes in the plating processes. The study indicated that the foam cathodes could be regenerated for reuse by acid chemical stripping. Panizza et al. (1999) suggested that in order to improve the process, more attention should be addressed to the filtration unit before the electrochemical cell, so as to reduce the concentration of phthalocyanine and sulphates in the effluent that were the major causes of low current efficiency and the poor quality of the copper deposit. The study also showed that the presence of chloride ions did not influence copper removal, and no evolution of toxic chlorine was detected because of the use of O_2-evolving anodes. This fact was confirmed as chloride concentration remained almost constant during the electrolysis.

9.2.5.3 Summary

Chen (2004) observed that electrochemical techniques have been used for wastewater treatment for over a century. More specifically, electrochemical processes have been utilised for many years to extract metals from aqueous wastes for reuse (Ali, 2011), and in the recent past electrochemical metal recovery has attracted increasing attention as compared to normal electrochemical metal removal (Jin and Zhang, 2020). There are a lot of electrochemical techniques being used for various purposes. However, this chapter focused particularly on electrocoagulation, electroflotation, and electrodeposition for the treatment and/or recovery of metals from wastewaters including AMD. According to Ali (2011), metal recovery via electrochemical techniques has the advantages that: (1) additional chemicals are not required, (2) selective metal recovery is possible taking into consideration thermodynamic and kinetic requirements of each species involved, (3) the metals can be recovered in their metallic form, and (4) the processes tend to operate at low temperature and pressure, specifically at room temperature and pressure of 25°C and 1 atm. In addition, electrochemical methods have attracted considerable attention because of their environmental compatibility, high efficiency, selectivity, versatility, feasibility, and cost effectiveness via the employment of the green redox reagent "electron" (Meunier et al., 2006; Jin and Zhang, 2020). Despite several advantages, there are also many bottlenecks associated with electrochemical metal recovery such as the concentration polarisation of dilute metal ions, the production of dendrites and spongy deposits, the sluggish kinetics of ions transportation, and side-reactions from the hydrogen evolution and oxygen reduction reactions (Jin and Zhang, 2020).

9.3 Recovery of Water

Water is one of the most important substances on earth, such that all plants and animals need water to survive (Simate, 2015). In other words, water supports life (EU Directive, 2000). Furthermore, water is at the centre of economic and social development (Goswami and Bisht, 2017). Though fresh water is considered a renewable resource through a continuous cycle of evaporation, precipitation, and runoff – commonly referred to as the water cycle (Goswami and Bisht, 2017), in the recent past, the demand for water has exceeded the supply of water in many places (UNEP 2012). Indeed, water is now a scarce commodity that is fast becoming an issue of prime concern globally (Masindi et al., 2019). Therefore, it is important to find alternative sources of water in order to meet the demand. Most importantly, developing additional sources of water supply such as the non-conventional ones is vital; and AMD is one such a non-conventional source for water supply.

In fact, studies in many countries including South Africa have reported that there are large volumes of AMD produced (Bologo et al., 2012; Masindi, 2016; Masindi et al., 2018), which, without doubt, can be used as an alternative source of water for various applications. In fact, the removal of heavy metals from AMD using the technologies discussed in the previous sections simultaneously recovers water (Simate and Ndlovu, 2014). However, the processes have several disadvantages which include pH dependence, which means that the removal of the mixture of heavy metals cannot be achieved at a single pH level (Simate and Ndlovu, 2014). Therefore, it is imperative that other technologies which are not pH dependent are pursued. In this regard, there have been significant advances by scientists and engineers to recover water from AMD. Therefore, this chapter will focus on conventional methods (e.g., electrodialysis, nanofiltration, reverse osmosis, etc.) and recently developed alternative technologies (e.g., membrane distillation, etc.).

9.3.1 Recovery of Water Using Membrane Technology

9.3.1.1 Concepts of Membrane Technology

Membranes are considered as thin semi-permeable sheets, usually made of biological, synthetic, or polymeric materials, which act as a selective barrier (Salas, 2017). In general, a membrane is a thin barrier that permits selective mass transport (Mortazavi, 2008). This is a vital property that the membrane techniques exploit, as it allows only specific components of the solution to permeate through the membrane freely while impeding the permeation of other components (Salas, 2017). Transport of permeating species through the membrane matrix is achieved by the application of a driving force across the membrane which provides a basis for the classification of membrane-separation processes (Mortazavi, 2008). Indeed, the classification is based on the type of driving force that drives mass transport across the membrane such as mechanical (pressure), concentration (chemical potential), temperature, or electrical potential (Porter, 1989; Mortazavi, 2008). Table 9.4 shows the

TABLE 9.4

Driving Forces and the Related Membrane Separation Processes

Driving Force	Membrane
Pressure difference	Microfiltration, ultrafiltration, nanofiltration, reverse osmosis or hyper filtration
Chemical potential difference	Pervaporation, per-traction, dialysis, gas separation vapour permeation, liquid membranes
Electrical potential difference	Electrodialysis, membrane electrophoresis, membrane electrolysis
Temperature difference	Membrane distillation

Source: Abhanga et al., 2013.

driving forces and the related membrane-separation processes. The characteristics of the feed solution and the desired permeate quality also dictate the choice of the membrane process, the membrane type, and module design and configuration (ITRC, 2010b).

The most broadly applied membrane technology processes are pressure driven which include reverse osmosis (RO), nanofiltration (NF), ultrafiltration (UF), and microfiltration (MF). A brief summary of the comparison amongst the four classes of pressure-driven membrane-separation processes is given in Table 9.5.

Membrane-separation processes have become a viable alternative to other physical methods of separation (Mortazavi, 2008; ITRC, 2010b). Some of their

TABLE 9.5

Comparison amongst the Four Classes of Pressure-Driven Membrane Techniques

	Reverse Osmosis	Nanofiltration	Ultrafiltration	Microfiltration
Membrane	Asymmetric	Asymmetric	Asymmetric	Asymmetric Symmetric
Thin film thickness	1 μm 150 μm	1 μm 150 μm	1 μm 150–250 μm	1–150 μm
Rejection	High and low molecular weight compounds, NaCl, glucose, amino acids	High molecular weight compounds, mono-, di-, and oligosaccharides, polyvalent ions	Macromolecules, proteins, polysaccharides, vira	Particles, clay, bacteria
Applications	Ultrapure water; desalination	Removal of (multivalent) ions and relatively small organics	Removal of macromolecules, bacteria, viruses	Classification; pre-treatment; removal of bacteria
Membrane materials	Cellulose acetate, thin film	Cellulose acetate, thin film	Ceramic, polysulphonic, poly vinylidene flouride, cellulose acetate, thin film	Ceramic, polysulphonic, poly vinylidene flouride, cellulose acetate
Pore size	<0.002 μm	<0.002 μm	0.02–0.2 μm	0.02–4 μm
Module configuration	Tubular, spiral wound, plate-and-frame	Tubular spiral wound, plate-and-frame	Tubular hollow fibre spiral wound, plate-and-frame	Tubular, hollow fibre
Operating pressure	15–150 bar	5–35 bar	1–10 bar	<2 bar
Separation mechanism	Diffusion + exclusion	Diffusion + exclusion	Sieving	Sieving
Permeate flux	Low	Medium	High	High

Sources: van der Bruggen, 2003; Mortazavi, 2008.

benefits include relatively low energy consumption, possibility of separation of both organic and inorganic substances, process selectivity, low chemical consumption, and operation at room temperature (García et al., 2013; Agboola, 2019). In addition, membrane processes are capable of continuous operation, require low capital, and their footprint is low (García et al., 2013).

The membrane technology can be used to treat AMD due to their capability in removing suspended solids, dissolved organic compounds, metallic ions, microbes, and multivalent ions from wastewater (Agboola, 2019). However, AMD treatment by membrane technology was very rare, in the past, owing to the moderately high cost of the membrane and high membrane fouling because of the susceptibility of membrane systems to high-salt concentrations found in AMD (Mortazavi, 2008; Agboola, 2019). Nevertheless, membrane technology processes are currently playing a crucial role as new state of the art methods used in treating AMD (Agboola, 2019), and recent studies on the treatment of AMD using semi-permeable membranes can be found in several literature (Mortazavi, 2008; ITRC, 2010b; Ambiado et al., 2017; Salas, 2017). In fact, various studies have shown successful application of various pressure-driven membranes for the treatment of AMD (Mortazavi, 2008).

9.3.1.2 Selected Typical Studies of Pressure-Driven Membrane Technology

The increasing demand for clean and portable water in countries that have or had mining industries has stimulated research and/or utilisation of membrane-separation processes for water and wastewater purification aimed at the use/reuse of water resources such as AMD. This section discusses selected examples of the use of membrane technology from published literature for the treatment of AMD with the view of recovering water using various classes of pressure-driven membranes. In fact, individual and/or combinations of MF, UF, NF, and RO membrane processes have been extensively studied for the recovery of both water and metals (Ahn et al., 1999; Garba et al., 1999; Zhong et al., 2007).

The aim of a study by Andrade et al. (2017) was to investigate the use of NF and RO for gold mining effluent treatment in order to obtain water for industrial reuse. Two effluents from a gold mining company were studied, i.e., an effluent from a sulphuric acid production plant and the water from the calcined dam. The effluent from gold mining was first pre-treated for each test performed. The pre-treatment used a commercial submerged polyvinylidene difluoride (PVDF) based UF membrane with an average pore diameter of 0.04 μm and a filtration area of 0.047 m^2 operating at a pressure of 0.7 bar. For the NF and RO filtration tests a bench scale unit was used. The RO membranes were TFCHR and BW30, while the NF membranes were MPF34, NF90, and NF270. The NF and RO filtration took place at a fixed pressure of 10 bar, feed flow rate of 2.4 L/min (corresponding to a cross-flow velocity on the surface of the membrane of 1.9 m/s and Reynolds number of 840), while the permeate was continuously removed and the concentrate returned to the supply tank. Before every test, all membranes were washed in two consecutive ultrasound baths for 20 min each; the first containing citric acid solution at pH 2.5 and the second containing 0.1% NaOH solution. After chemical cleaning, the membranes were flushed with distilled water.

A study by Andrade et al. (2017) concluded that in an effluent in which the major contaminants are bivalent ions, such as effluent from gold mining with high concentrations of sulphate, calcium and magnesium, effluent treatment with NF membranes was more effective than with RO membranes because they allow permeate fluxes of 7 to 12 times higher and compatible retention efficiencies. In the study, NF90 showed the best performance, and subsequently NF was declared to be a suitable treatment for gold mining effluent at an estimated cost of US$0.83/m³. However, it was observed that retention efficiencies of NF90 were similar to those of RO membranes except that the permeate fluxes obtained were 7 to 12 times higher for NF90. The study found that a feed pH of 5.0 provided both greater permeate flux and higher retention efficiencies. It was noted that slightly above the membrane isoelectric point (IEP) (pH 4.3), at pH 5.0, the membrane has a small negative charge and, therefore, it repels anions, which prevents fouling formation. It was observed that permeate flux decreased linearly as the permeate recovery rate increased. Additionally, there was a significant increase in conductivity for recovery rate above 40%. In other words, at a recovery rate above 40%, there was a significant decrease in permeate quality. Consequently, a permeate recovery rate of 40% was selected as the ideal for this process.

Three commercial NF (NF99, DK, and GE) membranes were employed in a laboratory scale study by Al-Zoubi et al. (2010) to investigate their performance in handling AMD collected from a copper mine in Chile. Table 9.6 shows the properties of the AMD solution used in the study. The pressure of the membrane in the laboratory scale test cell was set at 20 and 30 bar

TABLE 9.6

Properties of the Original Acid Mine Drainage and the Standard Concentration for Potable Water According to the World Health Organisation

Metal	Concentration (ppm) in AMD Solution	Standard Concentration for Potable Water (ppm)
Aluminium(Al)	1139.0	0.20
Sulphate (SO$_4$)	14 337	250
Calcium (Ca)	325.9	40
Copper (Cu)	2298.0	2.0
Iron (Fe)	627.5	0.200
Manganese (Mn)	224.5	0.050
Magnesium (Mg)	630.60	20
Sodium (Na)	6.89	200
Potassium (K)	4.31	12

Source: Al-Zoubi et al., 2010

for both the original and concentrated AMD. The concentrated AMD was obtained from RO of the original AMD until 50% of permeate (pure) water were removed and the total ion content was accordingly raised by a factor of 2.

The results of the study showed that DK and NF99 NF membranes successfully treated AMD with a very high rejection (>98%) of all divalent cations and anions for both 20 and 30 bar, which confirms that the maximum applied pressure on the NF membrane cell should not exceed 20 bar. The results indicate the suitability of NF membranes in treating AMD in a more environmentally friendly process. Moreover, the results showed that NF membranes are capable of reducing the heavy metals concentration found in AMD to low levels that are accepted by many international organisations especially for industrial and agricultural use. Detailed analysis of the results showed that the DK membrane is preferable for high concentration of AMD, while NF99 is used, when high permeate flux is required. The GE had the lowest rejection and permeate flux at all investigated conditions indicating its inappropriateness in treating AMD solution.

Andalaft et al. (2018) analysed, evaluated, and modelled the behaviour of two commercial spiral-wound NF membranes (NF270 and NF90) in the treatment of real AMD. A 150-L sample of AMD was made available by a large copper mining company in central Chile. The sample was micro-filtered with a 0.10-μm cut-off ceramic membrane to remove any existing colloidal solids. The results revealed that both membranes showed operational differences regarding the effect of pH, temperature, continuous operation, concentration polarisation, and fouling. However, the suitability of NF in treating AMD was confirmed, as the two tested membranes succeeded at removing almost all of the dissolved ions. Specifically, both NF90 and NF270 indicated a high rejection of divalent ions (~100%). However, the permeate flux modelling detected the occurrence of surface fouling on the NF90 membrane by colloid particles due to its elevated roughness, a theory supported by the elevated physically reversible resistance in the resistance model. The modelling of the permeate flux at various recoveries enabled the researchers to distinguish between the decline in permeate flux resulting from the increase in osmotic pressure and that caused by membrane fouling. As a result, the study was able to identify the maximum design recovery at which fouling begins to occur (~75%). Following this value, gypsum ($CaSO_4 \cdot 2H_2O$) was considered to be the only foulant, and its potential fouling on the tail elements of an industrial system could be controlled with antiscalants. Furthermore, it was also noted that adjustments of pH in the continuous tests reduced the precipitation of aluminium, and gypsum was identified as the main compound generating a significant decrease in the permeate flow in the membrane.

Sierra et al. (2013) studied the treatment of AMD from an abandoned mercury mine in Spain using NF270-2540 NF membrane (Filmtec). The sediment geochemistry and the origin of acid waters were analysed in order to understand the geochemical factors involved in NF. The study showed that 99% of the dissolved solids such as As, Fe, and Al could be removed. In particular, the researchers found that NF has the capability of rejecting up to 99% of aluminium, arsenic and iron content, and 97% of sulphate content. The studied NF membrane was found to be effective in the treatment of metallic acid drainage contaminants, even at low pH and low pressures, which, without doubt, can decrease the costs associated with the process at industrial scale.

A comparative study of the performance of the commercially available polyamide ultra-low-pressure RO (RE-4040-BL) and NF (DK4040F) processes for AMD treatment and reuse was investigated by Zhong et al. (2007). Samples of AMD for the study were collected from the Dong Gua Shan copper mine in Anhui province, China. The evident property of the AMD was the high conductivity that resulted from various ions such as sodium, chloride, sulphate, nitrate, heavy metal, and many others. Effects of operation pressure, pH, temperature, water recovery efficiency, and operation time on ultra-low-pressure RO and NF performance were studied. The experimental results showed that the removal efficiency of total conductivity of AMD by NF was low, although the removal efficiency of the heavy metal ions was very high while ultra-low-pressure RO effectively removed both total conductivity and heavy metal ions from AMD. In addition, the results of the study showed that the rejection increased with an increase in feed pressure and decreased with an increase in feed temperature, and it was dependent on the pH of the feed. The pH was found to influence the rejection of heavy metals because the charge property of the surface of polyamide ultra-low-pressure RO and NF membrane changes and heavy metal ions are capable of interacting with OH$^-$ ions and precipitate onto the membrane surface. The effect of the feed temperature on the permeate flux was very sensitive, and the removal efficiency kept almost constant below 25°C while it began to decrease above 25°C. The results suggested that with the rejections of heavy metals and total conductivity being greater than 97% and 96%, respectively, for the ultra-low-pressure RO, it showed that the membrane was suitable for the recovery of heavy metals and reclaiming of water from AMD. Compared to the ultra-low-pressure RO, the NF process had the capability of removing close to 90% of the ions in the AMD with 48% reduction of the total conductivity.

Haan et al. (2018) investigated the potential of different membrane modules (UF membrane with the nominal molecular weight cut-off of 10 kDa, UF membrane with molecular weight cut-off of 5 kDa, and RO membrane) in treating mine water collected from three different places in Selangor, Malaysia. The UF membranes were made from polyethersulphone whilst RO was made from polyamide thin film composite with 95% MgSO$_4$ and 99.4% NaCl rejection. The results indicated that different membrane modules demonstrated different extents of removal efficiency, depending on the rejection capability of the membranes, which in turn was predominantly governed by the membrane morphology. Given its tight membrane structure, the RO membrane module had the best performance in producing permeate water of an excellent quality. However, RO produced lower permeate flux due to the smaller pore size of the membrane module. Unfortunately, none of the water produced from the three studied single-membrane systems was able to meet the standard for residential use. Therefore, an integrated membrane filtration system with three stages of membrane filtration is required so as to further improve the water quality which could eventually lead to the attainment of the standard of water required for residential use.

In one study, Vaclav and Eva (2005) used RO technique to recover water of good quality from three different sources of AMD. Günther and Mey (2008) evaluated different water treatment technologies including the membrane technology with the view of treating AMD from a coal mine. The results showed that only the biological sulphate removal process and RO membranes could be used to produce a high water recovery, whilst being cost effective. Bhagwan (2012) studied the use of a high recovery precipitating reverse osmosis (HiPRO) process for the recovery of low salinity water from mine waters. The main advantage of the developed process is that it makes use of RO to concentrate the water and produce supersaturated brine from which the salts can be released in a simple precipitation process. Furthermore, Bhagwan (2012) states that the technology offers other key advantages including (1) very high recovery, (2) simple system configuration, (3) low operating and capital costs, (4) easy operation, and (5) minimum waste.

9.3.2 Recovery of Water Using Other Membrane Technologies

As shown in Table 9.4, there are several other types of membrane technologies based on other driving forces instead of pressure. This section of Chapter 9 discusses and gives typical examples of the use of two very important membranes in the treatment of AMD – electrodialysis and membrane distillation (Abhanga et al., 2013).

9.3.2.1 Concepts of Electrodialysis and Membrane Distillation Technologies

Amongst the different technologies, electrodialysis has been tested and proven to be an effective technology for water recovery from acidic solutions (Wisniewski and Wisniewska, 1999; Cifuentes et al., 2006; Agrawal and Sahu, 2009; Cifuentes et al., 2009). Electrodialysis is a membrane process where the driving force is electrochemical and ions are transported across a water swollen ion-exchange membrane under the influence of electrical potential (Mortazavi, 2008). Rodrigues et al. (2008) define electrodialysis as a membrane-separation process based on the selective migration of aqueous ions through ion-exchange membrane as a result of an electrical driving force. In this technology, the direction of transport and rate of each ion depends on its charge, mobility, solution conductivity, relative concentrations, applied voltage, etc. (Rodrigues et al., 2008; Benvenut et al., 2013). Electrodialysis provides a means of selective separation of anions and cations and is considered as a clean technology which does not require the addition of chemicals; it can be operated in continuous mode and allows obtaining profitable by-products (Mortazavi, 2008; Martí-Calatayud et al., 2013).

Membrane distillation process is an over five-decade old technique which is emerging commercially and is currently being explored for various applications including concentrating acid and recover fresh water from acidic waste solutions, desalination, water and wastewater treatment, removal of volatile organic compounds, and food processing (Tomaszewska, 2000; Tomaszewska et al., 2001; Alkhudhiri et al., 2012; Kesieme et al., 2012; Camacho, 2013; Shirazi et al., 2013; Simate and Ndlovu, 2014; Kesieme, 2015). Furthermore, Kesieme et al. (2012) confirm that membrane distillation has been used to recover water and concentrate acid and metal values from mining wastewater and process solutions. Ideally, the process combines both the conventional distillation and membrane-separation processes (Shirazi et al., 2014a).

Membrane distillation is a thermally driven separation process that utilises hydrophobic, microporous membranes as a contactor (Shirazi et al., 2014b). The driving force in the membrane distillation is the vapour pressure difference induced by the temperature difference across the hydrophobic membrane (Alkhudhiri et al., 2012). In other words, the temperature difference existing across the membrane results in a vapour pressure difference, thus vapour molecules are transported from the high-vapour pressure side to the low-vapour pressure side through the pores of the membrane (Tomaszewska, 2000). The hydrophobicity of the membrane prevents the transport of liquid across the pores of the partition while water vapour can be transported from the warm side, and condensing at the cold surface (Kesieme et al., 2012). Since the separation mechanism is based on the vapour/liquid equilibrium, it means that the component with the highest partial pressure will exhibit the highest permeation rate (Tomaszewska, 2000). Moreover, mass transfer in membrane distillation is controlled by three basic mechanisms – Knudsen diffusion, Poiseuille flow (viscous flow), and molecular diffusion (Alkhudhiri et al., 2012). Hull and Zodrow (2017) explain the operation of membrane distillation in simple terms as follows for the case of AMD. In membrane distillation, heated AMD is separated from a cooled distillate by a hydrophobic, water-excluding membrane. Due to the fact that water only passes through the membrane in the vapour phase, non-volatile sulphates and heavy metals are retained in the concentrate stream.

Membrane distillation processes have several configurations which are as follows (Kesieme et al., 2012): (1) direct contact membrane distillation (DCMD), (2) air gap membrane distillation (AGMD), (3) sweeping gas membrane distillation (SGMD), and (4) vacuum membrane distillation (VMD). The different membrane distillation configurations are represented in Figure 9.4 (Foureaux et al., 2020).

Among these configurations, DCMD is the most widely used because it is convenient to set up, consumes relatively low energy, gives high water flux (Kesieme et al., 2012), and has 100% (theoretical) macromolecules rejection

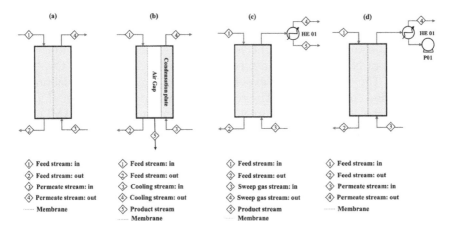

FIGURE 9.4
Different membrane distillation configurations (Foureaux et al., 2020). (a) direct contact membrane distillation (DCMD); (b) air gap membrane distillation (AGMD); (c) sweep gas membrane distillation (SGMD); and (d) vacuum membrane distillation (VMD). HE01: Heat exchanger. P01: Vacuum pump. Hot and cold side represented as red and blue streams, respectively.

ability, in addition to non-volatile compounds and inorganic ions (Foureaux et al., 2020). Membrane distillation processes, in general, have several advantages including the following: (i) the ability of 100% (theoretical) rejection of non-volatile compounds and inorganic ions; (ii) the possibility of operating at a relatively low temperature when compared to the conventional distillation process; (iii) lower operating pressures when compared to the classical membrane-separation processes that have the pressure gradient as the driving force and (iv) less influence of the fouling phenomenon due to less chemical interaction between process solution and membrane surface (Wang and Chung, 2015; Manna and Pal, 2016; Biniaz et al., 2019; Foureaux et al., 2020). Moreover, membrane-separation processes can operate without pre-treatment since it is able to treat a high solute concentration in the feed stream and it is not so sensitive to polarisation concentration effects like other classical membrane-separation processes (Alkhudhiri and Hilal, 2018; Biniaz et al., 2019; Foureaux et al., 2020).

9.3.2.2 Selected Typical Studies of Electrodialysis and Membrane Distillation Technologies

In a study of the use of electrodialysis, Buzzi et al. (2013) investigated the possibility of employing the technique to treat AMD generated by the mining of coal for water recovery purposes. The AMD samples that were used in the study were collected from a carboniferous area in Criciúma, Brazil at different locations throughout the carboniferous area to represent different situations where AMD is generated. After collection, any solids in each of the six AMD samples were allowed to settle. Thereafter, each sample was filtered through a membrane of 0.45-μm pores before the samples were chemically characterised for the following parameters: pH, conductivity, and levels of Na^+, K^+, Mg^{2+}, Ca^{2+}, Fe^{3+}, Cu^{2+}, Zn^{2+}, Mn^{2+}, Fe^{2+}, F^-, Cl^-, NO_3^-, and SO_4^{2-}. The electrodialysis experiments were conducted in a laboratory cell with five compartments. The results of the study by Buzzi et al. (2013) showed that electrodialysis is suitable for recovering water from AMD, with contaminant (e.g., Fe, Al, Mn, Pb, Zn, Cu, etc.) removal efficiencies that are greater than 97%. However, the precipitation of iron at the surface of the cation-exchange membrane constitutes a problem for the system because it causes a blockage of the membrane through the scaling phenomenon, which reduces the process efficiency. It is proposed that the elimination of iron from AMD prior to electrodialysis process would facilitate smooth recovery of water from the AMD.

In a 2011 study, Buzzi et al. conducted experiments to ascertain the possibility of using MF membrane, as pre-treatment, followed by electrodialysis technique for the treatment of AMD aimed at obtaining water for reuse. The AMD sample used in this experiment was collected from a carboniferous area in Criciúma in Brazil. The pluvial drainage from the coal was a more diluted drainage when compared to the drainage from the percolation of deposits of wastewater in the same carboniferous area. The AMD sample in the study was analysed for the following chemical compositions: pH, conductivity, total dissolved solids, Na^+, K^+, Mg^{2+}, Ca^{2+}, Fe^{3+}, Cu^{2+}, Zn^{2+}, Mn^{2+}, Fe^{2+}, F^-, Cl^-, NO_3^- and SO_4^{2-}. One pre-treatment with MF was necessary to prevent fouling and scaling of the electrodialysis, so the entire AMD sample used in the experiment was filtered through a membrane of 0.45 μm before the experiments. Electrodialysis tests were conducted in a laboratory cell with five compartments. The results showed that pre-treatment of AMD with MF membrane combined with electrodialysis was efficient to extract over 97.5% of cations and anions after 55 h when 2.6 mA·cm^{-2} current density was applied. The results indicated that water could be recovered by electrodialysis from AMD after a single pre-treatment with a MF membrane.

Zheng et al. (2015) used a laboratory scale electrodialysis system with an effective area of 88 cm^2 to remove copper and cyanide in simulated and real gold mine effluents. The electrodialysis stack consisted of two electrodes made of a titanium plate coated with ruthenium, five anion-exchange membranes, and six cation-exchange membranes. The gold mine effluent with a concentration of 47 mg/L and a cyanide concentration of 242 mg/L was provided by Zhaoyuan gold smelter plant (Shandong, China), whereas the simulated solutions were made from CuCN and NaCN. The concentrations of total copper and cyanide in simulated solutions, in mg/L, were as follows: sample 1 (Cu = 23.5, CN = 121), sample 2 (Cu = 47, CN = 242), sample 3 (Cu = 70.5, CN = 363), and sample 4 (Cu = 94, CN = 484). The effects of applied voltage, initial concentration, and flux rate on the removal rate of copper and cyanide were investigated. The results showed that the highest copper (99.41%) and cyanide (99.83%) removal rates were achieved under the following conditions: applied voltage of 25 V, initial concentration for copper and cyanide of 47 mg L^{-1} and 242 mg L^{-1} (sample 2 and real AMD), and a flux rate of 4.17 mL s^{-1}. The results of the study also showed that the lowest concentrations of copper (0.44 mg L^{-1}), cyanide (0.48 mg L^{-1}), and zinc (0.34 mg L^{-1}) found in the treated effluent water were all below regulatory limits (copper, cyanide < 0.5 mg L^{-1}, zinc < 2.0 mg L^{-1}) which makes it suitable for reuse. In addition, ion-exchange membrane fouling was studied and the results showed the presence of CuCN, $[Cu(CN)_3]^{2-}$, $Cu(OH)_2$, and $Zn(OH)_2$ in the precipitate, and the fouling of anion-exchange membranes could be decreased significantly via pH adjustment.

The recovery of water has also been studied in other acidic systems containing various metals (Wisniewski and Wisniewska, 1999; Agrawal and Sahu, 2009; Cifuentes et al., 2009; Benvenut et al., 2013). In all these studies, electrodialysis has been found to be an effective method for water recovery. In fact, membrane modules on average have a water recovery rate ranging from 50% to 80% (West et al., 2011). However, the only major disadvantage is that electrodialysis membrane technologies can be very expensive in applications where the wastewater contains elevated hardness and sulphate near gypsum saturation (West et al., 2011). The high cost of operation is due to the large reagent requirement for upstream pre-treatments, high power demand, and expensive disposal options for the concentrated brine solution produced by the membranes. At a time when responsible energy and environmental practices are under the spotlight, technologies with high energy use and that produce a large volume of wastewater are becoming less appealing.

In a study published almost a decade ago, Kesieme et al. (2014) claimed to have carried out experiments for the first time using DCMD for acid and water recovery from a real leach solution generated by a hydrometallurgical plant. In other words, the aim of the study conducted by Kesieme et al. (2014) was to assess the opportunity of using DCMD to recover fresh water and acids from real acid leach solutions generated from hydrometallurgical plants. Two different real leach solutions containing HCl or H_2SO_4 were obtained from a Jervois Mining process plant in Melbourne, Australia. The membranes used were flat sheet polytetrafluoroethylene (PTFE) supported on polypropylene scrim backing. The membranes had an active area of 0.0169 m^2, pore size of 0.45 μm. A cartridge filter with filtration size of 0.5 μm was used on the hot loop to collect precipitated matter prior to entering the MD module. The flow rate into the hot and cold sides of the module was 900 mL/min. The feed temperature was 60°C and the cold temperature was maintained at 20°C. For the solution containing non-volatile substances only water vapour was transferred across the membrane and the non-volatile compounds such as H_2SO_4 were retained by the membrane. Solutions containing volatiles compounds such as HCl and water vapour could pass through the membrane as permeate. The vapour (or permeate) was condensed directly into the solution (distillate) in which HCl was dissolved. The HCl flux was calculated from the material balance of HCl in the distillate collected every hour taking into account the changes in volume and the acid concentration in the distillate. The water flux was calculated based on Equation 9.16, and recovery was calculated as shown in Equation 9.17.

$$F_{water} = \frac{\text{mass of permeate (kg)}}{\text{effective membrane area (m}^2) \times \text{operating time (hours)}} \quad (9.16)$$

$$\text{Recovery (\%)} = \frac{\text{permeate produced (L)} \times 100}{\text{initial feed volume (L)}} \quad (9.17)$$

The results of the test work by Kesieme et al. (2014) showed that fluxes were within the range of 18–33 kg/m²/h and 15–35 kg/m²/h for the H_2SO_4 and HCl systems, respectively. In the H_2SO_4 leach system, the final concentration of free acid in the sample solution increased on the concentrate side of the DCMD system from 1.04 M up to 4.60 M. The sulphate separation efficiency was over 99.9% and overall water recovery exceeded 80%. In the HCl leach system, HCl vapour passed through the membrane from the feed side to the permeate side. The concentration of HCl captured in the permeate side was about 1.10 M leaving behind only 0.41 M in the feed from the initial concentration of 2.13 M. In all the experiments, salt rejection was > 99.9%. The results of this study clearly showed that DCMD was viable for high recovery of high water quality. The concentrated H_2SO_4 and metals remaining in the feed may be selectively recovered using solvent extraction. The HCl can be recovered for reuse using only DCMD.

In another study by Kesieme and Aral (2015) which acted as a follow-up to the 2014 study, an assessment of the potential and opportunities for DCMD to concentrate H_2SO_4 and recover fresh water from acidic process solutions was conducted. The study was also aimed at identifying how membrane distillation can work in combination with solvent extraction in the mineral processing industry for acid recovery. Table 9.7 shows the acid and metal compositions to the membrane distillation and solvent extraction systems used in the study. Experiments were conducted in DCMD mode to confirm the viability of membrane distillation to concentrate a 4-L synthetic acidic waste solution and to recover fresh water. The membrane had an active area of 0.0169 m² with pore size of 0.45 μm, and a cartridge filter with filtration size of 0.5 μm was used on the hot loop to collect precipitated matter prior to entering the membrane distillation module. The flow rate into the hot and cold sides of the module was 900 mL/min. The temperature on the hot side of the membrane distillation module was 60°C and the cold side temperature was 20°C. Permeate build-up was measured by the accumulated mass of water in the permeate tank.

The organic system consisting of 50% tris-2-ethylhexylamine (TEHA) and 10% ShellSol A150 (a 100% aromatic diluent) in octanol was used in the

TABLE 9.7

The Acid and Metal Compositions to the Membrane Distillation and Solvent Extraction Systems

Species	Feed to the Membrane Distillation System	Feed to the Solvent Extraction System Mimicking Concentrate from Membrane Distillation
Acid (M)	0.850	2.450
Aluminium (g/L)	0.056	0.250
Cobalt (g/L)	0.071	0.289
Copper (g/L)	0.269	1.040
Calcium (g/L)	0.218	0.307
Iron (g/L)	2.780	11.350
Magnesium (g/L)	0.050	0.220
Manganese (g/L)	0.002	0.008
Nickel (g/L)	0.065	0.259

Source: Kesieme and Aral, 2015.

solvent extraction system. The feed composition mimicking an acidic process solution after concentration using membrane distillation (see Table 9.7) was made by dissolving AR grade 245 g/L H_2SO_4 and sulphates of metals including Fe, Ni, Zn, Mg, Co, and Cu in distilled water. All batch solvent extraction tests were carried out in 100-mL hexagonal glass vessels immersed in a temperature-controlled water bath. The solution temperature was maintained at the desired temperature ($\pm1°C$) during testing. For the acid extraction test, the organic system was mixed with concentrated solution at an A/O ratio of 1:2 and a temperature of 22°C. The loaded organic solution was stripped twice at O/A ratios of 2:1 and 1:5 at 60°C. The raffinate and the loaded strip liquors were titrated to determine acid concentrations for extraction and mass balance calculations.

The results of the experiments by Kesieme and Aral (2015) confirmed that the membrane distillation was capable of concentrating H_2SO_4 and recover fresh water from process acidic solutions. The DCMD experiment showed that H_2SO_4 was concentrated from 0.85 to 4.44 M whereas the water recovery exceeded 80%. The sulphate and metal separation efficiency was >99.99%. After recovery of water with DCMD, over 80% of H_2SO_4 was extracted in the solvent extraction system in a single contact from the waste solution (i.e., the concentrated solution from the membrane distillation) containing 245 g/L H_2SO_4 and metals with various concentrations. After three stages of successive extraction, nearly 99% of acid was extracted, leaving only 2.4 g/L H_2SO_4 in the raffinate. The extracted acid was easily stripped from the loaded organic solution using water at 60°C. After scrubbing the loaded organic solution at an O/A ratio of 10 and 22°C, 98–100% of entrained metals were removed in a single contact with only 4.5% acid lost in the loaded scrub liquor. It was also found that the phase disengagement time was in the range of 2–4 min for both extraction and stripping which indicates a reasonable fast phase separation. In summary, the results from the study showed that membrane distillation and solvent extraction can be applied to recover acid and fresh water for reuse and metal values from mining and acidic process solutions.

Kang et al. (2019) investigated the techno-economic feasibility of using membrane distillation to recover clean water from AMD employing both renewable and non-renewable energy sources. The bench-scale set-up was used in batch tests in the study to establish whether membrane distillation is a viable technology to treat AMD. The bench-scale set-up was later modified to an open-loop continuous mode operation. Five membranes (two polypropylene, two polytetrafluoroethylene, and one polyvinylidene fluoride) that showed very high water flux of >35 kg/hm^2 and high liquid entry pressures of 25–40 psig were used in the bench-scale set-up. The bench-scale set-up employed simulated AMD solution made up from soluble metal sulphates (Cd, Cr, Fe, and Zn), sodium arsenate, sodium selenate, and adjusted to pH 2 with dilute sulphuric acid. The best of the five membrane distillation identified in the bench-scale set-up was investigated in an open-loop continuous process. In an open-loop continuous process, fresh simulated AMD was introduced into a tank and concentrated AMD was withdrawn at a predetermined rate so as to maintain a constant volume of the AMD feed in the tank, while purified water (i.e., the distillate) was also taken out at a predetermined rate in order to maintain it at a constant volume in the distillate tank.

Amongst all commercial membranes tested by Kang et al. (2019), 0.45-μm pore polypropylene membrane exhibited the highest water flux (˜62 kg/hm^2) and achieved 90% water recovery under optimal open-loop continuous membrane distillation conditions employing a realistic simulated AMD feed (total dissolved solids = 900 mg/L, pH = 2.4). However, the results indicated that the polypropylene membrane would need to be replaced once every 150 days under real process conditions. The cost analysis of the membrane distillation plant was performed for two major components of cost (i.e., capital cost and annual operating cost), and for four different energy sources, including local utility, photovoltaic, solar thermal, wind, and natural gas. The capital cost was defined as the cost associated with plant construction, process equipment purchases, and installation charges. The operating costs included amortisation, fixed charges, operating and maintenance costs, and membrane replacement costs. The economic analysis indicated pipelined natural gas and local electricity to be the most economical energy sources for heating AMD water and resulted in the total treatment costs of $0.476/m^3 and $0.607/m^3 of AMD, respectively.

In a study by Ryu et al. (2019), the performance of the following systems was evaluated: (1) natural and modified (heat treated) zeolite for heavy metal removal from AMD, (2) submerged DCMD for producing water for reuse from AMD, and (3) integrated submerged DCMD/sorption system for simultaneously removing heavy metals and producing water for reuse from AMD. The synthetic AMD solution used in the study was prepared by dissolving analytical grade $CaSO_4$, $MgSO_4 \cdot (3H_2O)$, $NaOH$, $FeO(OH)$, $Fe(SO_4) \cdot 7H_2O$, $ZnSO_4 \cdot 7H_2O$, $CuSO_4 \cdot 5H_2O$, $Al_2(SO_4)_3 \cdot 18H_2O$, and $Ni(NO_3)_2 \cdot 6H_2O$ in Milli-Q water. The pH of the solution was adjusted using concentrated H_2SO_4 (10 M). The natural zeolite in powder form (particle size < 75 mm) used in the experiments had a bulk density of $2.7g/cm^3$ and was mainly composed of clinoptilolite (~85 wt%) with minor quantities of quartz and mordenite (~15 wt%). Heat treatment method carried out at four different temperatures of 300, 400, 500 and 600°C for 24 h was used to potentially enhance the performance of natural zeolite. Heat treatment was chosen as it requires no additional chemicals and complex modification processes. The set-up of the DCMD consisted of a double-walled feed tank containing AMD solution with a submerged hollow fibre membrane made of polyvinylidene fluoride. The membrane pore size, inner and outer diameters, wall thickness, and contact angle were 0.1 µm, 0.7 mm, 1.2 mm, 250 µm, and 106 ± 2°, respectively. The membrane module was made of 18 fibres of 0.2-m length (active membrane area of 0.0136 m^2). The outer wall of the double-walled feed tank was circulated with heated water connected to a heating system, thus enabling an AMD feed solution to be maintained at a temperature of 55.0 ± 0.5°C. The permeate solution was maintained at 22.0 ± 0.5°C using a cooling system.

The results of a study by Ryu et al. (2019) showed that modified (heat treated) zeolite achieved 26–30% higher removal rate of heavy metals compared to natural untreated zeolite. Heavy metal sorption by heat treated zeolite followed the order of Fe > Al > Zn > Cu > Ni and the data fitted well to Langmuir and pseudo second-order kinetics model. A slight pH adjustment from 2 to 4 significantly increased Fe and Al removal rate (close to 100%) due to a combination of sorption and partial precipitation. An integrated system of submerged DCMD with zeolite for AMD treatment enabled to achieve 50% water recovery in 30 h. The integrated system provided a favourable condition for zeolite to be used in powder form with full contact time. Likewise, heavy metal removal from AMD by zeolite, specifically Fe and Al, mitigated membrane fouling on the surface of the hollow fibre submerged membrane. The integrated system produced fresh water of high quality while concentrating sulphuric acid and valuable heavy metals (Cu, Zn, and Ni).

9.3.3 Selected Commercially Developed Projects for Recovery of Water from Acid Mine Drainage

This section discusses a few selected commercially developed projects that are either in operation, being piloted or under evaluation. These processes were also discussed by Simate and Ndlovu (2014) in detail.

The Council for Scientific and Industrial Research (CSIR) of South developed the CSIR ABC (alkali-barium-calcium) process. The CSIR ABC desalination process, developed for AMD neutralisation and the removal of total dissolved solids from 2 600 to 360 mg/L, was demonstrated at a pilot plant in 2010. This precipitation process developed by CSIR uses barium carbonate to

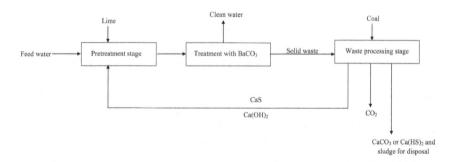

FIGURE 9.5
Process flow diagram for the CSIR-ABC process. (From Maree et al., 2012; Simate and Ndlovu, 2014.)

precipitate dissolved sulphate from AMD. It consists of the following three stages: pre-treatment, treatment with barium carbonate, and the waste processing stage as shown in Figure 9.5 (de Beer at al., 2010; Maree et al., 2012). In the pre-treatment stage the feed water is treated with CaS, $Ca(HS)_2$, or $Ca(OH)_2$ to remove free acid and metals. Ideally, metals are precipitated to low values as either hydroxides or sulphides, depending on the precipitation agent used. During this stage, the sulphate content is lowered from about 4500 mg/L to 1250 mg/L (de Beer et al., 2010; Maree et al., 2012). In the water treatment stage, $BaCO_3$ is added into the water thus producing barium sulphate as the solid waste and clean water. The alkalinity of the calcium bicarbonate-rich water was reduced from 1000 to 110 mg/L (as $CaCO_3$). The water treatment stage is integrated with a sludge processing stage to recover the alkali, barium, and calcium (ABC) from the sludge through reduction in a coal-fired kiln. In this process, good quality water containing less than 100 mg/L of sulphate was obtained in a cost-effective way from polluted mine water (de Beer et al., 2010; Maree et al., 2012). This process is a strong candidate for cost-effective treatment of mine water. However, the major limitation of this technology is the amount of sludge produced which is expensive to dispose. High capital and operating costs associated with the thermal reduction of waste to produce CaS, gypsum, and other solids for disposal also make the technology less cost-effective.

THIOPAQ process developed by PAQUES company is a biotechnological process which uses two distinct microbiological populations and stages (Boonstra et al., 1999): (1) conversion of sulphate to sulphide by using hydrogen gas (from the conversion of ethanol/butanol to acetate and hydrogen) as the electron donor and precipitation of metal sulphides, and (2) conversion of any excess hydrogen sulphide produced to elemental sulphur, using sulphide-oxidizing bacteria. In this way sulphate is removed from AMD to produce water of reusable quality. This process has lost attractiveness over the years due to an increase in the price of ethanol and butanol which are energy sources for the process (Boonstra et al., 1999).

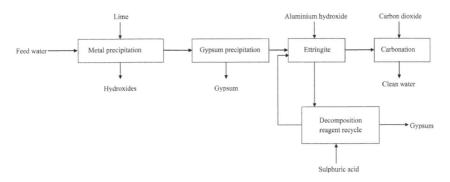

FIGURE 9.6
SAVMIN process flow diagram. (From Smith, 1999; Sibiliski, 2001; INAP, 2003; Simate and Ndlovu, 2014.)

The SAVMIM process was developed by the Council for Mineral Technology (Mintek) of South to treat polluted mine water (Smith, 1999; Sibiliski, 2001; Naidoo, 2018). The process was patented by Mintek in 1998 (Naidoo, 2018). Figure 9.6 shows the process flow diagram, and the main process stages are as also discussed below (Smith, 1999; Sibiliski, 2001; INAP, 2003; Simate and Ndlovu, 2014).

Stage 1 – Metal precipitation: Using lime, the pH of the feed water is raised to between 12.0 and 12.3 to precipitate metals (trace) and magnesium.

Stage 2 – Gypsum "de-supersaturation": Using gypsum seed crystals, gypsum is precipitated from the supersaturated solution and removed.

Stage 3 – Ettringite precipitation: Using aluminium hydroxide, dissolved calcium and sulphate are removed from the solution by the precipitation of ettringite (a calcium-aluminium sulphate mineral).

Stages 4 and 5 – Recycling of aluminium hydroxide: Using sulphuric acid, the ettringite slurry from stage 3 is decomposed at pH 6.5 in a solution supersaturated with gypsum (no precipitation). The resulting aluminium hydroxide is recycled to the third stage and the solution that is supersaturated with gypsum is contacted with seed crystals (stage 2) to precipitate and remove gypsum. The remaining solution saturated with gypsum is recycled.

Stage 6 – Carbonation and calcite precipitation: Using carbon dioxide, the pH of the solution from the third process stage (pH 11.2–12.4) is lowered to precipitate and remove calcite.

The end products of the SAVMIN process are potable water and a number of potentially saleable by-products (metal hydroxides, gypsum, and calcite). One of the major advantages of SAVMIN process is that high quality products can be obtained (Smith, 1999). A major disadvantage of this process is the vast amount of sludge produced which is expensive to dispose (Smith, 1999; INAP, 2003).

A process developed by Aveng Water called HiPRO (high-pressure reverse osmosis) process is capable of consistently achieving greater than 97% water recovery (Aveng Water, 2009). The final products from this process are portable water, a liquid brine solution (less than 3% of the total feed), and solid waste. The solid waste products are saleable grades of calcium sulphate and less pure calcium sulphate and metal sulphates (Aveng Water, 2009). The main disadvantages of this process are that it produces waste brine and sludge which are expensive to dispose.

9.3.4 Proposed Integrated Processes and Technologies for Recovery of Water from Acid Mine Drainage

In a study by Simate and Ndlovu (2014) it was suggested that the best way to working towards a sustainable solution of the AMD challenge is to take a business approach and consider the integration of existing technologies as well as technologies under development so as to come up with a solution that has the potential to address the problem in a holistic and sustainable manner. In other words, coupling different processes together as two- or three-stage processes would be more appropriate. Such measures would possibly include the integration of both the active and passive water treatment systems. In fact, previous studies have shown that combinations of physico-chemical treatments may be able to partially or completely remove some organic and inorganic contaminants (Dobias, 1993; Harrelkas et al., 2009). The first proposed option is to couple the AMD fuel cell with the cyclic electrowinning/precipitation (CEP) process. The first part of the process consists of the fuel cell where ferrous iron is completely removed through oxidation to insoluble Fe(III), forming a precipitate in the bottom of the anode chamber and on the anode electrode (Cheng et al., 2007). The iron contained in the precipitate or sludge could be marketed as a pigment for paint (Hedin, 2003), cosmetics, and possibly other uses. The electricity produced could be used to supplement the power in the electrowinning stage. The second part of the integrated process would consist of the CEP process. A pH swing (using NaOH or H_2SO_4) would be applied to the water coming from the fuel cell so as to precipitate the heavy metals such as cadmium, nickel, copper, etc. The precipitation and re-dissolution of metals would be repeated until the concentration of the heavy metals (<100 ppm) has reached a point where electrowinning can be efficiently done (Brown University, 2011). In electrowinning stage, heavy metal ions are converted using electric current to stable metal ions which can be recovered and separated from water. However, the metal barren solution would still contain high amount of sulphate ions. The solution could be concentrated and then reacted with sodium monochromate (from roasting and leaching of chromite ore) to form a mixture of sodium dichromate and sodium sulphate. This mixture can then be recovered and separated from the water. Depending on the purity, the product can be marketed as a fertiliser or as a metal finish and other uses. The proposed integrated process flow diagram is shown in Figure 9.7.

The second proposed integrated process by Simate and Ndlovu (2014) is shown in Figure 9.8. In this process the AMD is fed to the first part of the

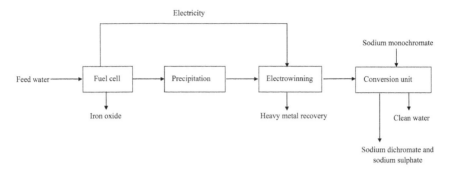

FIGURE 9.7
First proposed integrated process for the recovery of water from acid mine drainage. (From Simate and Ndlovu, 2014.)

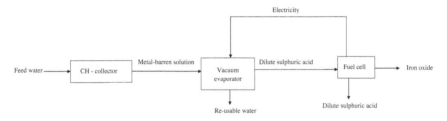

FIGURE 9.8
Second proposed integrated process for the recovery of water from acid mine drainage. (From Simate and Ndlovu, 2014.)

process in which the CH-collector (e.g., amino bisphosphonate adsorbent) adsorbs some of the heavy metals directly from the wastewater. The resulting solution is then passed through a vacuum evaporator where the water component is vaporised thus producing re-usable water while sulphuric acid remains in solution. A vacuum evaporator is used because it has the advantage of producing a large separation factor in the sulphuric acid/water system (Nleya et al., 2016). The remaining solution from the vacuum evaporator that contains dilute sulphuric acid and some remaining heavy metals such as ferrous iron is fed to a fuel cell that produces electricity, iron oxide and metal-barren dilute sulphuric acid.

9.4 Recovery of Acid

9.4.1 Introduction

The AMD is inherently acidic due to high concentrations of sulphuric acid (Simate and Ndlovu, 2014). Two predominant theories, (1) Arrhenius theory, and (2) Brønsted-Lowry theory, give two interrelated definitions of an acid and a base (Razzaq and Khudair, 2018). An Arrhenius acid is any species that increases the concentration of H^+ in aqueous solution. An Arrhenius base is any species that increases the concentration of OH^- in aqueous solution. In the Brønsted-Lowry definition, acids are proton donors, and bases are proton acceptors.

The predominantly acidic nature of AMD has made it to be extremely corrosive and polluting in nature (Agrawal and Sahu, 2009). The hazards associated with acidic pH in AMD on the environment and health have already been discussed in Chapter 5. Therefore, there is a need to develop techniques to recover acid from AMD and/or minimise its environmental and health effects. Actually, over the past half a century, meticulous efforts have been aimed at remediating AMD through acid removal so as to reduce the impact of the acidic water on the environment and produce water suitable for reuse (Johnson and Hallberg, 2005; Simate and Ndlovu, 2014; Nleya et al., 2016; Nleya, 2016). Indeed, apart from the production of reusable water and saleable metals, the recovery of sulphuric acid from AMD would also be used to offset its treatment costs (Simate and Ndlovu, 2014). Table 9.8 is a summary of the methods used to recover sulphuric acid from various wastewater solutions including AMD showing the recoveries, advantages, and disadvantages of the processes (Nleya et al., 2016). Furthermore, a study by Nleya et al. (2016) critically evaluated the technical and economic feasibilities of the processes in Table 9.8 for application to AMD, and the results of the study are given in Table 9.9. Based on the technical and economic feasibility results in Table 9.9, the freeze crystallisation and acid retardation processes are expected to be the most suitable technologies for acid recovery from wastewater solutions.

9.4.2 Selected Typical Studies of Recovery of Acid from Acid Mine Drainage

Several studies on processes shown in Table 9.8 have been conducted in the past and their fundamentals are well documented (Etter and Langill, 2006; Kim, 2006; Özdemir et al., 2006; Tjus et al., 2006; Agrawal and Sahu, 2009). This section only gives typical studies of recovery of acid from AMD only.

A study by Nleya (2016) focused on the removal of toxic heavy metals as well as the recovery of acid using low-cost adsorbents and acid retardation process, respectively. In the first part of the study, the adsorption efficiencies of zeolite and bentonite were found to be less than 50% for most metal ions, which were lower compared to the 90% efficiency obtained with cassava peel biomass. The second aspect of the study involved testing the feasibility of using a process known as acid retardation to recover sulphuric acid from metal-barren AMD. Acid retardation can be defined as a process which employs ion-exchange resins to selectively adsorb acids from solution while dissolved metal salts are rejected (Nleya, 2016). The term retardation emanates from the fact that the preferential adsorption of the acid causes

TABLE 9.8

Summary of Methods Used for Sulphuric Acid Recovery

Method	Solution Content	H_2SO_4 Recovery	Advantages	Disadvantages
Rectification	H_2SO_4, nitro compounds	R = 98.3%	Recovery of high-purity acid	High energy consumption High operating cost
Diffusion dialysis	H_2SO_4, Al H_2SO_4, Fe, V H_2SO_4, Ni H_2SO_4, rare earth sulphates	R = 82–90% R = 84% R = 80% R = 70–80%	High acid recovery Low pay back period Strong salt rejection	Not efficient at low acid concentration
Electrodialysis	H_2SO_4, Ni H_2SO_4, Fe H_2SO_4, Cu, Sb, As H_2SO_4, Fe, Na	R = 80–90% R = 90% R = Up to 99% –	Clean acid product Reduced solid waste for disposal	High operating cost Membrane fouling
Acid retardation	H_2SO_4, Fe H_2SO_4, Fe H_2SO_4, Ni	R = 74–96% R = 96% R = 70–95%	Low operating cost High acid recovery Small equipment size and space	Increases product volume High consumption of fresh water Dilute acid product
Crystallisation	H_2SO_4, Fe		Low cost Reduced waste for disposal	Risk of scale formation in crystalliser. Increased energy consumption
Solvent extraction	H_2SO_4, Cu H_2SO_4, Fe, Mn H_2SO_4, Zn	E = 75–79% E = 90% E = 90%	Can manage great volumes of solutions with high content of toxic solutes Clean acid product Only physical separation High throughput with compact equipment	Chemicals used are hazardous Pre-treatment is required to remove impurities Difficulties in stripping from Cyanex 923 Co-extraction of Fe and Zn

R = recovery, E = percentage extraction
Source: Nleya et al., 2016.

the movement of the acid on the resin bed to be retarded (slowed down), relative to the movement of the salts, resulting in the separation of the two entities (Sheedy, 1998; Sheedy and Parujen, 2012). The process is reversible and the acid can be recovered by water elution. In the study by Nleya (2016), sulphuric acid recovery from the metal barren solution was evaluated using Dowex MSA-1 ion-exchange resins. A column with inside diameter of 1.0 cm and 30 cm height was used in the tests. Sufficient resins were placed into the column to the required height. The AMD solution was fed in an upward

TABLE 9.9

A Summary of Capital and Operating Expenditures of Proposed Process Routes

Cost (US$)	Recti-fication	Diffusion Dialysis	Electro-dialysis	Solvent Extraction	Crystal-lisation	Acid Retardation
Fixed capital cost	1 193 165	381 780	1 523 478	442 956	407 268	307 973
Total estimated CAPEX	1 372 140	439 047	1752 000	509 399	468 368	354 169
Total estimated annual OPEX	427 789	247 752	266 809	125 124	154 798	124 810
Estimated annual revenue	≈9800	≈9800	≈9800	≈9800	≈10 320	≈9800

Source: Nleya et al., 2016.

flow direction and samples were collected from the top of the column at different time intervals, and analysed for metal, sulphate, and acid content. The results of the study showed that sulphuric acid can be recovered by the resins via the acid retardation process and could subsequently be upgraded to near market values of up to 70% sulphuric acid using an evaporator. Water of reusable quality could also be obtained during the acid upgrade process. Ideally, the process was stopped at approximately 70% acid (when approximately 98% of the water was evaporated) concentration, because almost no sulphuric acid could be detected in the water vapour up to an acid concentration of 70% (Nleya, 2016). An economic evaluation of the proposed process also showed that it was possible to obtain revenue from sulphuric acid which could be used to offset some of the operational costs in AMD remediation processes.

Martí-Calatayud et al. (2013) studied the recovery of sulphuric acid from AMD using an electrodialysis cell with three compartments. The ion-exchange membranes used in the study were heterogeneous HDX membranes (provided by Hidrodex®). The anion-exchange membrane (AEM, HDX 200) contained quaternary amine groups that were attached to the membrane matrix. The cation-exchange membrane (CEM, HDX 100) was charged with sulphonic acid groups and had a similar morphology to that of HDX 200. Both membranes had remarkably high ion-exchange capacities, which were 1.8 and 2.0 mmol/g for the AEM and the CEM, respectively (Buzzi et al., 2013). The structure of both membranes was reinforced with two nylon fabrics with the function of increasing their mechanical stability. The composition of AMD varied substantially depending on the source from which samples were collected. However, the AMD solution with the

TABLE 9.10

Composition of the Original Source of Acid Mine Drainage and the Synthetic
Solutions Used in the Electrodialysis Experiments

Solution	Fe(III) (mol L^{-1})	Na(I) (mol L^{-1})	SO$_4^{2-}$ (mol L^{-1})	pH
Acid mine drainage source	0.037	0.017	0.082	2.48
Synthetic solution: 0.02-M Fe$_2$(SO$_4$)$_3$ + 0.01-M Na$_2$SO$_4$	0.040	0.020	0.070	1.68

Source: Martí-Calatayud et al., 2013.

highest concentration of sulphates was selected as a basis for the study, since
the principal aim of the work was the recovery of sulphuric acid from AMD.
Synthetic solutions with a composition approximate to that of the original
AMD solution were prepared by mixing 0.02-M Fe$_2$(SO$_4$)$_3$ and 0.01-M Na$_2$SO$_4$.
Distilled water was used to prepare the synthetic solutions. The content of
the most concentrated species in the original AMD source is summarised
in Table 9.10, together with the concentrations and pH value of the synthetic
solutions.

The results of the study conducted by Martí-Calatayud et al. (2013) showed
that the recovery of sulphuric acid from AMD can be achieved by means of
an electrodialysis cell. Significant increases in sulphuric acid concentration
were obtained with the proposed scheme consisting of a three-compartment
electrodialysis cell with cation-exchange membrane and anion-exchange
membrane. An effective recovery of sulphuric acid free from Fe(III) species
was obtained in the anodic compartment as a result of the co-ion exclusion
mechanism in the membranes. The difference in the pH and pSO$_4^{2-}$ values
between the membrane phase and the external electrolyte promoted the
dissociation of complex species inside the membranes. This phenomenon
impeded the transport of Fe(III) and sulphates in the form of complex ions
towards the anodic and cathodic compartments, respectively. The current
efficiency values of the anion-exchange membrane at different current densities were approximately constant with time. However, the increase in the
recovery of acid decreased as the current increased. This result is explained
by the shift in the equilibrium at the membrane/solution interface as more
SO$_4^{2-}$ ions cross the anionic membrane and by the enhancement of the dissociation of water when the limiting current density is exceeded. The main
limitation of the process was related to an abrupt increase in the cell voltage due to the formation of precipitates at the surface of the cation-exchange
membrane.

In a study by Kesieme and Aral (2015) that has already been discussed
under the recovery of water from AMD (Section 9.3.2.2), the researchers
studied the potential and opportunities for DCMD to concentrate H$_2$SO$_4$
and recover fresh water from acidic process solutions. The study was also

aimed at identifying how membrane distillation can work in combination with solvent extraction in the mineral processing industry for acid recovery. After recovering water through the membrane distillation unit, the remaining concentrated acidic solution (see Table 9.7) was processed using solvent extraction so as to recover sulphuric acid. The organic system consisting of 50% TEHA and 10% ShellSol A150 (a 100% aromatic diluent) in octanol was used in the solvent extraction system. In the solvent extraction tests, the organic system was mixed with concentrated acidic solution at an A/O ratio of 1:2 and a temperature of 22°C. The loaded organic solution was stripped twice at O/A ratios of 2:1 and 1:5 at 60°C. The raffinate and the loaded strip liquors were titrated to determine acid concentrations for extraction and mass balance calculations. The results indicated that over 80% of H_2SO_4 was extracted in the solvent extraction system in a single contact from the waste solution (i.e., the concentrated solution from the membrane distillation) containing 245 g/L H_2SO_4 and metals with various concentrations. After three stages of successive extraction, nearly 99% of acid was extracted, leaving only 2.4 g/L H_2SO_4 in the raffinate. The extracted acid was easily stripped from the loaded organic solution using water at 60°C. After scrubbing the loaded organic solution at an O/A ratio of 10 and 22°C, 98–100% of entrained metals were removed in a single contact with only 4.5% acid lost in the loaded scrub liquor.

Three different types of NF membranes, namely, (1) a poly(piperazinamide) active layer (NF270), (2) a double active layer (poly(piperazinamide)/proprietary polyamide) (Desal DL), and (3) a sulphonated poly(ethersulphone) active layer (HydraCoRe 70pHT), were evaluated by López et al. (2019a) for the recovery of sulphuric acid from acidic mine waters and simultaneously increasing the concentration of valuable elements for further valorisation after the removal of iron. The NF270 and Desal DL membranes had the same top active layer based on a semi-aromatic poly(piperazineamide). However, Desal DL incorporates an additional proprietary second layer that approaches a tight UF and that was made of a material comparable to a polyamide. Both NF270 and Desal DL membranes possessed ionogenic amine ($R-NH_2$) and carboxylic ($R-COOH$) groups, which were responsible for the membrane charge. The IEPs for the two membranes were 2.5 and 4.0, respectively. The HydraCoRe 70pHT incorporated a sulphonated polyethersulphone as the active layer on a standard thin film composite membrane structure with polysulphone and polyester on the backside. The membrane charge was caused by the presence of sulphonic groups ($R-SO_3H$). Experiments were performed with a synthetic solution simulating the supernatant of a pre-treated acidic mine water from La Poderosa Mine at the Iberian Belt (Huelva, Spain). The synthetic solution was prepared by dissolving appropriate amounts of the metal-sulphate, nitrate, chloride and oxide salts in sulphuric acid.

The results of the study by López et al. (2019a) showed that NF technology offers a good chance to recover acids in the permeate and, at the same time, to concentrate metals in the retentate when treating acidic mine waters. Among the different membranes tested, NF270 showed the best performance, as it yielded negative H^+ rejections (i.e., permeate was more acidic than the feed solution), high metal rejections (>98%), and higher trans-membrane fluxes. The separation was highly influenced by the feed composition and the membrane chemistry of the active layer. Indeed, the membrane chemistry of the active layer (nature and acid-base membrane properties) and structure (single/double layer) were found to be strong parameters in the membrane-separation performance. The effect of the composition of active layer was observed when the same solution was filtered with different kinds of membranes. The composition of the active layer mainly influenced the solvent transport across the membrane and the superficial charge of the membrane. For example, polyamide membranes (NF270 and Desal DL) exhibited a positively charged surface leading to low anion rejections and high metal rejections. However, sulphonated polyethersulphone membrane (HydraCoRe 70pHT) was expected to exhibit a negative surface charge, leading to lower cation rejections and higher anion rejections than the other two membranes. Moreover, the fact that NF270 and Desal DL reached negative rejections of H^+, it made them suitable NF membranes to remove acidity from the feed solution. It must be noted that an increase in the pH in the feed solution, due to the negative rejections of H^+, could lead to a decrease in operational costs if any alkaline reagent is added downstream.

When the solution-electro-diffusion model coupled with reactive transport was applied to the study by López et al. (2019a), it was able to fit ions rejections properly by determining the membrane permeance values for each ion. Permeance values for different species of a given element were in agreement with the dielectric exclusion phenomenon. When the values of each element are compared, they could give information of the membrane charge. For instance, NF270 membrane permeances for Fe species followed the trend: $Fe(SO_4)_2^{2-}$ > $FeSO_4^+$ > $FeHSO_4^{2+}$ > Fe^{3+}, while for HydraCoRe 70pHT followed $FeSO_4^+$ > $Fe(SO_4)_2^{2-}$ > $FeHSO_4^{2+}$ > Fe^{3+}. The fact that $Fe(SO_4)_2^{2-}$ was the most permeable ion of Fe species for NF270 suggested that the membrane presents a negative charge at pH 1. As for the HydraCoRe 70pHT membrane, the fastest ion was $FeSO_4^+$ due to a negatively charged surface. Moreover, membrane permeance values were found to depend not only on the salt composition, but also on the total concentration (ionic strength).

López et al. (2019b) investigated the performance of a semi-aromatic poly(piperazine amide) NF membrane (NF270) in the treatment of streams generated from the off-gases treatment step of copper metallurgical industry. The streams contained a mixture of $H_2SO_4/HCl/H_3AsO_4$ and metallic species (Fe, Cu, Zn, Ni, Co, Cd) and alkaline metals (Na, K, Ca, Mg). The membrane performance was evaluated in terms of acid recovery and metal ions rejection taking into account their aqueous speciation in strong acid media. Transport of acids and metallic species through the membrane was evaluated under three different total acidity scenarios with pH values from 0.2 to 0.7 and modelled according to solution-electro-diffusion model coupled with reactive transport to determine the membrane permeances to species. The transport behaviour implications of both fully dissociated strong acids (H_2SO_4 and HCl) and weak acids (H_3AsO_4) with non-dissociated forms in the working conditions (0.2 < pH < 0.7) were critically evaluated in detail. The experiments were carried out in a cross-flow experimental set-up with flat-sheet membranes (0.014 m^2) placed in a test cell (GE SEPA™ CF II) with a spacer-filled feed channel. The set-up had a needle and a by-pass valve which allowed the variation of the cross-flow velocity and the trans-membrane pressure. The feed solution was kept in a thermostatic 30 L tank at a constant temperature ($25 \pm 2°C$) and was pumped into the membrane cell by a high-pressure diaphragm pump. The two outputs of the cell (permeate and concentrate) were recycled back to the tank to keep the same composition in the feed solution.

The experimental data of the study by López et al. (2019b) demonstrated that it was possible to recover strong acids (H_2SO_4, HCl) from hydrometallurgical streams by using a semi-aromatic poly(piperazine amide) membrane. The NF270 exhibited a positive surface charge at pH < 1.0, which favoured the transport of anions, while impeded the transport of metallic species that were present as cations. With the different solutions tested, the membrane exhibited negative chloride rejections and moderate sulphate rejections. Design of processes using more than one NF stage may allow the recovery of up to 90% of the total content of the strong acids. In addition, the membrane favoured the transport of non-metallic species (As) due to their presence as a non-charged species (H_3AsO_4) and not limited by the dielectric and Donnan exclusion. As a result, the levels of As are likely to limit the application of the technology. However, a pre-treatment stage if applied, such as using a reducing agent (e.g., H_2S or $S_2O_3^{2-}$) to obtain As(III) and then precipitate As as $As_2S_3(s)$ or as a mixture of S(s) and $As_2O_3(s)$, will maximise the recovery of the total content of the strong acids. The solution-electro-diffusion model coupled with reactive transport fitted the experimental ions rejections properly, and the calculated membrane permeances could be used to design stages in full-scale applications.

A comprehensive study by Nleya et al. (2016), which was briefly mentioned in Section 9.4.1, proposed a number of flowsheets in which acid could be recovered from AMD. These flowsheets are integrated to pre-existing technologies (see Table 9.8) that are used to recover various acids from other industrial waste streams. A first possible flow diagram for the recovery of sulphuric acid from AMD through rectification is shown in Figure 9.9. Rectification also known as slow distillation is a promising process for the recovery of high-purity sulphuric acid from waste acid solutions (Song et al., 2013). The rectification process which works by separating mixtures based on differences in volatilities of components in a boiling liquid mixture can concentrate and purify products in one step (Qian et al., 2011). It can be seen from Figure 9.9 that large quantities of heat energy are required for both pre-heating and rectification processes. In the rectification unit, the water

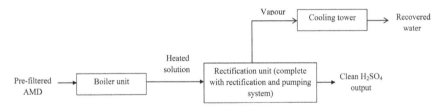

FIGURE 9.9
Proposed flow diagram for the recovery of sulphuric acid from acid mine drainage using the rectification process. (From Nleya et al., 2016.)

component is vaporised while sulphuric acid remains in solution and is recovered separately. Nleya et al. (2016) argued that when the rectification process is considered in the context of AMD, it might not be a suitable recovery process because the sulphuric acid content in AMD might be too low for any substantial economic benefits.

Figure 9.10 and Figure 9.11 are proposed flow diagrams for the recovery of acid from AMD using diffusion dialysis and electrodialysis, respectively. The concept of electrodialysis has been extensively discussed in Section 9.3.2.1. Diffusion dialysis makes use of a series of anion-exchange membranes to selectively attract the anion in the acid while electrodialysis uses an electric field to allow ions of one electrical charge to enter and pass through

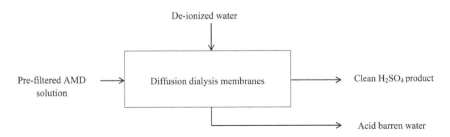

FIGURE 9.10
Proposed flow diagram for the recovery of sulphuric acid from acid mine drainage using diffusion dialysis method. (From Nleya et al., 2016.)

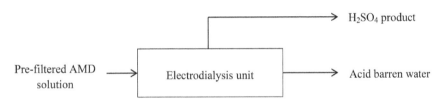

FIGURE 9.11
Proposed flow diagram for the recovery of sulphuric acid from acid mine drainage using electrodialysis method. (From Nleya et al., 2016.)

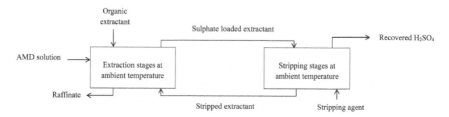

FIGURE 9.12

Proposed flow diagram for the recovery of sulphuric acid from acid mine drainage using solvent extraction method. (From Nleya et al., 2016.)

(perm-selectivity) (Nleya et al., 2016). In both flow diagrams (Figure 9.10 and Figure 9.11), the pre-filtered AMD solution is passed through a membrane unit where a clean acid and an acid barren water product can be obtained. In general, it can be seen that most of the membrane-separation processes are environmentally attractive. Apart from significantly reducing the solid waste, high-purity acid product can be obtained in most cases.

Figure 9.12 shows the proposed flowsheet for the recovery of sulphuric acid from AMD using solvent extraction method (Nleya et al., 2016). Solvent extraction process has long been used in the recycling and/or recovery of waste acids (Gottliebsen et al., 2000; Agrawal et al., 2008; Haghshenas et al., 2009; Shin et al., 2009), and is it a promising technology that can be extended to the recovery of sulphuric acid from AMD (Nleya et al., 2016). It is a clean and proven technology (Gottliebsen et al., 2000). The organic extractants tested have good selectivity for the acid; hence high recoveries can be expected (Gottliebsen et al., 2000; Agrawal et al., 2008).

Another promising technique for sulphuric acid recovery from AMD is freeze crystallisation. In addition to the acid, AMD also contains high quantities of ferrous ions which can also be recovered as crystals, purified and marketed as a commodity using the freeze crystallisation technique. A proposed flow diagram for the recovery of sulphuric acid via freeze crystallisation is shown in Figure 9.13. In the flow diagram, pre-filtered AMD solution is chilled in a heat exchanger unit using the cold sulphuric acid product. After the solution is pre-chilled, it enters the reactor where it is agitated and

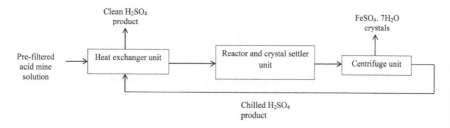

FIGURE 9.13

Proposed flow diagram for the recovery of sulphuric acid from acid mine drainage using freeze crystallisation technology. (From Nleya et al., 2016.)

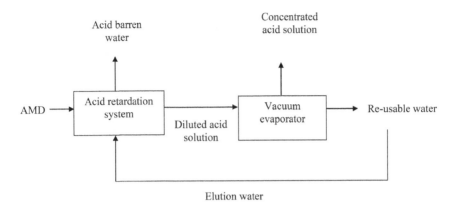

FIGURE 9.14
Proposed acid retardation set-up for sulphuric acid recovery from acid mine drainage. (From Nleya et al., 2016.)

chilled further until the ferrous sulphate heptahydrate crystals are formed. The settled crystalline solution is pumped to the centrifuge, where the crystalline product and the sulphuric acid solution are separated. The cooled sulphuric acid product is pumped back to the primary heat exchanger where it cools the incoming AMD solution and then collected for storage.

In addition to a study by Nleya (2016) that tested the feasibility of using acid retardation technique to recover sulphuric acid from metal-barren AMD, Nleya et al. (2016) proposed a flowsheet for the application of acid retardation process in the recovery of sulphuric acid from AMD as shown in Figure 9.14. In the proposed flow diagram a dilute stream of acid is obtained in the acid retardation process which can then be concentrated using a vacuum evaporator. Water of a quality suitable for recycling back to the system can also be obtained. A vacuum evaporator has the advantage of producing a large separation factor in the sulphuric acid/water system, and that it reduces the boiling point of the mixture which, subsequently, minimises the cost of heating (Nleya et al., 2016).

Simate and Ndlovu (2014) proposed an integrated process shown in Figure 9.15 that starts off with an AMD fuel cell where iron would be

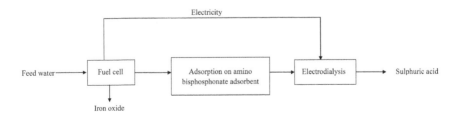

FIGURE 9.15
Proposed integrated process for the production of sulphuric acid from acid mine drainage. (From Simate and Ndlovu, 2014.)

precipitated as iron oxide at the anode. The water that contains most of the heavy metals would then be pumped to the adsorption circuit where the CH-collector (e.g., amino bisphosphonate adsorbent) would adsorb heavy metals directly from the wastewater (Turhanen and Vepsäläinen, 2013). The resulting water that is barren of metals is then pumped to the electrodialysis circuit where a more pure form of sulphuric acid and other residual metals are recovered (Martí-Calatayud et al., 2013). The electrodialysis process efficiency is expected to be high as metals which might cause membrane fouling would have been removed in the fuel cell and by the CH-collector. The products from this integrated process are expected to be iron oxide which could be sold as a pigment for paint (Hedin, 2003), cosmetics, and possibly other uses. The electricity produced from the fuel cell can act as a power source for the electrodialysis. The sulphuric acid also produced could be used in the leaching of metal ores.

9.5 Generation of Electricity from Acid Mine Drainage

9.5.1 Introduction

Over the years, microbial fuel cells (MFCs) have emerged as a promising yet challenging technology for converting organic waste including low-strength wastewaters and lignocellulosic biomass into electricity through the metabolic activities of the microorganisms that act as catalysts (Pant et al., 2010; Simate et al., 2011). In other words, MFC-based wastewater systems employ bioelectrochemical catalytic activities of microbes to produce electricity from the oxidation of organic and in some cases inorganic substrates present in urban sewage, agricultural, dairy, food, and industrial wastewaters (Gude, 2016). In principle, the MFC can enable the simultaneous treatment of wastewater such as AMD while generating electricity from organic matter in the wastewater. In other words, the MFCs treat wastewater and generate electricity at the same time (Bennetto, 1984; Habermann and Pommer, 1991). The MFC is a combined system with anaerobic and aerobic characteristics (Simate et al., 2011). The MFCs are designed for anaerobic treatment by bacteria in the solution near the anode, with the cathode exposed to oxygen (or an alternative chemical electron acceptor). Electrons released by bacterial oxidation of the organic matter are transferred through the external circuit to the cathode where they combine with oxygen to form water (Feng et al., 2008). It is noted that a combination of anaerobic-aerobic process can be constructed using a double-chamber MFC, in which effluent of anode chamber could be used directly as the influent of the cathode chamber so as to be treated further under aerobic condition to improve wastewater treatment efficiency (Wen et al., 2010).

The MFC is considered to be a promising and sustainable technology that would meet increasing energy needs, especially when using wastewaters as substrates (Du et al., 2007; Lu et al., 2009). Since it results in the production of electricity and clean water as final products, it may offset the operational costs of wastewater treatment plants (Lu et al., 2009).

9.5.2 Selected Typical Studies of the Generation of Electricity from Acid Mine Drainage

A study by Cheng et al. (2007) used fuel cell technologies to generate electricity while removing iron from the AMD. The AMD fuel cell (AMD-FC) was constructed from two plastic (Plexiglas) cylindrical chambers each 2 cm long by 3 cm in diameter separated by an anion-exchange membrane. The anode was a carbon cloth (non-wet-proofed) with a projected surface area of 7 cm^2 (one side). The cathode electrode was made by applying platinum (0.5 mg/cm^2) to a commercially available carbon cloth (30 wt% wet-proofed). The membrane was held between the two cylindrical chambers with a rubber O-ring to prevent leakage. The anode electrode was located at the end of one chamber and covered with a plastic end plate (5 × 5 × 0.6 cm). The cathode electrode was placed at the end of another chamber and covered with another end plate with a centre hole (3 cm in diameter), with the platinum-catalyst side facing the solution and another side facing the air (Figure 9.16). Platinum wires (1 mm in diameter) were used to connect both electrodes and used as terminals of the cell. The reactor was operated in open circuit mode for 0.5 h before connecting an external resistor (1000 Ω) to measure electricity generation.

The AMD-FC operated in fed-batch mode generated a maximum power density of 290 mW/m^2 at a Coulombic efficiency greater than 97%; and electricity generation was reported to be abiotic in nature. In the fuel cell system, ferrous iron (Fe^{2+}) was oxidised to ferric iron (Fe^{3+}) at the anode, and oxygen from the air was reduced to water at the cathode. In the best possible way, ferrous iron was completely removed through oxidation to insoluble Fe^{3+}, forming a precipitate at the bottom of the anode chamber and on the anode electrode. Several factors were examined to determine their effect on operation, including pH, ferrous iron concentration, and solution chemistry. Optimum conditions were reported at a pH of 6.3 and a ferrous iron concentration above 0.0036 M. These results suggested that fuel cell technologies can be used not only for treating AMD through removal of metals from solution, but also for producing useful products such as electricity and recoverable metals.

In a related study, Cheng et al. (2011) showed that fuel cell technologies are not only used for simultaneous treatment of AMD and power generation, but also they can generate useful products such as iron oxide particles having sizes appropriate for use in pigments and other applications. In the study, Cheng et al. (2011) used AMD-FC technique to generate spherical

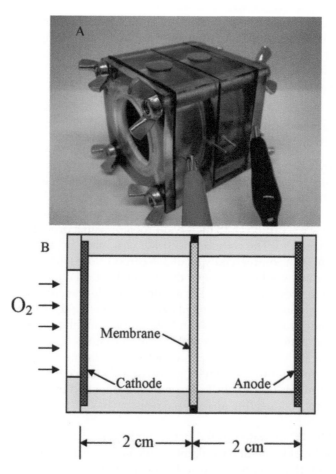

FIGURE 9.16
Laboratory scale prototype (A) and schematic (B) fuel cell system used to generate electricity. (From Cheng et al., 2007.)

nanoparticles of iron oxide that, upon drying, were transformed to goethite (a-FeOOH). This approach, therefore, provided a relatively straightforward way to generate a product that has commercial value. In other words, the results provided a method that could easily produce iron oxide particles that are essentially used in pigments and other products.

Hai et al. (2016) coupled membrane-free MFC with permeable reactive barrier (PRB) to treat AMD and generate electricity. The MFC-PRB system was carried out by employing parallel acrylic material columns, which were separated by a plate with a centre hole (3 cm inside diameter). The exterior chamber was used as PRB packed with corn cob media and inoculated with sulphate-reducing bacteria, and the cathode electrode was placed at the end of an exterior chamber and covered with another end plate. The inner chamber was directly used as an anode area that was filled with excess sewage sludge. In general, AMD lacks organic matter, therefore, additional organic substrate such as sewage sludge would need to be added so that it serves as microbial carbon sources for AMD treatment by MFCs (Jiang et al., 2009; Zhang et al., 2012; Peng et al., 2017). Sewage sludge is a by-product of biological wastewater treatment that requires treatment and disposal, but it contains high concentrations of organic matter, mainly protein and carbohydrate (Jiang et al., 2009; Zhang et al., 2012; Peng et al., 2017). The anode and cathode electrodes were made from a piece of 43.4-cm carbon rod and carbon felt without any pre-treatment and which were connected through a 1000 Ω resistor. The MFC-PRB system was continuously fed with synthetic AMD in a down flow mode using multiport peristaltic pumps, and it was operated for five periods at room temperature of $25 \pm 3°C$. The results showed that the MFC-PRB could continuously generate electricity from AMD, and the average sulphate removal rates of 51.2%, 39.8%, and 33.1% were obtained in effluents of 1000, 2000, and 3000 mg/L, respectively. High Cu^{2+}, Pb^{2+}, and Zn^{2+} removal efficiencies (99.5%) were also obtained during the operation, with most of the results being in the range of 0.01–0.05 mg/L which are far below the discharge level required by the Chinese government legislation of 0.5 mg/L, for example.

Peng et al. (2017) made use of the MFC to remove metals and sulphate from AMD using sewage sludge organics and simultaneously generated electricity. A total of six identical MFC reactors consisting of a vertical cylinder built using plexiglass were operated simultaneously and the reactors were closed during operation. To start up the MFC, 300-mL sludge was used to inoculate the MFC containing 700-mL AMD. To accelerate microbial growth, sodium acetate was added once into the reactors with an initial concentration of 2.0 g/L at the beginning of the start-up phase. After start-up, experiments were conducted in fed-batch mode at room temperature ($25 \pm 2°C$), and the reactors were replenished with fresh sludge and AMD every 10 days to initiate a new cycle. The results showed that under anaerobic conditions, 71.2% sulphate (from 2100 to 605 mg/L), 99.7% heavy metals, and 51.6% total chemical oxygen demand were removed at an electrode spacing of 4 cm and a sludge concentration of 30% (v/v) after 10-day treatment. A maximum power density of 51.3 mW/m^2 was obtained. Approximately 79.5% of the dissipated sulphate was converted to elemental sulphur or polysulphides. The sulphide concentration was kept below 20 mg/L. The concentrations of heavy metals were in the range of 0.02–0.06 mg/L in the effluent, which were far below the levels required by the Chinese government legislation. This study was one of many studies that showed the potential of synchronous degradation of residual sludge and treatment of AMD with electricity harvesting.

Lefebvre et al. (2012) investigated the bioelectrochemical treatment of AMD dominated by iron using acetate solution (as substrate component of the reactive mixture) in the anode compartment and simulated AMD ($FeCl_3$ solution) in the cathode compartment. The study was performed with two

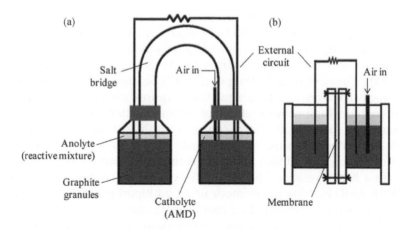

FIGURE 9.17
Schematic diagram of the (a) salt bridge and (b) membrane-microbial fuel cells. (From Lefebvre et al., 2012.)

different MFC designs: a salt bridge MFC and a membrane MFC. Overall, the salt bridge design – despite its rudimentary architecture – was intended to demonstrate the principle and feasibility of AMD bioelectrochemical treatment, while the membrane design was meant to improve iron recovery, thus demonstrating the potential of the technology. Figure 9.17 shows the two types of dual-chamber MFCs used in the study which only differed in the nature of the separator. The reactive mixture – used as the anolyte – consisted of a solution of nutrients, minerals, and vitamins, to which sodium acetate was added as the carbon source and electron donor (substrate). An artificial AMD, with 500 mg/L of Fe^{3+}, prepared using ferric chloride hexahydrate ($FeCl_3 \cdot 6H_2O$), was used as the catholyte. The selected Fe^{3+} concentrations corresponded to values found in natural AMD. The pH of the artificial AMD was left unadjusted at 2.4 ± 0.1. The experiment was carried out in batch mode. At the start of a batch test, the anode and the cathode chambers of the MFCs were filled with fresh solutions of anolyte and catholyte and the batch test was considered completed when the voltage recorded over an external resistance of 5 Ω dropped below 0.2 mV. The anode chamber was kept anaerobic throughout the batch testing, while the cathode chamber was constantly aerated using an aquarium air pump connected to an air diffuser.

Based on the findings from the study by Lefebvre et al. (2012), the AMD showed a potential to generate substantial amount of power (up to 8.6 ± 2.3 Wm^{-3}) in an MFC, which could help reduce the costs of full-scale bioelectrochemical treatment of AMD dominated with iron. In this study, Fe^{3+} was reduced to Fe^{2+} at the cathode of the MFC, followed by Fe^{2+} reoxidation and precipitation as oxy(hydroxi)des. In a broader perspective, the treatment process developed in this study could be attractive as a sustainable alternative for the treatment of AMD with high iron concentration.

This could involve the precipitation of iron prior to other challenging metals (e.g., Mn) or the co-precipitation of Fe-oxy(hydroxi)des with other dissolved metals in AMD. The optimum conditions were found at a charge of 662 Coulombs, which was achieved within 7 days at an acetate concentration of 1.6 g/L in a membrane MFC. This caused the pH to rise to 7.9 and resulted in iron removal of 99%. Treated effluent met the pH discharge limits of 6.5–9.0.

9.6 Production of Water Treatment Raw Materials

This section gives an overview of the developments and applications of AMD generated materials and/or AMD itself in the water treatment and purification processes. In particular, it covers advances in the use of AMD generated materials and/or AMD itself as adsorbents and coagulants or flocculants.

9.6.1 Production of Adsorbents

9.6.1.1 Introduction

A significant number of technologies for water treatment and purification have been developed. Of particular interest is the adsorption technology that offers several advantages for water treatment and purification because it can be operated at different scenarios besides its easy use, flexibility, versatile design, low-energy requirements, and cost-effectiveness (Bonilla-Petriciolet et al., 2017). The process exploits the ability of certain solids termed adsorbents to preferentially concentrate specific substances from the solution onto their surfaces (Bazrafshan et al., 2016). In this manner, the components which could be in either gases or liquid solutions can be separated from each other (Bazrafshan et al., 2016). There is a wide range of adsorbents used in adsorption processes and amongst them is sludge generated from AMD. The AMD sludge is produced from the treatment of AMD containing high amount of dissolved metals (Devi and Saroha, 2017). Sludge-based adsorbents are widely used for the removal of various pollutants from water and wastewater systems (Devi and Saroha, 2017). In fact, over the past years concerted efforts have been made to investigate the potential to use AMD sludge as adsorbents instead of disposing it into landfills at significant costs (Simate and Ndlovu, 2014).

9.6.1.2 Selected Typical Studies of the Use of Acid Mine Drainage Sludge as Adsorbents

In view of a wide range of research studies that accomplished taking out phosphorous from wastewaters using iron and/or aluminium hydroxide sludges, Wei et al. (2008) hypothesised that the AMD sludge containing a mixture of iron and aluminium hydroxide precipitates would be a suitable medium for the adsorption of dissolved orthophosphate from solution. In other words, sludges produced by the neutralisation of AMD are suitable for phosphorus sequestration as it is composed of aluminium and iron hydrous oxides, the same chemical forms produced when alum or ferric chloride is added to wastewater at near neutral pH. In addition, research by Sibrell et al. (2009) and Sibrell and Tucker (2012) has shown that dried AMD sludge or residuals can be used as a low-cost adsorbent to efficiently remove phosphorus from agricultural and municipal wastewaters. The phosphorus that has been adsorbed by the AMD sludge can later be stripped from the sludge and recycled into fertiliser; and the mine drainage sludge can be regenerated and reused for a number of additional applications.

A study by Wang et al. (2014) evaluated the application of AMD sludge, coal fly ash, and lignite as low-cost adsorbents for the removal of phosphate from dairy wastewater. In order to develop an insight into the potential deployment of the three low-cost adsorbents, batch adsorption isotherms and column experiments were conducted to investigate the breakthrough curves and cumulative removal under steady flow-through conditions. Both batch and column results indicated that fly ash exhibited the highest phosphate adsorption capacity followed by AMD sludge, while lignite had negligible adsorption capacity. These results are supported by the surface characteristics of the adsorbents that showed a significant presence of crystalline/amorphous Fe/Al/Si/Ca-based minerals and large surface areas for AMD sludge and fly ash. In other words, the differences were attributed to the physicochemical properties of the three low-cost adsorbents.

Studies by Wei and Viadero (2007a) investigated the utilisation of AMD sludge for the removal of Congo Red, an azo dye, from wastewater. Basically, the main goal was to ascertain the value-added use of AMD sludge as a low-cost waste material. The batch studies were carried out to examine the adsorption of Congo Red at different pH, temperature, dye concentration, contact time, and adsorbent dosage. A pre-coat filtration study was also conducted to test if the dye could be removed during continuous filtration by a pre-coated AMD sludge layer. Pre-coat filtration experiments were conducted using Buchner funnel filtration apparatus. First, a predetermined amount of wet AMD sludge (20–40 mL) was diluted to about 100 mL by adding deionised water. Then, the AMD sludge solution was poured onto the Whatman No. 2 filter paper in the Buchner funnel and a vacuum (75 ± 3 kPa) was applied. Once the AMD sludge was coated onto the filter paper to form a sludge layer and about 50 mL of filtrate was observed in the graduated cylinder, Congo Red solution (100, 150, or 200 mg/L) was poured into the funnel carefully until it was almost full. Based on batch adsorption and pre-coat filtration studies, sludge from AMD treatment successfully removed dissolved Congo Red from solutions by adsorption. Furthermore, it was noted that pre-coat filtration continuously removed Congo Red from aqueous solution with success, and dye removal mechanism via pre-coat filtration was found to be adsorption. It was observed that percent dye removal decreased with an increase in pH and the decrease was rapid when pH was above 8.5. The results also showed that the AMD sludge was effective in removing Congo Red over a broad temperature range. Through metal leaching tests, it was observed that most metals associated with AMD sludge remained insoluble when adsorption occurred at pH 6–10.

Studies by Edwards and Benjamin (1989) have also shown that the iron ferrihydrite component of AMD sludge from lime treatment plants can be used as a highly effective adsorbent for the removal of metals from water streams. Similarly, metal hydroxide sludge has also been used to remove carcinogenic dyes from wastewater (Netpradit et al., 2003).

The aim of a study by Cui et al. (2013) was to determine the applicability of coal mine drainage sludge for a wastewater treatment process. More specifically, the objective of the study was to elucidate the mechanism for Zn(II) removal using coal mine drainage sludge pre-treated at either 25 or 550°C. The study compared the AMD generated sludge with materials such as goethite, hematite, and calcite. Zeta potential analysis showed that coal mine drainage sludge dried at 25°C and coal mine drainage sludge dried at 550°C had a much lower IEP of pH than either goethite or calcite, which are the main constituents of coal mine drainage sludge. This indicates that the negatively charged anion (sulphate) was incorporated into the structural networks and adsorbed on the surface of coal mine drainage sludge via outer-sphere complexation. The study found that the removal of Zn(II) by coal mine drainage sludge was found to be primarily caused by sulphate-complexed iron (oxy)hydroxide and calcite. In particular, the electrostatic attraction of the negatively charged functional group, $FeOH-SO_4^{2-}$, to the dissolved Zn(II) provided high removal efficiencies over a wide pH range. Thermodynamic modelling and Fourier transform infrared spectroscopy (FT-IR) demonstrated that $ZnSO_4$ was the dominant species in the pH range of 3–7 as the sulphate complexes with the hydroxyl groups, whereas the precipitation of Zn(II) as $ZnCO_3$ or $Zn_5(CO_3)_2(OH)_6$ through the dissolution of calcite was the dominant mechanism in the pH range 7–9.6. From all of the analytical and batch test results observed, the study concluded that coal mine drainage sludge was a promising adsorbent for the treatment of various types of wastewater, including industrial wastewater and AMD, particularly, wastewater with a low pH. In terms of economic feasibility, the preparation of coal mine drainage sludge by drying at a low temperature was found to be favourable since there was no significant effect on the drying temperature.

9.6.2 Production of Coagulants and Flocculants

9.6.2.1 Introduction

Coagulation and flocculation have remained the most widely used processes for water and wastewater treatment (Simate et al., 2012; Simate, 2012). As a result, coagulation and flocculation are considered as the two key steps which often determine finished water quality (Zeta-Meter Inc., 1993). In water treatment plants, coagulation and flocculation are usually accomplished by the addition of trivalent metallic salts such as aluminium sulphate ($Al_2(SO_4)_3$) or ferric chloride ($FeCl_3$) (Simate, 2012).

According to Mwewa et al. (2019), there is a huge potential to recover alternative coagulants from AMD for water treatment. This is because two of the major constituents of AMD, iron and aluminium, can be recovered and engineered to function as coagulants (or AMD itself can be used directly for coagulation and/or flocculation). In fact, the high concentration of Fe and Al in AMD, as high as 5 000 mg/L for Fe and 500 mg/L for Al, has led to studies that have focused on developing an understanding of AMD's potential reuse as a coagulant in wastewater treatment. Section 9.6.2.2 gives selected typical examples of such studies.

9.6.2.2 Selected Typical Studies of the Use of Acid Mine Drainage as Coagulants and Flocculants

In a study by Mwewa et al. (2019), the application of an AMD-derived poly-alumino-ferric sulphate (AMD-PAFS) coagulant from coal AMD using chemical precipitation between pH 5.0 and 7.0 was evaluated. The efficiency of the AMD-PAFS was compared with conventional PFS coagulant in the treatment of brewery wastewater for turbidity, chemical oxygen demand and total dissolved solids removal. The effect of the coagulants on the electric conductivity of the wastewater was also evaluated. The results indicated that the recovery of Fe and Al from coal generated AMD at pH 5.0 was 99.9% for Fe and 94.7% for Al. With an increased pH of up to 7.0, the overall Al recovery increased to 99.1%. Although Al precipitation was 99.1% at pH 7.0, the precipitate formed at pH 5.0 was chosen for coagulant production due to the reduced chances of co-precipitation with other impurities should they exist in substantially higher concentrations. Dissolution of precipitate in 5.0% (w/w) sulphuric acid produced a coagulant containing 89.5% Fe and 10.0% Al. The coagulant produced had comparable characteristics to the commercially produced PFS coagulant. The subsequent brewery wastewater treatment tests showed that the AMD-derived coagulant was as effective as the conventional coagulants in the removal of chemical oxygen demand and turbidity. The total dissolved solids increased only slightly with an increase in coagulant dose for both the AMD-PAFS and PFS coagulants. In general, the increase in total dissolved solids is due to an increase in the number of solute particles or ions as a result of coagulant addition. From the results, it was clearly shown that the electric conductivity of the brewery wastewater increased as the dose of the coagulants increased. The initial conductivity of the original brewery wastewater sample was 3 510 µS/cm, but it was increased to 4 010 and 4 110 µS/cm for AMD-PAFS and PFS coagulants, respectively. The sporadic rise in electric conductivity observed in all the samples tested could be due to the presence of the dissolved ions in the wastewater coupled with the dissolved ions of the coagulants and the pH regulator (NaOH). The study by Mwewa et al. (2019) showed that the process can be easily integrated in existing AMD treatment plants, which would provide revenue and thereby subsidise the treatment costs. Furthermore, the issues associated with disposal of the voluminous sludge could be avoided, as the coagulant recovery would reduce the sludge volume by 95.0%.

Salama et al. (2015) developed a novel application using AMD directly for coagulation and/or flocculation of two morphologically different microalgae species for biomass recovery. There is no doubt the study showed that the environmentally recalcitrant AMD can be used as an effective flocculating agent for microalgae biomass. The effect of AMD dosage, microalgal cell density, and media pH (7 and 9) on flocculation efficiency was investigated. The AMD and microalgal biomass were mixed for 2 min at 500 rpm in a 200-mL glass beaker, followed by slow mixing for 10 min at 100 rpm to promote aggregation. Subsequently, the culture was transferred into a 100-mL gravimetric cylinder. A liquid sample of 5 mL was then collected at 2 cm below the surface of treated microalgae in gravimetric cylinder for optical density analysis to monitor microalgal settling. The study confirmed that AMD, as metal ions rich natural source (iron and aluminium ions), could be an effective option for harvesting of microalgal biomass. In the study, positively charged Fe(III) and Al(III) hydroxides in AMD were rapidly formed leading to the destabilisation of microalgal suspension, and thus making AMD useful for coagulation/flocculation of different microalgal species (*Chlorella vulgaris* and *Scenedesmus obliquus*) within 20 min at initial suspension pHs of 7 and 9. The flocculation efficiency using AMD at the optimal conditions was 89% and 93% for *S. obliquus* and *C. vulgaris*, respectively. Scanning electron microscope with energy-dispersive X-ray (SEM/EDX) micrograph of the microalgae aggregates revealed that the sweeping floc formation was the dominating mechanism.

Lopes et al. (2011) used AMD directly as a coagulant for the treatment of sewage wastewater. This study was based on the fact that when AMD is rich in iron in the form of Fe^{3+} (and secondarily Al^{+3}), it can be used as a coagulant. In addition, the AMD can also be applied in wastewater treatment as Fenton's reaction when it is rich in iron in the form of Fe^{2+}. The sewage wastewater treatment tests were carried out in a standard jar test apparatus using two different methodologies (i.e., simple coagulation and Fenton's reaction). For simple coagulation, 15 mL of AMD was added to 1 L of sewage, which provided a concentration of 264 mg/L Fe and 38 mg/L Al. The solution was adjusted to a pH of 3.5 and mixed for 1 h. Thereafter, the pH was adjusted to 9.0 with a 2-M NaOH solution for metals precipitation. Sludge settling was carried out for 1 h in Imhoff's cone. Thereafter, the sludge was filtered, dried for 24 h at 60°C, and weighed. In the case of Fenton's reaction, the same amount of AMD and sludge were used as in the simple coagulation test thus obtaining the same Fe and Al concentration. The solution was initially mixed for 1 h before adjusting its pH to 3.5 with a 2-M HCl solution followed by the addition of 1.25 mL/L of 35% (w/v) H_2O_2. Fenton's reaction was conducted over 3 h. Thereafter, the mixture was adjusted to pH of 9.0 with a 2-M NaOH solution for metals precipitation. The rest of the procedure was similar to the simple coagulation test. In both cases, the treated water was analysed by considering the following parameters: pH, total solids, suspended solids, dissolved solids, settleable solids, thermotolerant coliforms, total coliforms, total Fe, total Al, chemical oxygen demand, biological oxygen demand, total Kjeldahl nitrogen, phosphorus, and sulphates. The results showed that the AMD, especially the most concentrated ones (in terms of iron), can be used in wastewater treatment. The treatment by simple coagulation allowed significant reductions in the sewage levels of suspended solids, organic matter, and phosphorous. The treatment by Fenton's reaction allowed better results in terms of the final concentration of suspended solids, organic matter, phosphorous and also promoted the disinfection of the treated water.

Another study that used AMD directly as a coagulant and compared it with ferric chloride in the treatment of municipal wastewater was performed by Rao et al. (1992). The jar tests were conducted using a 6 place Phipps and Bird jar test apparatus. When ferric chloride or AMD was added as a sole coagulant to each jar, rapid mixing was done for 2 min at 100 rpm, followed by flocculation for 20 min at 30 rpm, and sedimentation for 30 min. Effluent samples (supernatants) were then analysed for turbidity, pH, suspended solids, total phosphorus, and heavy metals. The study showed that the AMD was as effective as the commercial ferric chloride coagulant for the removal of suspended solids/turbidity, and phosphorus. In some cases, on an equivalent Fe^{3+} dose basis, AMD was even more effective than commercially available ferric chloride due to the presence of Al^{3+} in AMD. However, as a result of the presence of other heavy metals, especially Zn as in the case of a study by Rao et al. (1992), it will be necessary to pre-treat the AMD before use as coagulants with domestic wastes so as to allow discharge of the treated effluent into typical receiving water bodies. The initial experimental work by Rao et al. (1992) led to other studies having been conducted to recover ferric sulphate coagulant by reacting the ferric hydroxide precipitate formed from AMD at pH 3.5–3.6 with sulphuric acid (Rao et al., 1992). In addition, the use of dodecylamine surfactant to avoid co-precipitation of other metals improved the purity of the precipitate, and thus the recovered coagulant was effective in municipal wastewater treatment and compared favourably with conventional coagulants (Rao et al., 1992).

Menezes et al. (2009) produced a ferric sulphate rich solution from acidic coal mine drainage and was assessed for its applicability as a coagulant. Iron was recovered from the AMD by an oxidation/selective precipitation process. Initially, the AMD was aerated for 24 h at pH 2.5–3.0 to convert all of the Fe^{2+} to Fe^{3+}. Thereafter, the pH of the solution was increased to and maintained at 3.8 ± 0.1, with the addition of 4-N NaOH solution, in order to precipitate the iron as ferrichydroxide/oxyhydroxide, which was further separated from the AMD by centrifugation at 3000 rpm. The final precipitate was dissolved in sulphuric acid to achieve a clear solution, which was used as a chemical coagulant. The coagulation procedure was carried out using a 1000 mL of raw water sample from a lake using a conventional jar test apparatus. The sludge generated was filtered using a quantitative filter paper, dried, and weighed. The treated water was analysed for: pH, suspended solids, turbidity, colour, conductivity, metals (Fe, Al, Mn, Zn, Cu, Cr, Cd, Pb, and As), hardness, and sulphate. Overall, the study showed that by precipitating the iron at pH 3.8, followed by dissolution in sulphuric acid, a coagulant consisting of 12.4% iron and 1.3% aluminium was able to be produced. This coagulant production process could reduce the overall volume of sludge by 70% (Menezes et al., 2009). The raw water treatment tests proved that the AMD generated coagulant was as efficient as the conventional chemical coagulants used in water treatment plants.

9.7 Utilisation of Acid Mine Drainage Sludge

9.7.1 Introduction

As stated in a number of chapters in this book, AMD has always been one of the mining problems that is difficult to avoid. The AMD that has not been treated before it is discarded into the water bodies can cause serious negative impacts to the environment because of its low pH values and higher content of heavy metals (Amanda and Moersidik, 2019) as discussed in Chapter 5. As extensively discussed in Chapter 7, the treatment of AMD can be carried out by active and passive methods. Unfortunately, both AMD treatment methods produce sludge with various compositions depending on the treatment type, the use of lime, and the quality of the water treated (Amanda and Moersidik, 2019). The objective of this section of the chapter is to discuss and illustrate the potential of AMD sludge as a valuable material. One of the applications of sludge – adsorbents – has already been discussed in great detail in Section 9.6.1 of this chapter; and the rest of the other pertinent applications of AMD sludge are discussed in Section 9.7.2. Simate and Ndlovu (2014), Ndlovu et al. (2017), and Rakotonimaro et al. (2017) also highlighted some of these applications in their publications.

9.7.2 Selected Typical Studies of the Reuse of Acid Mine Drainage Sludge

9.7.2.1 Production of Fertiliser

Fertiliser is considered as a material that is produced in order to supply elements, in a readily available form, that are known to be essential for plant growth and development. Zinck (2006) argues that low metal content sludges, such as sludges from coal mining operations, which have been found to have excess alkalinity present in the sludge, can be utilised to raise soil pH. A study by Dobbie et al. (2005) investigated the use of phosphorus-enriched ochre (hydrous iron oxide sludge) as a phosphorus fertiliser. The AMD sludge according to Sibrell et al. (2009) is a waste product produced by the neutralisation of AMD and consists mainly of the same metal hydroxides used in traditional wastewater treatment for the removal of phosphorus. Ideally, when ochre (or simply AMD sludge) is used for remediation of wastewaters, it adsorbs phosphorus (in the form of inorganic phosphate) from the solution (Heal et al., 2003; Shepherd, 2017) effectively which makes the resulting phosphate-enriched ochre a potential phosphorus fertiliser (Dobbie et al., 2005). In the study by Dobbie et al. (2005), pot and field experiments were set up to assess performance and environmental acceptability of ochre as a fertiliser, using grass and barley as test crops, as well as birch and spruce tree seedlings. It was noted from the study that applying phosphate-saturated ochre as a fertiliser increases the phosphorus status of soils and has a useful liming effect. Phosphate-saturated ochre is also less water-soluble than conventional phosphorus fertiliser, thus reducing the potential for diffuse pollution from agricultural land. The results in the study by Dobbie et al. (2005) also showed that the ochre caused no metal contamination, but other AMD sludge sources would need to be monitored to ensure that they do not contain undesirable concentrations of metals. The slow release of phosphorus from phosphate-saturated ochre means that less frequent applications would be required than when using conventional phosphorus fertiliser. It was found that all crops studied grew well comparatively when using fertiliser with phosphate-saturated ochre or with conventional phosphorus fertiliser.

Heal et al. (2004) state that when the phosphorus removal capacity of ochre is finally exhausted after removal of phosphorus from wastewater such as agricultural runoff and sewage effluent, the "spent" material will require removal and disposal, and a more sustainable alternative to landfill disposal is to recycle the phosphorus as a fertiliser. Pot experiments and field trials comparing barley and grass grown in soils amended with phosphorus-saturated ochre with the plants grown with conventional phosphorus fertiliser studied by Heal et al. (2004) showed that ochre additions improved soil fertility and increased the pH of the soil whilst the same crop yields were maintained similar to conventional fertiliser. At the end of the growing season, the results also showed that there was more phosphorus available in the ochre-amended soil than in soil treated with conventional fertiliser, indicating that phosphorus-saturated ochre had a further desirable property of acting as a slow-release fertiliser, thus reducing the need for future phosphorus fertiliser applications.

Several other studies also suggested that AMD sludge has the potential to adsorb phosphorus from agricultural wastewaters for possible use as a fertiliser (Adler and Sibrell, 2003; Fenton et al., 2009).

9.7.2.2 *Production of Iron Pigments*

Studies have shown that the sludge obtained from AMD can be considered for the production of inorganic pigments (Hedin, 1998; Hedin, 2003; Marcello et al., 2008; Michalková et al., 2013) and magnetic particles like ferrites (Wang et al., 1996). Indeed, a range of products with various purities, phase compositions, and properties, including surface properties, can be synthesised from AMD depending on the reaction conditions (Michalková et al., 2013). To produce commercially usable iron oxides as raw material for production of pigments, additives to ceramics, etc., treatment of AMD using a two-step selective precipitation process was developed (Hedin, 1998; Hedin, 2003). The two-step process that uses magnesium oxide and sodium hydroxide resulted in the ferrous and ferric oxyhydroxide sludge that can be thermally transformed to basic ferric pigment. A study by Hedin (2003), however, indicated that while the end product was of high quality, the costs associated with the processing made the materials more costly to produce than mined oxides although this may be offset when considering the high cost of hydrous ferric oxide disposal.

A study by Bernardin et al. (2006) used acid drainage mud from a coal mine to produce ceramic pigments. The raw material was collected at an effluent treatment station, and the mud was dried (105°C, 8 h), ground (250 µm) and calcined (~1 250°C). The calcined pigment was then micronised (D_{50}~2 µm). After calcination and micronisation, mineralogical analyses (XRD) were used to determine the pigment structure at 1 250°C. Finally, the pigments were mixed with transparent glaze and fired in a laboratory roller kiln (1130°C, 5 min). The results showed that the drainage residue can be used as a pigment only when mixed with pure oxides or as part of a commercial pigment. When used alone, the residue pigment presented a faded brown colour, inadequate for ceramic glazes. However, when mixed with transparent glaze, the residue pigment had better results, but was still poor compared with a commercial pigment. Nevertheless, the AMD residue could be used in other ceramic applications, as filler for brick pastes and other ceramic products. Most importantly, the study has shown that the residue can be eliminated from the environment through the production of pigments. A similar study was carried out by Marcello et al. (2008) who investigated the use of hydrous ferric oxides from active coal mine drainage treatment as pigment within ceramic tile glaze. Favourable results similar to the ones obtained by Bernardin et al. (2006) were also obtained when the ferrous hydrous oxides were blended with an industrial standard pigment.

Research by Cheng et al. (2007) and Cheng et al. (2011) has also shown that fuel cell technologies are not only used for simultaneous AMD treatment and power generation, but also generate iron oxide particles having sizes appropriate for use as pigments and other applications. As already discussed in Section 9.5, a fuel cell called an AMD fuel cell based on an MFC was developed and used during the studies. During the AMD treatment process in the studies, ferrous iron was oxidised in the anode chamber under anoxic conditions, while oxygen was reduced to water at the cathode. Ferrous iron was completely removed through oxidation to insoluble ferric iron and precipitated at the bottom of the anode chamber. The particle diameter of the iron oxides could be controlled by varying the conditions in the fuel cell, especially current density, pH and initial ferrous iron concentration. Upon drying, the iron oxide particles were then transformed to goethite (α-FeOOH).

Silva et al. (2019) evaluated several processes for purifying iron sludge from AMD so as to obtain a yellow pigment (goethite) of good quality. The study optimised the process for precipitating iron (III) selectively by assessing three variables (the reagent, the number of washes, and the separation method). Two alkaline agents (sodium hydroxide or sodium bicarbonate) with different neutralisation powers and two processes for solid-liquid separation (filtration or centrifugation) were used. In other words, the experiments were carried out by causing precipitation with strong (NaOH) and weak ($NaHCO_3$) bases and removing other metals from the sludge by washing and filtering the sludge or by centrifugation. The results of the study found that high quality goethite can be produced from AMD effluent provided that the process for recovering iron can remove the contaminants, particularly aluminium which adversely affects the growth of crystals, thereby preventing goethite from taking an acicular form, which is characteristic of pigment goethite. Basically, the results showed that the colour, type, and morphology of the compounds formed depended on the number of contaminants; and that the removal of various contaminants was strongly dependent on the type of reagent used and less dependent on the separation process and the number of washes. In other words, the purification results indicated that it was the kind of reagent which was mainly responsible for separating iron and aluminium during neutralisation process. When the reagent was $NaHCO_3$, 67% of the samples produced yellow pigment; whereas when the reagent was NaOH, 33% of the samples produced yellow pigment. The results clearly show that the weak base ($NaHCO_3$) prevents aluminium from contaminating the sludge during the precipitation process.

Lottermoser (2011) states that reuse of mine wastes allows their beneficial application, whereas recycling extracts resource ingredients or converts wastes into valuable products. In the study by Lottermoser (2011) various reuse and recycling options that have been proposed for mine wastes by numerous researchers were listed. Extraction of hydrous ferric oxides for paint pigments and extraction of manganese for pottery glaze were considered as two of the reuse and recycling options for AMD sludge.

9.7.2.3 Building and Construction Related Materials

When AMD is treated, dewatered and dried, the resulting sludge is composed largely of inorganic components that are suitable for use in building materials such as in the manufacture of cement (Simate and Ndlovu, 2014;

Michael, 2016; Ndlovu et al., 2017; Rakotonimaro et al., 2017). Basically, many of the constituents of sludge are the same as those used in cement manufacturing (Simate and Ndlovu, 2014; Michael, 2016; Ndlovu et al., 2017). For example, calcite, gypsum, silica, Al, Fe, and Mn are common raw materials for cement (Simate and Ndlovu, 2014; Ndlovu et al., 2017). Therefore, the components that make up AMD treatment sludge such as gypsum, calcite, and ferrihydrite can be utilised as raw materials in the manufacture of construction materials and other products (Simate and Ndlovu, 2014; Ndlovu et al., 2017). According to Michael (2016), calcium, iron, and aluminium are three of the four principal components of Portland cement; therefore, the use of AMD sludge as a feedstock for the manufacture of cement could have both economical and environmental benefits. Some studies have actually suggested that sludge can replace up to as much as 30% Portland cement in blended cement (Tay and Show, 1994); thus in such cases, sludge is expected to lower the use of binders (Rakotonimaro et al., 2017).

In some studies it was observed that the high aluminium content in sludge produced from the treatment of acidic drainage at coal and gold mines could be used for the production of aluminous cement (Lubarski et al., 1996). The production of bricks by adding sludges of various compositions of inorganic components has also been studied by several researchers (Benzaazoua et al., 1999; Rouf and Hossain, 2003; Weng et al., 2003; Benzaazoua et al., 2006; Mahzuz et al., 2009; Hassan et al., 2014). The studies showed that the addition of sludge coupled with high curing temperatures produced bricks of high quality. Some of the studies found that the bricks manufactured with the addition of sludge had high comprehensive strength compared to normal clay bricks (Rouf and Hossain, 2003). However, though there was less arsenic release by leaching from bricks, the presence of arsenic in sludge does not produce good quality bricks and hence sludge that contains arsenic in large quantities is not preferable for manufacturing bricks (Mahzuz et al., 2009). There is no doubt that sludge containing inorganic materials in reasonable quantities can be utilised in building and construction related materials and thus reduce mining of raw materials for production of building material and construction materials (Simate and Ndlovu, 2014; Ndlovu et al., 2017).

9.7.2.4 Material for Carbon Dioxide Sequestration

The increasing CO_2 concentration in the Earth's atmosphere, mainly caused by fossil fuel combustion, has led to concerns about global warming (Montes-Hernandez et al., 2008). Without drastic market, technological, and societal changes, CO_2 concentrations are projected to increase to alarming levels in the near future (Feely et al., 2004; Olajire, 2013). It is, therefore, paramount that carbon capture and sequestration are instituted if meaningful CO_2 reduction is to be achieved.

Among various technologies for capturing and storing CO_2, mineral carbonation technology has been found to be a potentially attractive sequestration technology for permanent and safe storage of CO_2 (Olajire, 2013). Mineral carbonation technology is a process whereby CO_2 is chemically reacted with calcium and/or magnesium containing minerals to form stable carbonate materials which do not incur any long-term liability or monitoring commitments (Olajire, 2013). In other words, mineral carbonation technology stores CO_2 by reacting natural minerals and industrial by-products containing a lot of calcium or magnesium with CO_2 and subsequently forming carbonate minerals (Lee et al., 2016). In general, the mineral carbonation process consists of extracting reactive calcium or magnesium from the raw materials, and, thereafter, the leached calcium or magnesium ions react with CO_2 to form the carbonate minerals in high pH conditions (Lee et al., 2016).

According to Zinck (2006), the same mechanism that generates CO_2 during the production of lime can be utilised to sequester CO_2. In this regard, CO_2 gas can react with AMD treatment sludges and iron-rich metallurgical residues to produce solid calcium, magnesium, and iron carbonates while stabilizing the sludge/residue and its impurities. Furthermore, the extraction of calcium ions from the raw materials will not be necessary if neutralised mine drainage is utilised because calcium ions are already present in the AMD solution (or sludge) through the reaction between AMD and dissolved lime that is added during the AMD neutralisation process. In comparison with other common mineral carbonation processes, carbonation utilizing the neutralisation process of mine drainage does not need pre-treatment and any additional facilities to sequester CO_2. There is also an additional advantage of short treatment time and the process can be carried out at ambient temperature and pressure.

A study by Lee et al. (2016) demonstrated the concept of using AMD sludge for CO_2 sequestration at laboratory scale on both synthetic and real AMD. In the study, hydrated lime, as used in the process of neutralisation, was added to adjust the pH of AMD solution and evaluated its feasibility as a probable technology for CO_2 sequestration through carbonation. In the first step, hydrated lime was added to the mine drainage, and then the mine drainage was stirred for 5 min in order to increase its pH. Thereafter, CO_2 gas was injected into the mine drainage until the pH reached 8.3, which is the minimum pH level for the production of carbonate ions. To evaluate the efficiency of CO_2 injection, two sets of experiments were carried out: (1) a carbonation experiment in which CO_2 was injected, and (2) a non-carbonation experiment without any CO_2 injection. The results showed that as hydrated lime was added into the mine drainage the overall pH increased up to about 12. In the carbonation treatment, the pH decreased after CO_2 injection because CO_2 generates H^+ ions when dissolved in water (i.e., $CO_2 + H_2O \leftrightarrow H^+ + HCO_3^-$). In the non-carbonated study in which there was no CO_2 injection, the pH remained high at about 12. In the case of real AMD, 1 kg of mine drainage could retain CO_2 of up to 0.54 g through carbonation treatment using the neutralisation process. Undoubtedly, CO_2 sequestration using the neutralisation process of mine drainage can be considered as a positive technique in terms of sustainable development.

Merkel et al. (2005) developed a sustainable low risk concept – CDEAL – on how to incorporate CO_2 into the subsurface and thus exclude it from the atmosphere. However, the results of the concept do not seem to have been published. The overall goal of CDEAL was twofold: (1) reduce the CO_2 emissions into the atmosphere, and (2) rehabilitate contaminated, acidic mine waters using carbonation. As the storage of CO_2 would be in the form of carbonate, it would, therefore, be sustainable. Thus, in this case, CDEAL would have positively contributed to the reduction of greenhouse gas emissions by CO_2 sequestration. The details of the concept indicate that CDEAL would only use pre-treated mine water and mine water with elevated CaO-contents, where an excess of CaO existed and can be used to react with CO_2 to form $CaCO_3$. In addition, where iron-hydroxide sludge is available in great amounts in some parts of the open pit lakes and the waste rock piles, it could be used as a reacting material as well.

Unger-Lindig et al. (2010) conducted a study that investigated whether alkaline cations in both the deposited sludge and the pore water can be used to improve the alkalinity in lake water, when CO_2 was added. The batch test results showed that addition of low-density sludge to acidic water (from mining lake) increased the pH of the water and the injected CO_2 could be captured mainly in the form of metal-bicarbonate complexes. According to Rakotonimaro et al. (2017), this was associated with the availability of oxy-hydroxides contained in the AMD sludge. Rakotonimaro et al. (2017) state further that the advantage of sludge incorporation in an acidic pit lake, for CO_2 sequestration, is its capacity to be employed as a neutraliser because it still contains unreacted hydrated lime (or calcite if dried), thus reducing the acidity of the lake by up to 30% and stabilise the sludge itself. However, depending on the concentration of the elements and mineral solubility in the AMD sludge, the possibility of contaminant release into the surrounding environment is a serious risk (Rakotonimaro et al., 2017).

9.7.2.5 Stabilisation of Contaminated Soil

Soils may become contaminated by the accumulation of heavy metals and metalloids through emissions from the rapidly expanding industrial areas, mine tailings, disposal of high metal wastes, leaded gasoline and paints, application of fertilisers on land, animal manures, sewage sludge, pesticides, wastewater irrigation, coal combustion residues, spillage of petrochemicals, and atmospheric deposition (Khan et al., 2008; Zhang et al., 2010; Wuana and Okieimen, 2011). Immobilisation, soil washing, and phytoremediation techniques are frequently listed among the best demonstrated available technologies for remediation of heavy metal-contaminated soils (Wuana and Okieimen, 2011).

At the moment, research has shown that AMD sludge which is found in abundance contains lots of metal oxides (or hydroxides) that may be useful for heavy metal stabilisation in soils (Kim et al., 2014). For example, Tsang et al. (2013) explored the potential use of AMD sludge and carbonaceous materials (green waste compost, manure compost, and lignite) for minimizing the environmental risks of As and Cu in the soil. After 9-month soil incubation, significant sequestration of As and Cu in soil solution was accomplished by AMD sludge, on which adsorption and co-precipitation could take place. However, in a moderately aggressive environment, AMD sludge only suppressed the leachability of As, but not Cu. Therefore, the provision of compost and lignite augmented the simultaneous reduction of Cu leachability, probably via surface complexation with oxygen-containing functional groups. Under continuous acid leaching in column experiments, combined application of AMD sludge with compost proved more effective than AMD sludge with lignite. This was attributed to the larger amount of dissolved organic matter with aromatic moieties from lignite, which may have enhanced Cu and As mobility.

The main objective of a study by Lee et al. (2013) was to investigate the effectiveness of soil stabilisation treatments using waste resources such as calcined oyster shell and coal mine drainage sludge. The study focused, particularly, on the feasibility of the simultaneous stabilisation of As and other heavy metals using mixed stabilizing agents. Both batch and column-leaching tests were used as evaluation methods, and the free and easy movement (or mobility) of As and other heavy metals after stabilisation treatments was compared to that of the control samples. The overall results of both the batch and column tests indicated that a combination of calcined oyster shell and coal mine drainage sludge was effective for stabilizing As and the other heavy metals. More specifically, in the acid extraction experiments, after the batch wet-curing process, the stabilisation efficiencies of As, Pb, and Cu were more than 90%, compared to the control experiments. In addition, in the column tests, the stabilisation process successfully prevented the migration of contaminants by leachate infiltration into the lower part of the soil samples, which suggested the feasible application of oyster shell and coal mine drainage sludge waste resources for stabilisation of As and other heavy metals in the soil. A similar study to that of Lee et al. (2013) in which calcined oyster shell was combined with coal mine drainage sludge was performed by Moon et al. (2016). The results also showed a good retention of As (>93%), Cu (>99%), and Pb (>99%). However, it is noted that the calcined oyster shell wastes mixed with mine sludge might raise calcium content in the soil, thus inducing a potential increase of hardness in the surrounding water in case of any leaching (Rakotonimaro et al., 2017).

Kim et al. (2014) studied stabilisation of heavy metals in agricultural soils affected by the abandoned mine sites nearby using AMD sludge. The results indicated that AMD sludge could be applied to soil contaminated with heavy metals as an alternative for reducing heavy metal mobility and bioavailability. Ko et al. (2012) investigated the stability of arsenic in solution and soil using various additives, such as limestone, steel mill slag, granular ferric hydroxide, and AMD sludge. The total concentrations of arsenic in the area where soil used in the study was collected ranged up to 145 mg/kg. After the stabilisation tests, the removal percentages of dissolved As(III) and As(V) were found to differ depending on the additives employed. Approximately 80% and 40% of the As(V) and As(III), respectively, were removed with the use of steel mill slag. The addition of limestone had a lesser effect on the removal of arsenic from the solution. However, more than 99% of arsenic was removed from solution within 24 h when using granular ferric hydroxide and AMD sludge, and similar results were observed when the contaminated soils were stabilised using granular ferric hydroxide and AMD sludge. These results suggested that granular ferric hydroxide and AMD sludge may play a significant role on the arsenic stabilisation. Moreover, the result in the study by Ko et al. (2012) showed that AMD sludge can be used as a suitable additive for the stabilisation of arsenic.

9.7.2.6 Covers in the Prevention of Acid Mine Drainage

In Chapter 6, several techniques for preventing the generation of AMD have been discussed. One of the methods involves the creation of physical separation barriers for water and oxygen such as the use of dry covers that have a number of roles (Pozo-Antonio et al., 2014). The AMD sludge was given as an example of alternatives for covering materials in place of natural soil in the prevention of AMD generation (Demers et al., 2017). This section gives an in-depth discussion of studies where AMD sludges have been used as covers for acid-generating waste materials.

A number of studies have been performed by Demers et al. with the view of using AMD sludge to control AMD produced by tailings and waste rocks (Demers et al., 2015a,b; Demers et al., 2017). In the first study by Demers et al. (2015a), the possibility of using AMD treatment sludge as a cover component for controlling AMD generation by tailings and waste rocks was investigated in the laboratory. Column experiments were conducted in order to identify the potential mixtures that could reduce acid generation when placed over acid-generating tailings and waste rock. The sludge-waste rock mixtures, when placed on waste rock, were not able to limit the transport of gaseous oxygen. However, the tests on the use of sludge-waste rock mixtures placed over waste rock demonstrated the capacity of the sludge to neutralise part of acid generated by waste rocks. In addition, the sludge waste rock mixtures significantly reduced metal loading in the effluents. Covers made of sludge and tailings mixture were able to reduce the generation of copper, zinc, calcium, and sulphur from the tailings; and acidic conditions were not observed for any test that was conducted, including the control conditions. The second study by Demers et al. (2015b) involved field work in order to evaluate the effectiveness of waste rock-sludge and tailings-sludge mixtures. The field results showed that waste rock-sludge mixture placed over waste rock was able to reduce the generation of AMD from the waste rock, therefore, confirming laboratory results, and was able to produce a neutral effluent with low concentrations of dissolved metals. The tailings-sludge mixture placed over tailings, with an evaporation protection layer, maintained a high volumetric water content and reduced sulphide oxidation from the tailings as exhibited by a neutral effluent. The two studies by Demers et al. (2015a,b) highlighted two sludge valorisation options as follows: (1) the use of waste rock-sludge mixture to reduce, at least temporarily, acidity and metal loads from a waste rock pile effluent, and (2) the use of tailings-sludge mixtures as cover to limit oxygen transport towards acid-generating tailings.

Mbonimpa et al. (2016) investigated the geotechnical properties of silty soil–sludge mixtures as possible components for covers with capillary barrier effects to prevent AMD generation from mine waste. It must be noted that both the soil and sludge used in the study were non-acid-generating. The silty soil–sludge mixtures with β values of 10%, 15%, 20%, and 25% sludge (β = wet sludge mass/wet soil mass) were studied. Two water contents were considered for each of the mixture components: 175% and 200% for the sludge and 7.5% and 12.5% for the soil. The results indicated that adding up to 25% of sludge to a silty soil can provide a mixture with appropriate saturated hydraulic conductivity (about 10^{-5} cm/s) and water retention properties (air-entry value about 30 kPa) for the mixture to be used in the moisture retention layer for covers with capillary barrier effects. This understanding was based on the comparison with existing efficient covers with capillary barrier effects. The impact of sludge addition to the silty soil on freeze-thaw behaviour was relatively limited. Volumetric shrinkage at complete drying of the mixtures (worst-case scenario) increased with the sludge content, but shrinkage can be reduced by covering the mixture with a layer of coarse material (the drainage and protection layer of the covers with capillary barrier effect) to control evaporation. The study indicated that the sludge mass that can be reused in the moisture retention layer of covers with capillary barrier effects could be significant, which consequently reduces the mass (or volume) of soil required as well.

Previous research, as indicated already, showed that AMD sludge has geotechnical and geochemical properties that can be used in combination with silty soil to be part of covers (oxygen barriers) that can prevent AMD generation from waste rock and tailings impoundments (Mbonimpa et al., 2016). On this basis, Demers et al. (2017) studied the use of sludge as a replacement for a portion of natural soil used for cover systems. Mixtures of sludge and a natural silty soil were tested in the laboratory (for over 500 days) and in field experiments (4 years) as an oxygen barrier cover placed over acid-generating tailings and waste rock. Field cell experiments were conducted in order to validate the results obtained in the laboratory and to evaluate the effect of a protective sand-gravel layer placed over the sludge-soil mixture. Data from the test work included monitoring of leachate geochemical parameters (e.g., pH, conductivity, metal and sulphate content) and hydrogeological parameters (water content, suction, effluent flow rate). The results of the laboratory column experiments showed that the sludge-soil mixture was efficient to prevent AMD generation, as long as the cover maintains its integrity. The results for the four monitoring seasons, in the field experiments, indicated that the sludge-soil mixture covers are effective in limiting AMD generation and reduce dissolved metal concentration in the cells effluents. Volumetric water content and suction measurements confirmed that the covers maintained a high degree of saturation, which made them efficient oxygen barriers. The presence of the sand-gravel layer possibly helped to reduce evaporation.

It is well known that Portland cement is an effective, but expensive option for source control of AMD (Sephton and Webb, 2017); but the cost could be reduced by blending the cement with cheap waste materials such as fly ash from coal combustion and sludge produced from neutralisation of AMD with lime (Sephton et al., 2019). To test how the two additives affect cement performance in reducing AMD generation, a study was carried out by Sephton et al. (2019). In the study, blended cement slurries were applied to sulphidic waste rocks in leaching columns that were monitored for about a year. The study found that applications of fly ash-blended cements and AMD sludge-blended cements to acid-producing sulphidic waste rocks in leaching columns considerably reduced AMD generation, decreased acidity, metal and sulphate loads in column leachates by 80–95%, similar to the effects of unblended cements. The AMD sludge showed no evidence of releasing its adsorbed heavy metals, but the fly ash released some silicon, indicating that it is not chemically stable in the cement. The overall analysis of the long-term effectiveness of the blended cement applications shows that cement placed as a surface cap on top of the waste rock provides more value, because the slower cement dissolution rates ensure continued effectiveness for many years.

9.8 Production of Nanoparticles from Acid Mine Drainage

9.8.1 Introduction

Nanoparticles represent an active area of research and a techno-economic sector with full expansion for applications (Jeevanandam et al., 2018). Nanoparticles are solid particles ranging in size from 1 to 100 nm (Yah et al., 2012; Ansari et al., 2019). Of particular interest in this section of the chapter are metallic nanoparticles which are regarded as nano-sized metals with dimensions (length, width, thickness) within the size range of 1 to 100 nm (Kumar et al., 2018). In general, nanoparticles have gained prominence in technological advancements due to their tunable physicochemical characteristics such as melting point, wettability, electrical and thermal conductivity, catalytic activity, light absorption and scattering resulting in enhanced performance over their bulk counterparts (Jeevanandam et al., 2018). The subsequent section gives a number of examples of nano-metallic particles that have been generated from AMD.

9.8.2 Selected Typical Studies of Production of Nanoparticles from Acid Mine Drainage

Nanoscale zerovalent iron (nZVI) particles were investigated by Crane and Sapsford (2017) for the extent at which the particles could be used for the selective formation of Cu bearing nanoparticles from AMD. A methodology by Wang and Zhang (1997) was used to synthesise pure nZVI by dissolving 7.65 g of $FeSO_4 \cdot 7H_2O$ into 50 mL of Milli-Q water (>18.2 MΩ cm) and the pH was adjusted to 6.8 using 4-M NaOH. The NaOH was added drop-wise to avoid the formation of hydroxo-carbonyl complexes. The salts were reduced to nZVI by the addition of 3.0 g of $NaBH_4$. The nanoparticle products were separated by centrifugation, rinsed with absolute ethanol and then centrifuged again. This step was repeated three times. Thereafter, the nanoparticles were dried in a vacuum desiccator for 72 h and then stored in an argon filled (BOC, 99.998%) MBraun glovebox until required for use. Prior to conducting any nZVI-AMD experiments, the AMD that was collected from a disused open cast Cu-Pb-Zn mine in Wales (UK) was removed from the refrigerator and allowed to equilibrate at the ambient laboratory temperature of 20.0 ± 1°C for 24 h. All batch experiments comprised adding 200 mL of the AMD into 250-mL glass jars, and after the addition of nZVI, the batch experimental systems were sonicated for 120 s using an ultrasonic bath. Basically, batch experiments were conducted containing unbuffered (pH 2.67 at $t = 0$) and pH buffered (pH < 3.1) AMD which were exposed to nZVI of about 0.1–2.0 g/L. Each system was then sealed and placed on the benchtop. About 5 mL of aqueous-nZVI suspensions were periodically taken out using an auto-pipette. The extracted suspensions were centrifuged at 4000 rpm for 240 s after which the supernatant became clear (i.e., all of the nZVI were centrifuged to the bottom of the vial). The solid nZVI at the base of each centrifuge vial was collected and analysed using X-ray diffraction (XRD), energy dispersive spectroscopy (EDS), X-ray photoelectron spectroscopy (XPS), and high resolution transmission electron microscopy (HRTEM). The results of the study by Crane and Sapsford (2017) demonstrated that nZVI was selective for Cu, Cd, and Al removal (> 99.9% removal of all metals within 1 h when nZVI ≥ 1.0 g/L) from unbuffered AMD despite the co-existence of numerous other metals in the AMD such as Na, Ca, Mg, K, Mn, and Zn. Basically, the addition of nZVI concentrations of ≥1 g/L to the AMD leads to rapid and near total selective removal of Cu, Al, and Cd from the solution through a combination of mechanisms including cementation (for Cu), precipitation and sorption to corrosion products (for Al and Cd). The selectivity of nZVI for Cu can be further enhanced by the application of an acidic pH buffer due to the restriction of Zn, Cd sorption onto nZVI corrosion products along with the concurrent prevention of hydrolysis and precipitation of Al as $Al(OH)_3$. For example, an acidic pH buffer enabled a similar high Cu removal, but maximum removal of only < 1.5% and < 0.5% Cd and Al, respectively. The HRTEM-EDS confirmed the formation of discrete spherical nanoparticles comprised of up to 68% weight Cu, with a relatively narrow size distribution (typically 20–100-nm diameter). On the other hand, the XPS confirmed such nanoparticles as containing Cu°, with the Cu removal mechanism, therefore, likely to have been via cementation with Fe°. Overall the results demonstrate that nZVI is effective for the selective formation of Cu°-bearing nanoparticles from acidic wastewater. Therefore, this technique has proven to be a highly useful mechanism for the valorisation of Cu-bearing AMD, thereby unlocking a new economic incentive for AMD treatment. It is also noted that the Cu°-bearing nanoparticles from acidic wastewater such as AMD have a wide range of applications including catalysis, optics, electronics, and as antifungal/antibacterial agents.

Wei and Viadero (2007b) studied the recovery of ferric iron from AMD via an oxidation selective precipitation process and used it as a feed stock to synthesise magnetite nanoparticles by co-precipitation of ferric and ferrous iron at the pH of 9.5 under an inert atmosphere. Raw AMD pumped from abandoned underground coal mines in North Central West Virginia was collected and sealed in high-density polyethylene bottles. At the laboratory, the solids and debris in the water samples were removed by settling and the remaining suspended solids were removed by filtration through a 0.45-µm membrane. Thereafter, the samples were stored at 4°C prior to metal recovery experiments. Iron recovery from AMD was achieved by an oxidation selective precipitation process as described by Wei et al. (2005). Initially, raw AMD was oxidised with hydrogen peroxide. Thereafter, the pH of the solution was raised to 3.5–4.0 with the addition of 4-N sodium hydroxide solution, where iron was precipitated as ferric hydroxide/oxyhydroxide which was separated from AMD by centrifugation. The ferric hydroxide/oxyhydroxide solids recovered were resolubilised with sulphuric acid to achieve a clear solution, which was used as feed stock to synthesise magnetite nanoparticles. As discussed in a study by Wei et al. (2005), the supernatant from centrifugation was further neutralised with sodium hydroxide to pH 6.0–7.0 and an aluminium-rich precipitate was obtained after settling.

Wei and Viadero (2007b) state that the magnetite nanoparticles were synthesised through co-precipitation at room temperature, which required the presence of both ferric and ferrous iron at a ratio of 2:1. Resolubilised ferric iron from the iron recovery process was used as the ferric iron source for the generation of magnetite nanoparticles from AMD. A typical synthesis process was as follows: a solution of $[Fe^{3+}]:[Fe^{2+}] = 2:1$ of which the concentrations of $Fe^{3+}:Fe^{2+}$ ranged from 0.02 M:0.01 M, 0.04 M:0.02 M, to 0.08 M:0.04 M was prepared and mixed for 30 min under a $N_2(g)$ atmosphere to prevent oxidation by completely removing dissolved oxygen from solution as previously discussed by Kim et al. (2001). Thereafter, a solution of 6.4-M NH_4OH was added to raise the pH to 9.5. The crystals of magnetite were allowed to grow for 30 min, under vigorous mixing and N_2 bubbling. The black precipitate (magnetite) was then isolated from solution using an external superconducting high gradient magnetic separation system. The magnetite nanoparticles were then washed three times with deoxygenated deionised water until the pH was near neutral (~7.5). Lastly, the nanoparticles were vacuum-dried and later characterised. The synthesis test results from the study by Wei and Viadero (2007b) demonstrated that it was feasible to prepare magnetite nanoparticles via the co-precipitation method with the recovered ferric iron from AMD. Based on X-ray diffraction analysis, the iron oxide phase in the black precipitate was magnetite. Through scanning and transmission electron microscopic studies, it was demonstrated that most of the magnetite particles ranged from 10 to 15 nm and were spheroidal or cubic in shape. The results of this study have shown that the synthesis of magnetite nanoparticles from AMD with the iron recovered from AMD was technically feasible.

Silva et al. (2017) assessed the feasibility of iron recovery by selective precipitation and the synthesis of goethite particles for use as pigment. The AMD was collected from a drainage channel near a coal tailings deposit in the state of Santa Catarina (Brazil). Iron recovery was achieved by selective precipitation of 1 L of AMD at a pH of 3.6. In this study, the method used for iron precipitation was similar to the methods used by Wei et al. (2005) and Menezes et al. (2009). The pH of the AMD was increased and maintained at 3.6 ± 0.1, with the addition of a 4-M NaOH solution to precipitate the iron as ferric hydroxide/oxyhydroxide. The ferric hydroxide/oxyhydroxide precipitate was separated from the remaining solution by centrifugation at 3 000 rpm. After centrifugation, the ferric hydroxide/oxyhydroxide precipitate obtained at pH 3.6 was washed with distilled water at pH of 3.6 ± 0.1, re-suspended, and centrifuged, and the cycle was repeated three times. The final precipitate was dissolved in nitric acid so as to achieve a clear acidic iron solution which subsequently formed an iron-hexa-aqua-ion complex $[Fe(H_2O)_6]^{3+}$. The solution of an iron-hexa-aqua-ion complex was alkalised with potassium hydroxide to a pH of 12.0 so as to form ferrihydrite. Thereafter, the mixture was diluted with water and heated to 70°C for 60 min for goethite crystallisation. The synthesised goethite particles were prepared in two different forms: (1) as a paste – the goethite particles were centrifuged and prepared as a water suspension containing about 50% solids, and (2) as a powder – the goethite particles were dried at 60°C. The solids were further analysed for particle size (laser diffraction, in aqueous solution with 1% sodium polyacrylate dispersing agent), specific surface area (BET), crystalline compounds (X-ray diffraction) and elemental chemical composition (atomic absorption spectroscopy). Thereafter, the coloured pastes were produced with commercial white Portland cement in a water/cement ratio of 1:2.5 to which goethite powder was added at a powder/cement mass ratio of 1:10. The results showed that the study was able to recover iron from AMD using a selective precipitation process at pH 3.6, and the iron was used to synthetise goethite particles. Furthermore, the results showed that goethite particles were successfully produced. The goethite particles varied from 0.04 to 5.0 µm when produced as paste suspension and 0.04 to 25.0 µm when dried at 60°C and converted to a solid powder. The pigment was also successfully used in a 10% pigment/cement paste mixture to colour a white cement paste giving it a yellow ochre colour.

The study by Kwon et al. (2016) focused on the formation of iron oxide from AMD and used it for the adsorption of heavy metals. The AMD used in the study was collected from the abandoned mine in Korea. The iron oxide recovery system from the AMD included the flow equalisation tank, neutralisation tank, and the precipitation tank. Iron oxide was collected at the precipitation tank where the solution pH was maintained at 3.4 to reduce the precipitation of impurities (e.g., Al. Zn, Cu, etc.). The iron oxides were classified based on precipitation time of 1 h and 24 h. The collected precipitates were dried at 40°C for 24 h after centrifugation and were later characterised. The sorption capacity of the iron oxides was determined in a batch reactor for heavy metals (e.g., As, Pb, and Cu) removal. The concentration of each metal was 1 mmol/L, and the adsorbent to liquid ratio was 10 g/L. The adsorbent and heavy metal ions were mixed on an orbital shaker at 200 rpm for 4 h at room temperature. After sorption equilibrium the mixture was filtered with 0.45-μm filter and analysed for the remaining metal concentrations. The results showed that the iron oxide was successfully synthesised from the AMD solution by controlling solution pH. The synthesised iron oxides were found to be good adsorbents for heavy metals such as As, Cu, and Pb. However, the sorption characteristics were highly dependent on the type of iron oxide and solution pH. For example, iron oxide sampled after 1 h had higher sorption capacity for Pb and Cu than iron oxide sampled after 24 h. However, iron oxide obtained at 24 h had higher sorption capacity for As than iron oxide sampled after 1 h. The results also showed that the two iron oxides had similar particle distribution patterns, but iron oxide samples obtained at 24 h were of relatively bigger size.

In a study by Kefeni et al. (2015), the possibility of synthesizing magnetic nanoparticles with and without heat from pure chemicals and real AMD by co-precipitation method was explored. Analytical grade chemicals of purity ≥98% of $FeCl_3 \cdot 6H_2O$, $FeSO_4 \cdot 7H_2O$ and $Co(NO_3)_2 \cdot 6H_2O$ (Merck, Darmstadt, Germany) were used to synthesise Fe_3O_4 and $CoFe_2O_4$ magnetic nanoparticles by the co-precipitation method. For each experimental set-up for pure chemicals, the required amount of salts (Fe^{3+}/Fe^{2+} or Fe^{3+}/Co^{2+} at a mole ratio of 2:1) was weighed and dissolved by deionised water in 1.5-L volumetric flask and the pH of the solution was recorded. For each set, six replicates of 200 mL of the solution were measured separately and added to 250-mL beaker. For comparison purpose, pHs of paired samples were adjusted to 8.5 and 11.5 by using 5-N NaOH (aqueous); and in between 8.5 and 9 by using 25% NH_4OH (aqueous). For each set of a pair of samples, one was heated at 60°C and the other was not, whilst both samples were stirred continuously for 2 h. Thereafter, the samples were filtered and the precipitates were washed using deionised water until the pH of the filtrate was about 7. The precipitates were then dried at 105°C in the oven for 6 h to remove water and other volatile substances adhered to synthesised magnetic nanoparticles. Amongst the dried samples, four samples were selected and a small portion of each sample was heated at 500°C in muffle furnace for 3 h to examine the effect of high temperature. Finally, the synthesised magnetic nanoparticles were stored at room temperature for investigation of their capacity to remove metals from AMD.

In the same study by Kefeni et al. (2015), samples of simulated AMD were prepared from $FeCl_3 \cdot 6H_2O$, $FeSO_4 \cdot 7H_2O$, $CrCl_3 \cdot 6H_2O$, $Al(NO_3)_3 \cdot 9H_2O$, $Co(NO_3)_2 \cdot 6H_2O$, $MnSO_4 \cdot H_2O$, $ZnSO_4 \cdot 6H_2O$, and $Ni(NO_3)_2 \cdot 6H_2O$ (Merck, Darmstadt, Germany). The salts were weighed to achieve a desired 2:1 mol ratio of trivalent to bivalent ions present in the spinal structure of ferrite. For the real AMD that was collected, the types and concentrations of metals present in the AMD were measured, in order to determine the amount of AMD that is needed to be oxidised and mixed with fresh AMD so as to attain the 2:1 ratio of trivalent:divalent metals in the ferrite. However, only $Fe^{3+}:Fe^{2+}$ ratio was considered due to low concentration of other trace metals (trivalent or divalent) in the real AMD studied. For the oxidation of Fe^{2+} to Fe^{3+} about 1 L of fresh AMD was taken and added into 2.5-L plastic container, and its pH was adjusted to between pH 5 and 6 using either NaOH (aqueous) or NH_4OH (aqueous) and then aerated for 2 h using compressed air.

The results of the study by Kefeni et al. (2015) showed that it is possible to synthesise Fe_3O_4 and $CoFe_2O_4$ from their corresponding pure chemical binary salts. When the method was applied to simulated and real AMD, it showed that formation of well-crystalline magnetic nanoparticles at lower pH and temperature from both samples were hampered due to interference and combined effect of the metal cations. For example, these observations were reflected by the formation of Fe_3O_4, γ-Fe_2O_3, Mn_3O_4, MnO_2, and ZnO mixtures as major components from real AMD. However, increasing the pH and temperature increased crystalline size of synthesised magnetic nanoparticles. In other words, the results showed that the higher pH and temperature are favourable conditions for the formation of magnetic nanoparticles from AMD than the binary cations from the standard chemicals. Under all conditions, higher intensity and better resolution of XRD peaks of synthesised magnetic nanoparticle were obtained when NH_4OH (aqueous) was used for neutralisation than NaOH (aqueous). Generally, this study demonstrated the possibility of synthesizing magnetic nanoparticles from real AMD by optimizing pH, temperature, and string time. The results also indicated that treating AMD in the presence of magnetic nanoparticle seeds accelerated the formation of ferrite and resulted in increased magnetic moment of ferrite sludge. In general, this study demonstrated the possibility of converting environmental pollutants into commercially valuable chemicals.

9.9 Concluding Remarks

This chapter covered two elements that appear in the title –recovery and utilisation – of valuable materials from AMD. The AMD with its low pH (between 2 and 4) and high concentrations of hazardous and toxic elements and sulphate contents is considered as a very serious global environmental problem. For sustainability, the focus when dealing with AMD should be to prevent, minimise, reuse, and recycle. However, at present, there is no general solution to the problem of AMD, either through prevention or remediation. Nevertheless, the chapter has provided a detailed analysis of processes used for the recovery of valuable resources such as water, metals, pigments, and sulphuric acid from AMD. In addition, the utilisation of waste materials generated during the treatment of AMD such as AMD sludge in the manufacture of several products or use in processes, for example, has also been discussed. Actually, sustainable resource recovery and/ or utilisation is the key to manage the overburden of various waste entities of mining practices. More specifically, the chapter has clearly shown that a major water contamination problem and environmental hazard can be transformed into valuable resources which can meet the needs of a range of users, safely and reliably.

It is clear from the chapter that the recovery processes should not only target a specific material alone present in AMD, but should holistically consider all available valuable resources as well either concomitantly or separately. For example, metals, water, acid, pigments and sludge and other minor resources need to be recovered and valorised in order to make the process economically viable and resource efficient. In conclusion, the chapter has shown that the resource recovery and utilisation is a non-debatable holistic approach to environmental sustainability and AMD pollution reduction. More specifically, the chapter has clearly shown that the 3Rs – Reuse, Recycle, Recovery – are the most important strategies for dealing with AMD, and these are depicted in Figure 9.18. Reuse in the context of this chapter refers to the use or application of a significant component of the AMD such as water for a specific purpose after undergoing some treatment. Recycling is defined as the conversion of the entire waste such as AMD sludge into a

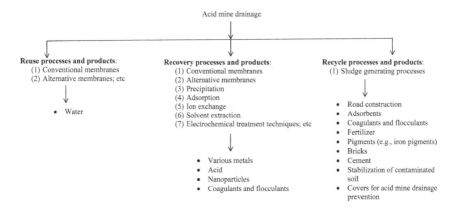

FIGURE 9.18
Reuse, recycle, and recovery processes of acid mine drainage and/or acid mine drainage waste residues.

new valuable product or application. Recovery on the other hand refers to the extraction of valuable resources or ingredients from the AMD with the aid of processing and/or reprocessing techniques.

References

Abhanga, R.M., Wanib, K.S., Patil, V.S., Pangarkara, B.L. and Parjanea, S.B. (2013). Nanofiltration for recovery of heavy metal ions from waste water – a review. International Journal of Research in Environmental Science and Technology 3(1), 29–34.

Adler, P.R. and Sibrell, P.L. (2003). Sequestration of phosphorus by acid mine drainage floc. Journal of Environmental Quality 32, 1122–1129.

Agboola, O. (2019). The role of membrane technology in acid mine water treatment: a review. Korean Journal of Chemical Engineering 36(9), 1389–1400.

Agrawal, A., Kumari, S. and Sahu, K.K. (2008). Liquid-liquid extraction of sulphuric acid from zinc bleed stream. Hydrometallurgy 92, 42–47.

Agrawal, A. and Sahu, K.K. (2009). An overview of the recovery of acid from spent acidic solutions from steel and electroplating industries. Journal of Hazardous Materials 171(1–3), 61–75.

Akbal, F. and Camcı, S. (2011). Copper, chromium and nickel removal from metal plating wastewater by electrocoagulation. Desalination 269(1), 214–222.

Akdeniz, Y. and Ulku, S. (2007). Microwave effect on ion-exchange and structure of clinoptilolite. Journal of Porous Materials 14, 55–60.

Ahn, K.H., Song, K.G., Cha, H.Y. and Yeom, I.T. (1999). Removal of ions in nickel electroplating rinse water using low-pressure nanofiltration. Desalination 122(1), 77–84.

Al-Rashdi, B., Somerfield, C. and Hilal, N. (2011). Heavy metals removal using adsorption and nanofiltration techniques. Separation and Purification Reviews 40, 209–259.

Alexandrova, L., Nedialkova, T. and Nishkov, I. (1994). Electroflotation of metal ions in waste water. International Journal of Mineral Processing 41(304), 285–294.

Ali, U.F.M. (2011). Electrochemical separation and purification of metals from waste electrical and electronic equipment (WEEE). PhD thesis. Imperial College of Science, Technology and Medicine, London, UK.

Alkhudhiri, A., Darwish, N. and Hilal, N. (2012). Membrane distillation: a comprehensive review. Desalination 287, 2–18.

Alkhudhiri, A. and Hilal, N. (2018). Membrane distillation – principles, applications, configurations, design, and implementation. In: Gude, V. G. (Editors), Emerging Technologies for Sustainable Desalination Handbook. Elsevier, Oxford, pp. 55–106.

Al-Zoubi, H., Rieger, A., Steinberger, P., Pelz, W., Haseneder, R. and Härtelet, G. (2010). Nanofiltration of acid mine drainage. Desalination and Water Treatment 21(1–3), 148–161.

Amanda, N. and Moersidik, S.S. (2019). Characterization of sludge generated from acid mine drainage treatment plants. Journal of Physics: Conference Series 1351(2019), 012113.

Ambiado, K., Bustos, C., Schwarz, A. and Bórquez, R. (2017). Membrane technology applied to acid mine drainage from copper mining. Water Science and Technology 75(3–4), 705–715.

Andalaft, J., Schwarz, A., Pino, L., Fuentes, P., Bórquez, R. and Aybar, M. (2018). Assessment and modeling of nanofiltration of acid mine drainage. Industrial and Engineering Chemistry Research 57, 14727–14739.

Andrade, L.H., Aguiar, A.O., Pires, W.L., Miranda, G.A., Teixeira, L.P.T., Almeida, G.C.C. and Amaral, M.C.S. (2017). Nanofiltration and reverse osmosis applied to gold mining effluent treatment and reuse. Brazilian Journal of Chemical Engineering. 34(1), 93–107.

Ansari, S.A.M.K., Ficiarà, E., Ruffinatti, F.A., Stura, I., Argenziano, M., Abollino, O., Cavalli, R., Guiot, C. and D'Agata, F. (2019). Magnetic iron oxide nanoparticles: synthesis, characterization and functionalization for biomedical applications in the central nervous system. Materials 12, 465.

Arpa, Ç., Başyilmaz, E., Bektaş, S., Genç, Ö. and Yürüm, Y. (2000). Cation exchange properties of low rank Turkish coals: removal of Hg, Cd and Pb from waste water. Fuel Processing Technology 68, 111–120.

Aveng Water (2009). Blueprint for treatment of AMD. Available from http://www.avengwater.co.za/news-room/press-releases/emalahleni-blueprint-treatment-amd [Accessed 16 April 2014].

Azimi, A., Azari, A., Rezakazemi, M. and Ansarpour, M. (2017). Removal of heavy metals from industrial wastewaters: a review. ChemBioEng Reviews 4, 37–59.

Barrer, R. (1978). Zeolites and Clay Minerals as Sorbents and Molecular Sieves. Academic Press Inc., London.

Bayer A.G. (2000). Removing Traces Treating Industrial Waste Water, Process Water and Ground-water with Lewatit Ion Exchange Resins. Bayer AG, Leverkusen.

Bazrafshan, E., Amirian, P., Mahvi, A.H. and Ansari-Moghaddam, A. (2016). Application of adsorption process for phenolic compounds removal from aqueous environments: a systematic review. Global NEST Journal 18(1), 146–163.

Bejan, D. and Bunce, N.J. (2015). Acid mine drainage: electrochemical approaches to prevention and remediation of acidity and toxic metals. Journal of Applied Electrochemistry 45, 1239–1254.

Bennetto, H.P. (1984). Microbial fuel cell. In: Michelson, A.M. and Bannister, J.V. (Editors), Life Chemistry Reports, Volume 2. Harwood Academic, London.

Benvenut, T., Rodrigues, M.A., Krapf, R.S., Bernardes, A.M. and Zoppas-Ferreira, J. (2013). Electrodialysis as an alternative for treatment of nickel electroplating effluent: water and salts recovery. In: 4th International Workshop, Advances in Cleaner Production, 22–24 May, São Paulo, Brazil.

Benzaazoua, M., Ouellet, J., Servant, S., Newman, P. and Verburg, R. (1999). Cementitious backfill with high sulfur content: physical, chemical, and mineralogical characterization. Cement and Concrete Research 29, 719–725.

Benzaazoua, M., Fiset, J.F., Bussière, B., Villeneuve, M. and Plante, B. (2006). Sludge recycling within cemented backfill: study of the mechanical and leachability properties. Minerals Engineering 19, 420–432.

Bernardin, A.M., Marcello, R.R., Peterson, M., Galato, S., Izidoro, G., Saulo, V. and Riella, H.G. (2006). Inorganic pigments obtained from coal mine drainage residues. In: Proceedings of the 8th World Congress on Ceramic Tile Quality, Volume 3. Official Chamber of Commerce, Industry and Navigation, Castellón, Spain, pp. 169–174.

Bessho, M., Markovic, R., Trujic, T. A., Bozic, D., Masuda, N. and Stevanovic, Z. (2017). Removal of dissolved metals from acid wastewater using organic polymer hydrogels. In: Wolkersdorfer, C., Sartz, L., Sillanpää, M. and Häkkinen, A. (Editors), Mine Water and Circular Economy. IMWA 2017, Lappeenranta, Finland, pp. 1080–1086.

Bessho, M., Markovic, R., Trujic, T. A., Masuda, N., Bozic, D. and Stevanovic, Z. (2019). Recovery of metals from acid mine drainage using organic polymer. In: Wolkersdorfer, C., Khayrulina, E., Polyakova, S. and Bogush, A. (Editors), Mine Water: Technological and Ecological Challenges. IMWA 2019, Perm, Russia, pp. 155–161

Bhagwan, J. (2012). Turning acid mine drainage water into drinking water: the eMalahleni water recycling project. Available at https://www3.epa.gov/region1/npdes/merrimackstation/pdfs/ar/AR-1530.pdf [Accessed 23 July 2020].

Biniaz, P., Ardekani, N.T., Makarem, M. and Rahimpour, M. (2019). Water and wastewater treatment systems by novel integrated membrane distillation. Chemical Engineering 3(8), 1–36.

Blais, J.F., Djedidi, Z., Cheikh, R.B., Tyagi, R.D. and Mercier, G. (2008). Metals precipitation from effluents: review. Practice Periodical of Hazardous, Toxic, and Radioactive Waste Management 12, 135–149.

Bogoczek, R. and Kociołek-Balawejder, E. (1988). Chemicznie aktywne kopolimery styreno-diwinylobenzenowe o siarkowych grupach funkcyjnych. Chemik 1, 10–16.

Bologo, V., Maree, J.P. and Zvinowanda, C.M. (2009). Treatment of acid mine drainage using magnesium hydroxide. Available at https://www.imwa.info/docs/imwa_2009/IMWA2009_Bologo.pdf [Accessed 13 July 2020].

Bologo, V., Maree, J.P. and Carlsson, F. (2012). Application of magnesium hydroxide and barium hydroxide for the removal of metals and sulphate from mine water. Water SA 38, 23–28.

Bolto, B.A. and Pawłowski, L. (1987). Wastewater Treatment by Ion-Exchange. E. and F.N. Spoon Ltd., London.

Bonilla-Petriciolet, A., Mendoza-Castillo, D.I. and Reynel- Ávila, H.E. (2017). Adsorption Processes for Water Treatment and Purification. Springer Nature, Cham, Switzerland.

Boonstra, J., van Lier, R., Janssen, G., Dijkman, H. and Buisman, C.J.N. (1999). Biological treatment of acid mine drainage. In: Amils, R. and Ballester A. (Editors), Biohydrometallurgy and the Environment Toward the Mining of the 21st Century, Volume 9. Elsevier, Amsterdam, pp. 559–567.

Botha, M., Bester, L. and Hardwick, E. (2009). Removal of uranium from mine water using ion exchange at DRIEFONTEIN mine. Available at http://www.imwa.info/docs/imwa_2009/IMWA2009_BothaHardwick.pdf [Accessed 24 May 2020].

Brown, L.M. (1996). Removal of heavy metals from water with microalgal resins I. process development. Available at https://www.usbr.gov/research/dwpr/reportpdfs/report074.pdf [Accessed 6 May 2020].

Butler, C.J., Green, A.M. and Chaffee, A.L. (2007). Remediation of mechanical thermal expression product waters using raw Latrobe Valley brown coals as adsorbents. Fuel 86, 1130–1138.

Buzzi, D.C., Viegas, L.S., Silvas, F.P.C., Espinosa, D.C.R., Rodrigues, M.A.S., Bernandes, A.M. and Tenório, J.A.S. (2011). The use of microfiltration and electrodialysis for treatment of acid mine drainage. In: Rüde, T.R., Freund, A. and

Wolkersdorfer, C. (Editors), 11th International Mine Water Association Congress, Mine Water – Managing the Challenges, Aachen, Germany, pp. 287–291.

Buzzi, D.C., Viegas, L.S., Rodrigues, M.A.S., Bernardes, A.M. and Tenório, J.A.S. (2013). Water recovery from acid mine drainage by electrodialysis. Minerals Engineering 4, 82–89.

Camacho, L.M., Dumée L., Zhang, J., Li, J., Duke, M., Gomez, J. and Gray, S. (2013). Advances in membrane distillation for water desalination and purification applications. Water 5, 94–196.

Campbell, D.A., Dalrymple, I.M. and Sunderlend, J.G. (1994). The electrochemical recovery of metals from effluent and process streams. Resources, Conservation and Recycling 10, 25–33.

Chartrand, M.M.G. and Bunce, N.J. (2003). Electrochemical remediation of acid mine drainage. Journal of Applied Electrochemistry 33, 259–264.

Chen, G. (2004). Electrochemical technologies in wastewater treatment. Separation and Purification Technology 38, 11– 41.

Cheng, S., Dempsey, B.A. and Logan, B.E. (2007). Electricity generation from synthetic acid-mine drainage (AMD) water using fuel cell technologies. Environmental Science and Technology 41(23), 8149–8153.

Cheng, S., Jang, J.H., Dempsey, B.A. and Logan, B.E. (2011). Efficient recovery of nano-sized iron oxide particles from synthetic acid-mine drainage (AMD) water using fuel cell technologies. Water Research 45(1), 303–307.

Churms, S.C. (1966). The effect of pH on the ion-exchange properties of hydrated alumina: part 1. Capacity and selectivity. South African Journal of Chemistry 19(2), 98–107.

Cifuentes, L., García, I., Ortiz, R. and Casas, J.M. (2006). The use of electrohydrolysis for the recovery of sulphuric acid from copper-containing solutions. Separation and Purification Technology 50(2), 167–174.

Cifuentes, L., García, I., Arriagada, P. and Casas, J.M. (2009). The use of electrodialysis for metal separation and water recovery from CuSO4–H2SO4–Fe solutions. Separation and Purification Technology 68(1), 105–108.

Clifford, D. (2004). Fundamentals of radium and uranium removal from drinking water supplies. Available at https://19january2017snapshot.epa.gov/sites/production/files/2015-09/documents/rads_treatment_dennis_clifford.pdf [Accessed 6 May 2020].

Crane, R.A. and Sapsford, D.J. (2018). Selective formation of copper nanoparticles from acid mine drainage using nanoscale zerovalent iron particles. Journal of Hazardous Materials 347, 252–265.

Cui, M., Jang, M., Cannon, F.S., Na, S., Khim, J. and Park, J.K. (2013). Removal of dissolved Zn(II) using coal mine drainage sludge: implications for acidic wastewater treatment. Journal of Environmental Management 116, 107–112.

Dąbrowski, A., Hubicki, Z., Podkościelny, P. and Robens, E. (2004). Selective removal of the heavy metal ions from waters and industrial wastewaters by ion-exchange method. Chemosphere 56, 91–106.

da Mota, I.O., de Castro, J.A., Casqueira, R.G. and Oliveira Junior, A.G. (2015). Study of electroflotation method for treatment of wastewater from washing soil contaminated by heavy metals. Journal of Materials Research and Technology 4, 109–113.

de Beer, M., Maree, J.P., Wilsenach, J., Motaung, S., Bologo, L. and Radebe, V. (2010). Acid mine water reclamation using the ABC process. Available at http://researchspace.csir.co.za/dspace/bitstream/handle/10204/5137/De%20Beer2_2010.pdf?sequence=1&isAllowed=y [Accessed 31 May 2020].

Demers, I., Bouda, M., Mbonimpa, M., Benzaazoua, M., Bois, D. and Gagnon M. (2015a). Valorisation of acid mine drainage treatment sludge as remediation component to control acid generation from mine wastes, part 1: Material characterization and laboratory kinetic testing. Minerals Engineering 76, 109–116.

Demers, I., Bouda, M., Mbonimpa, M., Benzaazoua, M., Bois, D. and Gagnon, M. (2015b). Valorisation of acid mine drainage treatment sludge as remediation component to control acid generation from mine wastes, part 2: Field experimentation. Minerals Engineering 76, 117–125.

Demers, I., Mbonimpa, M., Benzaazoua, M., Bouda, M., Awoh, S., Lortie, S. and Gagnon, M. (2017). Use of acid mine drainage treatment sludge by combination with a natural soil as an oxygen barrier cover for mine waste reclamation: laboratory column tests and intermediate scale field tests. Minerals Engineering 107, 43–52.

Devi, P. and Saroha, A.K. (2017). Utilization of sludge based adsorbents for the removal of various pollutants: a review. Science of the Total Environment 578, 16–33.

Dhir, S. (2018). Biotechnological tools for remediation of acid mine drainage (removal of metals from wastewater and leachate). In: Prasad, M.N.V., Favas, P.J.C. and Maiti, S.K. (Editors), Bio-Geotechnologies for Mine Site Rehabilitation. Elsevier, London.

Dinardo, O., Kondos, P.D., Mackinnon, D.J., Mccready, R.G.L., Riveros, P.A. and Skaff, M. (1991). Study on metals recovery/recycling from acid mine drainage phase IA: literature survey. Available at http://mend-nedem.org/wp-content/uploads/2013/01/3.21.1a.pdf [Accessed 19 July 2020].

Dobbie, K.E., Heal, K.V. and Smith, K.A. (2005). Assessing the performance of phosphorus-saturated ochre as a fertiliser and its environmental acceptability. Soil Use Manage 21, 231–239.

Dobias, B. (1993). Coagulation and Flocculation. Marcel Dekker, New York.

Doerkar, N.V. and Tavlarides, L.L. (1998). An adsorption process for metal recovery from acid mine waste: The Berkeley Pit Problem. Environmental Progress 17(2), 120–125.

Dorfner, K. (1991). Ion-Exchangers. Walter de Gruyter, Berlin, New York.

Dold, B. (2010). Basic concepts in environmental geochemistry of sulphide mine waste management. In: Kumar, E.S. (Editor), Waste Management. In-Tech, Rijeka, pp. 173–198.

Drogui, P., Blais, J.F. and Mercier, G. (2007). Review of electrochemical technologies for environmental applications. Recent Patents on Engineering 1, 257–272.

Du, Z., Li, H., and Gu T. (2007). A state of the art review on microbial fuel cells: A promising technology for wastewater treatment and bioenergy. Biotechnology Advances 25, 464–482.

Dubpernel, G. (1978). Selected Topics in the History of Electrochemistry. Electrochemical Society, Princeton, NJ, p. 1.

Dudzińska, M.R. and Clifford, D.A. (1991). Anion exchange studies of lead-EDTA complexes. Reactive Polymers 16, 71–80.

Dudzińska, M.R. and Pawłowski, L. (1993). Anion exchange removal of heavy metal–EDTA complex. In: Dyer, A., Hudson, M.J. and Williams, P.A. (Editors), Ion Exchange Processes: Advances and Applications. Royal Society of Chemistry, Cambridge.

Duffus, J. H. (2002). Heavy metals – a meaningless term? Pure and Applied Chemistry 74(5), 793–807.

Dunaway, R. S. (1984). Heating Systems Specialist (AFSC 54750). Extension Course Institute, Air Training Command, University of Minnesota, Minneapolis, MN.

Dyer, A. (1988). An Introduction to Zeolite Molecular Sieves. John Wiley & Sons, Toronto.

Edwards, M. and Benjamin, M. (1989). Regeneration and reuse of iron hydroxide adsorbents in the treatment of metal-bearing wastes. Journal Water Pollution Control Federation 61(4), 481–490.

Englert, A.H. and Rubio, J. (2005). Characterisation and environmental application of a Chilean natural zeolite. International Journal of Mineral Processing 75, 21–29.

EPA (2014). Reference guide to treatment technologies for mining-influenced water. Available at https://nepis.epa.gov/Exe/ZyPDF.cgi/P100I4PB.PDF?Dockey= P100I4PB.PDF [Accessed 6 May 2020].

Erdem, E., Karapinar, N. and Donat, R. (2004). The removal of heavy metal cations by natural zeolite. Journal of Colloid and Interface Science 280, 309–314.

Etter, K. and Langill, P.D. (2006). Acid recycling-process and design notes on hot-dip galvanizing. Available at http://www.galvanizeit.org/images/uploads/ memberGalvanizingNotes/AcidRecycling,KurtEtter(GalvanizingNotes,2006 Feb).pdf [Accessed 14 March 2016].

EU Directive (2000). The EU Water Framework Directive. Available at https://ec. europa.eu/environment/pubs/pdf/factsheets/wfd/en.pdf [Accessed 7 July 2020].

Feely, R.A., Sabine, C.L., Lee, K., Berelson, W., Kleypas, J., Fabry, V.J. and Millero, F.J. (2004). Impact of anthropogenic CO_2 on the $CaCO_3$ system in the oceans. Science 305(5682), 362–366.

Feng, Y., Wang, X., Logan, B. E. and Lee, H. (2008). Brewery wastewater treatment using air-cathode microbial fuel cells. Applied Microbiology and Biotechnology 78, 873–880.

Feng, Y., Yang, L., Liu, J. and Logan, B. E. (2016). Electrochemical technologies for wastewater treatment and resource reclamation. Environmental Science: Water Research and Technology 2(5), 800–831.

Fenton, O., Healy, M.G. and Rodgers, M. (2009). Use of ochre from an abandoned metal mine in the south east of Ireland for phosphorus sequestration from dairy dirty water. Journal of Environmental Quality 38, 1120–1125.

Figueroa, L. and Wolkersdorfer, C. (2014). Electrochemical recovery of metals in mining influenced water: State of the art. In: Sui, W., Sun, Y. and Wang, C. (Editors), An Interdisciplinary Response to Mine Water Challenges. China University of Mining and Technology, Xuzhou, China, pp. 627–631.

Foureaux, A.F.S., Moreira, V.R., Lebron, Y.A.R., Santos, L.V.S. and Amaral, M.C.S. (2020). Direct contact membrane distillation as an alternative to the conventional methods for value-added compounds recovery from acidic effluents: a review. Separation and Purification Technology 236, 116251–116264.

Fu, F. and Wang, Q. (2011). Removal of heavy metal ions from wastewaters: a review. Journal of Environmental Management 92, 407–418.

Gharabaghi, M., Irannajad, M. and Azadmehr, A.R. (2012). Selective sulphide precipitation of heavy metals from acidic polymetallic aqueous solution by thioacetamide. Industrial and Engineering Chemistry Research 51, 954–963.

Gaikwad, R.W. and Gupta, D.V. (2008). Review on removal of heavy metals from acid mine drainage. Applied Ecology and Environmental Research 6(3), 81–98.

Gaikwad, R.W., Misal, S.A. and Gupta, D.V. (2009). Removal of copper ions from acid mine drainage (AMD) by ion exchange resins: Indion 820 and Indion 850. Journal of Applied Sciences in Environmental Sanitation 4(2), 133–140.

Gaikwad, R.W., Sapkal, R.S. and Sapkal, V.S. (2010). Removal of copper ions from acid mine drainage wastewater using ion exchange technique: factorial design analysis. Journal of Water Resource and Protection 2, 984–989.

Garba, Y., Taha, S., Gondrexon, N. and Dorange, G. (1999). Ion transport modelling through nanofiltration membranes. Journal of Membrane Science 160(2), 187–200.

García, V., Häyrynen, P., Landaburu-Aguirre, J., Pirilä, M., Keiski, R.L. and Urtiag, A. (2013). Purification techniques for the recovery of valuable compounds from acid mine drainage and cyanide tailings: application of green engineering principles. Journal of Chemical Technology and Biotechnology 89, 803–813.

Garland, R. (2011). Acid mine drainage – the chemistry. Quest 7(2), 50–52.

Geiger, A. and Cooper, J. (2010). Overview of airborne metals regulations, exposure limits, health effects, and contemporary research. Available at https://www3.epa.gov/ttnemc01/prelim/otm31appC.pdf [Accessed 4 May 2020].

Gordyatskaya, Y. (2017). Selective removal of copper and nickel ions from acid mine drainage. Available at https://www.longdom.org/conference-abstracts-files/2375-4397-C1-009-005.pdf [Accessed 23 May 2020].

Gorgievski, M., Božić, D., Stanković, V. and Bogdanović, G. (2009). Copper electrowinning from acid mine drainage: A case study from the closed mine "Cerovo". Journal of Hazardous Materials 170(2), 716–721.

Gottliebsen, K., Grinbaum, B., Chen, D. and Stevens, G.W. (2000). Recovery of sulfuric acid from copper tank house electrolyte bleeds. Hydrometallurgy 56, 293–307.

Goswami, K.B. and Bisht, P.S. (2017). The role of water resources in socio-economic development. International Journal for Research in Applied Science and Engineering Technology 5(12), 1669–1674.

Gude, V.G. (2016). Wastewater treatment in microbial fuel cells – an overview. Journal of Cleaner Production 122, 287–307.

Günther, P. and Mey, W. (2008). Selection of mine water treatment technologies for the eMalahleni (Witbank) water reclamation project. Available at http://www.ewisa.co.za/literature/files/122%20Gunther.pdf [Accessed 12 March 2016].

Haan, T.Y., Mohammad, A.W., Ramli, S., Sajab, M.S. and Mazuki, N.I.M. (2018). Potential of membrane technology for treatment and reuse of water from old mining lakes. Sains Malaysiana 47(11), 2887–2897.

Hai, T., Pu, W.C., Cai, C.F., Xu, J.P. and He, W.J. (2016). Remediation of acid mine drainage based on a novel coupled membrane-free microbial fuel cell with permeable reactive barrier system. Polish Journal of Environmental Studies 25(1), 107–112.

Habermann, W. and Pommer, E.H. (1991). Biological fuel cells with sulphide storage capacity. Applied Microbiology and Biotechnology 25(1), 128–133.

Haghshenas, D.F., Darvishi, D., Rafieipour, H., Alamdari, E.K. and Salardini, A.A. (2009). A comparison between TEHA and Cyanex 923 on the separation and the recovery of sulfuric acid from aqueous solutions. Hydrometallurgy 97, 173–179.

Hall, D.A. and Giglio, N.M. (2010). Architectural Graphic Standards for Residential Construction. John Wiley & Sons, Hoboken, NJ.

Hammer, D.A. (1997). Creating Freshwater Wetlands. CRC Press, Boca Raton, FL.

Hardwick, E. and Hardwick, J. (2016). An overview of the use of ion exchange to extract wealth from mine waters. In: Carsten, D. and Michael, P. (Editors), Mining Meets Water – Conflicts and Solutions. Proceedings IMWA 2016, Freiberg, Germany, pp. 1274–1279.

Harrelkas, F., Azizi, A., Yaacoubi, A., Benhammou, A. and Pons, M.N. (2009). Treatment of textile dye effluents using coagulation-flocculation coupled with membrane processes or adsorption on powdered activated carbon. Desalination 235, 330–339.

Hassan, K.M., Fukushi, K., Turikuzzaman, K. and Moniruzzaman, S.M. (2014). Effects of using arsenic-iron sludge wastes in brick making. Waste Management 34, 1072–1078.

Heal, K., Younger, P.L., Smith, K. Glendinning, S., Quinn, P. and Dobbie, K. (2003). Novel use of ochre from mine water treatment plants to reduce point and diffuse phosphorus pollution. Land Contamination and Reclamation 11, 145–152.

Heal, K.V., Younger, P.L., Smith, K.A., Quinn, P., Glendinning, S., Aumônier, J. and Dobbie, K.E. A. (2004). A sustainable use of ochre from mine water treatment plants for phosphorus removal and recycling. Available at https://www.imwa.info/docs/imwa_2004/IMWA2004_34_Heal.pdf [Accessed 22 June 2020].

Hedin, R. (1998). Potential recovery of iron oxides from coal mine drainage. In: Nineteenth Annual West Virginia Surface Mine Drainage Task Force Symposium, West Virginia University, Morgantown.

Hedin R.S. (2003). Recovery of marketable iron oxide from mine drainage in the USA. Land Contamination and Reclamation 11(2), 93–97.

Heidmann, I. and Calmano, W. (2008). Removal of Zn(II), Cu(II), Ni(II), Ag(I) and Cr(VI) present in aqueous solutions by aluminum electrocoagulation. Journal of Hazardous Materials 152, 934– 941.

Herrera, P., Uchiyama, H., Igarashi, T., Asakura, K. and Ochi, Y. (2007). Treatment of acid mine drainage through a ferrite formation process in central Hokkaido, Japan: evaluation of dissolved silica and aluminium interference in ferrite formation. Minerals Engineering 20, 1255–1260.

Hu, J., Chen, G. and Lo, I. M. C. (2006). Selective removal of heavy metals from industrial wastewater using maghemite nanoparticle: performance and mechanisms. Journal of Environmental Engineering 132(7), 709–715.

Hubicki, Z., Jakowicz, A. and Łodyga, A. (1999). Application of the ions from waters and sewages. In: Dąbrowski, A. (Editor), Adsorption and Environmental Protection, Studies in Surface Science and Catalysis, Volume 120. Elsevier, Amsterdam.

Hubicki, Z. and Jusiak, S. (1978). Selective removal of copper from ammonical wastewater on ion exchangers of various types. Materials Science 4, 17–21.

Hubicki, Z. and Kołodyńska, D. (2012). Selective removal of heavy metal ions from waters and waste waters using ion exchange methods. In: Kilislioğlu, A. (Editor), Ion Exchange Technologies. In-Tech, Croatia.

Hubicki, Z. and Pawłowski, L. (1986). Possibility of copper recovery from wastewater containing copper-ammine complexes. Environment Protection Engineering 12, 5–16.

Huisman, J.L., Schouten, G. and Schultz, C. (2006). Biologically produced sulphide for purification of process streams, effluent treatment and recovery of metals in the metal and mining industry. Hydrometallurgy 83, 106–113.

Hull, E.J. and Zodrow, K.R. (2017). Acid rock drainage treatment using membrane distillation: impacts of chemical-free pretreatment on scale formation, pore wetting, and product water quality. Environmental Science and Technology 51, 11928–11934.

Iijima, S. (1991). Helical microtubules of graphitic carbon. Nature 354, 56–58.

International Network for Acid Prevention (INAP) (2003). Treatment of sulphate in mine effluents. Available at http://www.inap.com.au/public_downloads/Research_Projects/Treatment_of_Sulphate_in_Mine_Effluents_-_Lorax_Report.pdf [Accessed 28 July 2020].

International Network for Acid Prevention (INAP) (2009). The global acid rock drainage (GARD) guide. Available at http://www.gardguide.com/images/5/5f/TheGlobalAcidRockDrainageGuide.pdf. [Accessed 13 July 2020].

ITRC (2010a). Mining waste treatment technology selection. Available at https://www.itrcweb.org/miningwaste-guidance/to_chem_precip.htm [Accessed 13 July 2020].

ITRC (2010b). Pressure-driven membrane separation technologies. Available at https://www.itrcweb.org/miningwaste-guidance/to_membrane_sep.htm [Accessed 24 May 2020].

Jeevanandam, J., Barhoum, A., Chan, Y.S., Dufresne, A. and Danquah, M.K. (2018). Review on nanoparticles and nanostructured materials: history, sources, toxicity and regulations. Beilstein Journal of Nanotechnology 9, 1050–1074.

Jenke, D.R. and Diebold, F.E. (1984). Electroprecipitation treament of acid mine wastewater. Water Research 18(7), 855–859.

Jiang, J., Zhao, Q., Zhang, J., Zhang, G. and Lee, D.J. (2009). Electricity generation from biotreatment of sewage sludge with microbial fuel cell. Bioresource Technology 100, 5808–5812.

Jin, W. and Zhang, Y. (2020). Sustainable electrochemical extraction of metal resources from waste streams: From removal to recovery. ACS Sustainable Chemistry and Engineering 8(12), 4693–4707.

Johnson, D.B. and Hallberg, K. B. (2005). Acid mine drainage remediation options: a review. Science of the Total Environment 338, 3–14.

Kabdaşlı, I., Arslan, T., Ölmez-Hancı, T., Arslan-Alaton, I. and Tünay, O. (2009). Complexing agent and heavy metal removals from metal plating effluent by electrocoagulation with stainless steel electrodes. Journal of Hazardous Materials 165, 838–845.

Kang, J., Guliants, V. and Lee, J.Y. (2019). Simultaneous concentration of dissolved solids and recovery of clean water from acid mine drainage employing membrane distillation. Available at https://s3-eu-west-1.amazonaws.com/itempdf74155353254prod/11347055/Simultaneous_Concentration_of_Dissolved_Solids_and_Recovery_of_Clean_Water_from_Acid_Mine_Drainage_Employing_Membrane_Di_v1.pdf [Accessed 29 May 2020].

Karabulut, S., Karabakan, A., Denizli, A. and Yürüm, Y. (2000). Batch removal of copper (II) and zinc (II) from aqueous solutions with low-rank Turkish coals. Separation and Purification Technology 18, 177–184.

Kaur, G., Hatton-Jones, B.W., Couperthwaite, S.J. and Millar, G.J. (2018). Alternative neutralisation materials for acid mine drainage treatment. Journal of Water Process Engineering 22, 46–58.

Khan, S., Cao, Q., Zheng, Y.M., Huang, Y.Z. and Zhu, Y.G. (2008). Health risks of heavy metals in contaminated soils and food crops irrigated with wastewater in Beijing, China. Environmental Pollution 152(3), 686–692.

Khelifa, A., Aoudj, S., Moulay, S., de Petris-Wery, M. (2013). A one-step electrochlorination/electroflotation process for the treatment of heavy metals wastewater in presence of EDTA. Chemical Engineering and Processing 70, 110–116.

Khelifa, A., Moulay, S. and Naceur, A.W. (2005). Treatment of metal finishing effluents by the electroflotation technique. Desalination 181, 27–33.

Kefeni, K.K., Msagati, T.M. and Mamba, B.B. (2015). Synthesis and characterization of magnetic nanoparticles and study their removal capacity of metals from acid mine drainage. Chemical Engineering Journal 276, 222–231.

Kefeni, K.K., Msagati, T.A.M. and Mamba, B.B. (2017). Acid mine drainage: prevention, treatment options, and resource recovery: a review. Journal of Cleaner Production 151, 475–493.

Kesieme, U.K. (2015). Mine waste water treatment and acid recovery using membrane distillation and solvent extraction. PhD thesis, Victoria University

Kesieme, U.K. and Aral, H. (2015). Application of membrane distillation and solvent extraction for water and acid recovery from acidic mining waste and process solutions. Journal of Environmental Chemical Engineering 3, 2050–2056.

Kesieme, U.K., Milne, N., Aral, H., Cheng, C.Y. and Duke, M. (2012). Novel application of membrane distillation for acid and water recovery from mining waste waters. In: McCullough, C. D., Lund, M.A. and Wyse, L. (Editors), International Mine Water Association Symposium 2012. Mine Water and the Environment, Red Hook, New York.

Kesieme, U.K., Milne, N., Cheng, C.Y., Aral, H. and Duke, M. (2014). Recovery of water and acid from leach solutions using direct contact membrane distillation. Water Science and Technology 69(4), 868–875.

Kim, D.K., Zhang, Y., Voit, W., Rao, K.V. and Muhammed, M. (2001). Synthesis and characterization of surfactant-coated superparamagnetic monodispersed iron oxide nanoparticles. Journal of Magnetism and Magnetic Materials 225, 30–36.

Kim, M.S., Min, H.G., Lee, B.J., Chang, S., Kim, J.G., Koo, N. Park, J.S. and Bak, G.I. (2014). The applicability of the acid mine drainage sludge in the heavy metal stabilization in soils. Korean Journal of Environmental Agriculture 33(2), 78–85.

Kim, K.J. (2006). Purification of phosphoric acid from waste acid etchant using layer melt crystallization. Chemical Engineering Technology 29(2), 271–276.

Ko, M.S., Kim, J.Y., Bang, S., Lee, J.S., Ko, J.I. and Kim, K.W. (2012). Stabilization of the As contaminated soil from the metal mining areas in Korea. Environmental Geochemistry and Health 34, 143–149.

Kraft, A. (2004). Electrochemical water treatment processes. Vom Wasser 102(3), 12–19.

Kuhr, J. H., Robertson, J.D., Lafferty, C.J., Wong, A.S. and Stalnaker, N.D. (1997). Ion exchange properties of a Western Kentucky low-rank coal. Energy and Fuels 11(2), 323–326.

Kumar, H.K., Venkatesh, N., Bhowmik, H. and Kuila. A. (2018). Metallic nanoparticle: a review. Biomedical Journal of Scientific and Technical Research 4, 3765.

Kwon, H.W., Kim, J.J., Ha, D.W. and Kim, Y.H. (2016). Formation of iron oxides from acid mine drainage and magnetic separation of the heavy metals adsorbed iron oxides. Progress in Superconductivity and Cryogenics 18(1), 28–32.

Ladeira, A.C.Q. and Gonçalves, C.R. (2017). Influence of anionic species on uranium separation from acid mine water using strong base resins. Journal of Hazardous Materials 148, 499–504.

Larsson, M., Nosrati, A., Kaur, S., Wagner, J., Baus, U. and Nydén, M. (2018). Copper removal from acid mine drainage-polluted water using glutaraldehyde-polyethyleneimine modified diatomaceous earth particles. Heliyon 4, e00520.

Lata, S., Singh, P.K. and Samadder, S.R. (2014). Regeneration of adsorbents and recovery of heavy metals: a review. International Journal of Environmental Science and Technology 12, 1461–1478.

Lee, K.Y., Moon, D.H., Lee, S.H., Kim, K.W., Cheong, K.H., Park, J.H., Ok, Y.S. and Chang, Y.Y. (2013). Simultaneous stabilization of arsenic, lead, and copper in contaminated soil using mixed waste resources. Environmental Earth Sciences 69, 1813–1820.

Lee, H.C., Min, K.W. and Seo, E.Y. (2016). A feasibility study on CO_2 sequestration using the neutralization process of acid mine drainage. Geosystem Engineering 2016, 1–9.

Lefebvre, O., Neculita, C.M., Yue, X. and Ng, H.Y. (2012). Bioelectrochemical treatment of acid mine drainage dominated with iron. Journal of Hazardous Materials 241, 411–417.

Lewis, A.E. (2010). Review of metal sulphide precipitation. Hydrometallurgy 104, 222–234.

Li, Y.-H., Wang, S., Zhang, X., Wei, J., Xu, C., Luan, Z. and Wu, D. (2003a). Adsorption of fluoride from water by aligned carbon nanotubes. Materials Research Bulletin 38(3), 469–476.

Li, Y.-H., Ding, J., Luan, Z., Di, Z., Zhu, Y., Xu, C, Wu, D. and Wei, B., (2003b). Competitive adsorption of Pb^{2+}, Cu^{2+} and Cd^{2+} ions from aqueous solutions by multiwalled carbon nanotubes. Carbon 41, 278–2792.

Li, Y.-H., Zhu, Y., Zhao, Y., Wu, D. and Luan, Z. (2006). Different morphologies of carbon nanotubes effect on the lead removal from aqueous solution. Diamond and Related Materials 15, 90–94.

Li, Y.-H., Zhao, Y.M., Hu, W.B., Ahmad, I., Zhu, Y.Q., Peng, J.X. and Luan, Z.K. (2007). Carbon nanotubes – the promising adsorbent in wastewater treatment. Journal of Physics: Conference Series 61, 698–702.

Liu, S., Zhang, X. and Wei, F. (2016). Recovery of dissolved metals from beneficiation wastewater by electrochemical oxidation. International Journal of Electrochemical Science 11, 7173–7181.

Lopes, F.A., Menezes, J.C. and Schneider, I.A. (2011) Acid Mine Drainage as Source of Iron for the Treatment of Sewage by Coagulation and Fenton's reaction. IMWA, Aachen, Germany.

López, J., Reig, M., Gibert, O. and Cortina, J.L. (2019a). Recovery of sulphuric acid and added value metals (Zn, Cu and rare earths) from acidic mine waters using nanofiltration membranes. Separation and Purification Technology 212, 180–190.

López, J., Reig, M., Gibert, O. and Cortina, J.L. (2019b). Increasing sustainability on the metallurgical industry by integration of membrane nanofiltration processes: acid recovery. Separation and Purification Technology, 226, 267–277.

Lottermoser, B.G. (2011). Recycling, reuse and rehabilitation of mine wastes. Elements 7, 405–410.

Lozet, J. and Mathieu, C. (1991). Dictionary of Soil Science, 2nd Edition. A. A. Balkema Publishers, Rotterdam.

Lu, C., Chung, Y.L. and Chang, K.F. (2005). Adsorption of trihalomethanes from water with carbon nanotubes. Water Research 39, 1183–1189.

Lu, N., Zhou, S.G., Zhuang, L., Zhnag, J.T. and Ni, J.R. (2009). Electricity generation from starch processing wastewater using microbial fuel cell technology. Biochemical Engineering Journal 43, 246–251.

Luptáková, A., Mačingová, E. and Jencárová, J. (2010). Application of bacterially produced hydrogen sulphide for selective precipitation of heavy metals. In: Václavíková, M., Vitale, K., Gallios, G.P. and Ivaničová, L. (Editors), Water Treatment Technologies for the Removal of High-Toxicity Pollutants. Springer, Netherlands, pp. 267–273.

Mačingová, E. and Luptáková, A. (2012). Recovery of metals from acid mine drainage. Chemical Engineering Transactions 28, 109–114.

Mačingová, E., Ubaldini, S. and Luptáková, A. (2016). Study of manganese removal in the process of mine water remediation. Journal of the Polish Mineral Engineering Society (January to June), 121–127.

Mahzuz, H.M.A., Alam, R., Alam, M.N., Basak, R. and Islam, M.S. (2009). Use of arsenic contaminated sludge in making ornamental bricks. International Journal of Environmental Science and Technology 6, 291–298.

Mamelkina, M.A., Tuunila, Ritva., Sillänpää, M. and Häkkinen, A. (2017). Electrocoagulation treatment of real mining waters and solid-liquid separation of solids formed. In: Wolkersdorfer, C., Sartz, L., Sillanpää, M. and Häkkinen, A (Editors), Mine Water and Circular Economy. IMWA 2017, Lappeenranta, Finland, pp. 1070–1075.

Manna, A.K. and Pal, P. (2016). Solar-driven flash vaporization membrane distillation for arsenic removal from groundwater: experimental investigation and analysis of performance parameters. Chemical Engineering and Processing – Process Intensification 99, 51–57.

Marcello, R.R., Galato, S., Peterson, M., Riella, H.G. and Bernardin, A.M. (2008). Inorganic pigments made from the recycling of coal mine drainage treatment sludge. Environmental Management 88, 1280–1284.

Maree, J.P., Wilsenach, J., Motaung, S., Bologo, L. and Radebe, V. (2012). Acid mine water reclamation using the ABC process. In: CSIR Natural Resources and the Environment, September 2010.

Marracino, J.M., Coeuret, F. and Langlois, S. (1987). A first investigation of flow-through porous electrodes made of metallic felts or foams. Electrochimica Acta 32(9), 1303–1309.

Martí-Calatayud, M.C., Buzzi, D.C., García-Gabaldón, M., Ortega, E., Bernardes, A.M., Tenório, J. A.S. and Pérez-Herranz, V. (2013). Sulfuric acid recovery from acid mine drainage by means of electrodialysis. Desalination 343, 120–127.

Masindi, V. (2016). A novel technology for neutralizing acidity and attenuating toxic chemical species from acid mine drainage using cryptocrystalline magnesite tailings. Journal of Water Process Engineering 10, 67–77.

Masindi, V., Madzivire, G. and Tekere, M. (2018). Reclamation of water and the synthesis of gypsum and limestone from acid mine drainage treatment process using a combination of pre-treated magnesite nanosheets, lime, and CO_2 bubbling. Water Resources and Industry 20, 1–14.

Masindi, V., Osman, M.S. and Shingwenyana, R. (2019). Valorization of acid mine drainage (AMD): A simplified approach to reclaim drinking water and synthesize valuable minerals – pilot study. Journal of Environmental Chemical Engineering 7, 103082.

Matlock, M.M., Howerton, B.S. and Atwood, D.A. (2001). Irreversible binding of mercury and lead with a newly designed multidentate ligand. Journal of Hazardous Materials 84, 73–82.

Matlock, M.M., Henke, K.R. and Atwood, D.A. (2002a). Effectiveness of commercial reagents for heavy metal removal from water with new insights for future chelate designs. Journal of Hazardous Materials 92(2), 129–142.

Matlock, M.M., Howerton, B.S. and Atwood, D.A. (2002b). Chemical precipitation of heavy metals from acid mine drainage. Water Research 36, 4757–4764.

Mbonimpa, M., Bouda, M., Demers, I., Benzaazoua, M., Bois, D. and Gagnon, M. (2016). Preliminary geotechnical assessment of the potential use of mixtures of soil and acid mine drainage neutralization sludge as materials for the moisture retention layer of covers with capillary barrier effects. Canadian Geotechnical Journal 53(5), 828–838.

McCauley, C.A. (2011). Assessment of passive treatment and biogeochemical reactors for ameliorating acid mine drainage at Stockton coal mine. PhD thesis, University of Canterbury, Australia.

Melling, J. and West, D.W. (1984). A comparative study of some chelating ion exchange resins for applications in hydrometallurgy. In: Naden, D. and Streat, M. (Editors), Ion Exchange Technology. Ellis Horwood Limited, Chichester.

Menezes, J.C.S.S., Silva, R.A., Arce, I.S. and Schneider, I.A.H. (2009). Production of a poly-ferric sulphate chemical coagulant by selective precipitation of iron from acidic coal mine drainage. Mine Water and the Environment 28(4), 311–314.

Merkel, B., Werner, F. and Wolkersdorfer, C. (2005). Carbon dioxide elimination by using acid mine lakes and calcium oxide suspensions (CDEAL). Available at https://www.wolkersdorfer.info/publication/pdf/Carbon_dioxide_elimination_by_using_acid_mine_lakes_and_calcium_oxide_suspensions_CDEAL.pdf [Accessed 26 June 2020].

Meunier, N., Blais, J.F., Lounes, M., Tyagi, R.D. and Sasseville, J.L. (2002). Different options for metal recovery after sludge decontamination at the Montreal Urban Community wastewater treatment plant. Water Science and Technology 46 (10), 33–41.

Meunier, N., Drogui, P., Montane, C., Hausler, R., Mercier, G. and Blais, J. (2006). Comparison between electrocoagulation and chemical precipitation for metals removal from acidic soil leachate. Journal of Hazardous Materials 137, 581–590.

Michael, S. (2016). Final report: The use of sludge generated by the neutralization of acid mine drainage in the cement industry. Available at https://cfpub.epa.gov/ncer_abstracts/index.cfm/fuseaction/display.abstractDetail/abstract/7939/report/F [Accessed 23 June 2020].

Michalková, E., Schwarz, M., Pulišová, P., Máša, B. and Sudovský, P. (2013). Metals recovery from acid mine drainage and possibilities for their utilization. Polish Journal of Environmental Studies 22(4), 1111–1118.

Mohan, D. and Chander, S. (2006). Removal and recovery of metal ions from acid mine drainage using lignite – a low cost sorbent. Journal of Hazardous Materials B 137, 1545–1553.

Mollah, M.Y.A., Morkovsky, P., Gomes, J.A.G., Kesmez, M., Parga, J. and Cocke, D.L. (2004). Fundamentals, present and future perspectives of electrocoagulation. Journal of Hazardous Materials 114, 199–210.

Montes-Hernandez, G., Perez-Lopez, R., Renard, F., Nieto, J.M. and Charlet, L. (2008). Mineral sequestration of CO_2 by aqueous carbonation of coal combustion fly-ash. Journal of Hazardous Materials 161(2–3), 1347–1354.

Moon, D.H., Cheong, K.H., Koutsospyros, A., Chang, Y.Y., Hyun, S., Ok, Y.S. and Park, J.H. (2016) Assessment of waste oyster shells and coal mine drainage sludge for the stabilization of As-, Pb-, and Cu-contaminated soil. Environmental Science and Pollution Research 23, 2362–2370.

Morris, C. (1992). Academic Press Dictionary of Science and Technology. Academic Press, San Diego, CA.

Mortazavi, S. (2008). Application of membrane separation technology to mitigation of mine effluent and acidic drainage. Available at http://mend-nedem.org/wp-content/uploads/2013/01/3.15.1.pdf [Accessed 24 May 2020].

Motsi, T., Rowson, N.A. and Simmons, M.J.H. (2009). Adsorption of heavy metals from acid mine drainage by natural zeolite. International Journal of Mineral Processing 92, 42–48.

Muddemann, T., Haupt, D., Sievers, M. and Kunz, U. (2019). Electrochemical reactors for wastewater treatment. ChemBioEng Reviews 6(5), 142–156.

Mulopo, J. (2015). Making sense of our mining wastes: removal of heavy metals from AMD using sulphidation media derived from waste gypsum. The Journal of the Southern African Institute of Mining and Metallurgy 115, 1193–1197.

Mwewa, B., Stopic, S., Ndlovu, S., Simate, G.S., Xakalashe, B. and Friedrich, B. (2019). Synthesis of poly-alumino-ferric sulphate coagulant from acid mine drainage by precipitation. Metals 2019 (9), 1166.

Naidoo, R., du Preez, K. and Govender-Ragubee, Y. (2018). Are we making progress in the treatment of acid mine drainage? In: Wolkersdorfer, C., Sartz, L., Weber, A., Burgess, J. and Tremblay, G. (Editors), 11th International Conference on Acid Rock Drainage – Risk to Opportunity, Pretoria, South Africa, pp. 215–220.

Naidu, G., Ryu, S., Thiruvenkatachari, R., Choi, Y., Jeong, S. and Vigneswaran, S. (2019). A critical review on remediation, reuse, and resource recovery from acid mine drainage. Environmental Pollution 247, 1110–1124.

Ñancucheo, I. and Johnson, D.B. (2012). Selective removal of transition metals from acidic mine waters by novel consortia of acidophilic sulfidogenic bacteria. Microbial Biotechnology 5(1), 34–44.

Nariyan, E., Sillanpää, M. and Wolkersdorfer, C. (2016). Cadmium removal from real mine water by electrocoagulation. In: Carsten, D. and Michael, P. (Editors), Mining Meets Water – Conflicts and Solutions. Proceedings of IMWA 2016, Freiberg, Germany, pp. 902–905.

Nariyan, E., Sillanpää, M. and Wolkersdorfer, C. (2017). Electrocoagulation treatment of mine water from the deepest working European metal mine – performance, isotherm and kinetic studies. Separation and Purification Technology 177, 363–373.

Nassaralla, C.L. (2001). Pyrometallurgy. In: Buschow, K.H.J., Flemings, M.C., Kramer, E.J., Veyssière, P., Cahn, R. W., Ilschner, B. and Mahajan, S. (Editors), Encyclopedia of Materials: Science and Technology, 2nd Edition. Elsevier, London.

Ndlovu, S., Simate, G.S. and Matinde, E. (2017). Waste Production and Utilization in the Metal Etraction Industry. CRC Press, Boca Raton, FL.

Netpradit, S., Thiravetyan, P. and Towprayoon, S. (2003). Application of 'waste' metal hydroxides for adsorption of azo reactive dyes. Water Research 37, 763–772.

Nleya, Y. (2016). Removal of toxic metals and recovery of acid from acid mine drainage using acid retardation and adsorption processes. MSc dissertation, University of the Witwatersrand, Johannesburg.

Nleya, Y. Simate, G.S. and Sehliselo, N. (2016). Sustainability assessment of the recovery and utilisation of acid from acid mine drainage. Journal of Cleaner Production 113, 17–27.

Nordstrom, D.K., Bowell, R.J., Campbell, K.M. and Alpers, C.N. (2017). Challenges in recovering resources from acid mine drainage. In: Wolkersdorfer, C., Sartz, L., Sillanpää, M. and Häkkinen, A. (Editors), Mine Water Circular Economy. IMWA, Lappeenranta, Finland, pp. 1138–1146.

Norgate, T.E. and Rankin, W.J. (2002). The role of metals in sustainable development. Green Processing 2002. Australasian Institute of Mining and Metallurgy, Cairns, Australia, May 2002, pp. 49–55.

Oliveira, L.C.A., Rios, R.V.R.A., Fabris, J.D., Sapag, K., Garg, V.K. and Lago, R.M. (2003). Clay-iron oxide magnetic composites for the adsorption of contaminants in water. Applied Clay Science 22(4), 169–177.

Oh, C., Han, Y.-S., Park, J.H., Bok, S., Cheong, Y., Yim, G. and Ji, S. (2016). Field application of selective precipitation for recovering Cu and Zn in drainage discharged from an operating Mine. Science of the Total Environment 557–558, 212–220.

Ohgushi, T. and Nagae, M. (2003). Quick activation of optimized zeolites with microwave heating and utilization of zeolite for reuseable desiccant. Journal of Porous Materials 10, 139–143.

Ohgushi, T. and Nagae, M. (2005). Durability of zeolite against repeated activation treatments with microwave heating. Journal of Porous Materials 12, 265–271.

Olajire, A.A. (2013). A review of mineral carbonation technology in sequestration of CO_2. Journal of Petroleum Science and Engineering 109, 364–392.

Olds, W.E., Tsang, D.C., Weber, P.A. and Weisener, C.G. (2013). Nickel and zinc removal from acid mine drainage: roles of sludge surface area and neutralising agents. Journal of Mining 2013, 1–5.

Oncel, M.S., Muhcu, A., Demirbas, E. and Kobya, M. (2013). A comparative study of chemical precipitation and electrocoagulation for treatment of coal acid drainage wastewater Journal of Environmental Chemical Engineering 1, 989–995.

Orescanin, V. and Kollar, R. (2012). A combined CaO/electrochemical treatment of the acid mine drainage from the "Robule" lake. Journal of Environmental Science and Health, Part A 47, 1186–1191.

Özdemir, T., Öztin, C. and Kincal, N.S. (2006). Treatment of waste pickling liquors: process synthesis and economic analysis. Chemical Engineering Communications 193, 548–563.

Panizza, M., Solisio, C. and Cerisola, G. (1999). Electrochemical remediation of copper (II) from an industrial effluent Part II: three-dimensional foam electrodes. Resources, Conservation and Recycling 27, 299–307.

Pant, D., van Bogaert, G., Diels, L. and Vanbroekhoven, K. (2010). A review of the substrates used in microbial fuel cells (MFCs) for sustainable energy production. Bioresource Technology 101, 1533–1543.

Park, S.M., Shin, S.Y., Yang, J.S., Ji, S.W. and Baek, K. (2015). Selective recovery of dissolved metals from mine drainage using electrochemical reactions. Electrochimica Acta 181, 248–254.

Parker, S.P. (1989). McGraw-Hill Dictionary of Scientific and Technical Terms, 4th Edition. McGrawHill, New York.

Parnell, D. (2019). Basic Principles of Metallurgy and Metalworking. Available at https://www.cedengineering.com/userfiles/Basic%20Principles%20of%20Metallurgy%20and%20Metalworking.pdf [Accessed 21 May 2020].

Patil, D.S., Chavan, S.M. and Oubagaranadin, J.U.K. (2016). A review of technologies for manganese removal from wastewaters. Journal of Environmental Chemical Engineering 4, 468–487.

Peng, X., Tang, T., Zhu, X., Jia, G., Ding, Y., Chen, Y., Yang, Y. and Tang, W. (2017). Remediation of acid mine drainage using microbial fuel cell based on sludge anaerobic fermentation. Environmental Technology 38, 2400–2409.

Petriláková, A. and Bálintová, M. (2011). Utilization of sorbents for heavy metals removal from acid mine drainage. Chemical Engineering Transactions 25, 339–344.

Ponder, S.M., Darab, J.G. and Mallouk, T.E. (2000). Remediation of Cr(VI) and Pb(II) aqueous solutions using supported nanoscale zerovalent iron. Environmental Science and Technology 34, 2564–2569.

Porter, M.C. (1989). Handbook of Industrial Membrane Technology. Noyes Publications, Park Ridge, NJ.

Pozo-Antonio, S., Puente-Luna, I., Lagüela-López, S. and Veiga-Ríos, M. (2014). Techniques to correct and prevent acid mine drainage: a review. Dyna 81(186), 73–80.

Qian, Q. L., Wang, H.X., Bai, P. and Yuan, G.Q. (2011). Effects of water on steam rectification in a packed column. Chemical Engineering Research and Design 89, 2560–2565.

Raju, G.B. and Khangaonkar, P.R. (1984). Electroflotation – a critical review. Transactions of the Indian Institute of Metals 37(1), 59–66.

Rakotonimaro, T.V., Neculita, C.M., Bussière, B., Benzaazoua, M. and Zagury, G.J. (2017). Recovery and reuse of sludge from active and passive treatment of mine drainage impacted waters: a review. Environmental Science and Pollution Research 24(1), 73–91.

Ram, L.C. and Masto, R.E. (2014). Fly ash for soil amelioration: a review on the influence of ash blending with inorganic and organic amendments. Earth-Science Reviews 128, 52–74.

Rao, G.P., Lu, C. and Su, F. (2007). Sorption of divalent metal ions from aqueous solution by carbon nanotubes: a review. Separation and Purification Technology 58, 224–231.

Rao, S.R., Gehr, R., Riendeau, M., Lu, D. and Finch, J.A. (1992). Acid mine drainage as a coagulant. Minerals Engineering 5, 1011–1020.

Razzaq, A.S. and Khudair, Z.J. (2018). Acid-base equilibria and pH calculations. Available at https://www.researchgate.net/publication/329538900_Acid-Base_Equilibria_pH_Calculations_Analytical_Chemistry [Accessed 21 May 2020].

Reddad, Z., Gerente, C., Andres, Y. and Le Cloirec P. (2002). Adsorption of several metal ions onto a low-cost biosorbent: kinetic and equilibrium studies. Environmental Science and Technology 36(9), 2067–2073.

Ríos, C.A., Williams, C.D. and Roberts, C.L. (2008). Removal of heavy metals from acid mine drainage (AMD) using coal fly ash, natural clinker and synthetic zeolites. Journal of Hazardous Materials 156, 23–35.

Rodrigues, M.A.S., Amado, F.D.R., Xavier, J.L.N., Streit, K.F., Bernardes, A.M. and Zoppas- Ferreira, J. (2008). Application of photoelectrochemical-electrodialysis treatment for the recovery and reuse of water from tannery effluents. Journal of Cleaner Production 16, 605–611.

Rodriguez, J., Stopić, S., Krause, G. and Friedrich. B. (2007). Feasibility assessment of electrocoagulation towards a new sustainable wastewater treatment. Environmental Science and Pollution Research 14(7), 477–482.

Rodríguez-Galán, M., Baena-Moreno, F.M., Vázquez, S., Arroyo-Torralvo, F., Vilches, L.F. and Zhang, Z. (2019). Remediation of acid mine drainage. Environmental Chemistry Letters 17, 1529– 1538.

Roto, P. (1998). Smelting and refining. In: Stellman, J.M. (Editor), Encyclopedia of Occupational Health and Safety: Chemical, Industries and Occupations, Volume III, 4th Edition, Geneva, 82.2

Rouf, A. and Hossain, D. (2003). Effects of using arsenic-iron sludge in brick making. Available at https://pdfs.semanticscholar.org/5218/6c9a4118e68c7ab8881ac45c4cbb2f9a74d0.pdf [Accessed 24 June 2020].

Ryu, S., Naidu, G., Johir, M.A.H., Choi, Y., Jeong, S. and Vigneswaran, S. (2019). Acid mine drainage treatment by integrated submerged membrane distillation-sorption system. Chemosphere 218, 955–965.

Salama, E.S., Kim, J.R., Ji, M.-K., Cho, D.W., Abou-Shanab, R.A., Kabra, A.N. and Jeon, B.H. (2015). Application of acid mine drainage for coagulation/flocculation of microalgal biomass. Bioresource Technology 186, 232–237.

Salas, B. S. (2017). Acid mine drainage treatment by nanofiltration membranes: impact of aluminium and iron concentration on membrane performance. Dissertation, Escola Tècnica Superior d'Enginyeria Industrial de Barcelona, Spain.

Sampaio, R., Timmers, R., Xu, Y., Keesman, K. and Lens, P. (2009). Selective precipitation of Cu from Zn in a pS controlled continuously stirred tank reactor. Journal of Hazardous Materials 165, 256–265.

Savage, N. and Diallo, M.S. (2005). Nanomaterials and water purification: opportunities and challenges. Journal of Nanoparticle Research 7, 331–342.

Schiewer, S. and Volesky, B. (1995). Modeling of the proton-metal ion exchange in biosorption. Environmental Science and Technology 29(12), 3049–3058.

Sethurajan, M. (2015). Metallurgical sludges, bio/leaching and heavy metals recovery (Zn, Cu). PhD thesis, Université Paris-Est, France.

Shafaei, A., Rezayee, M., Arami, M. and Nikazar, M. (2010). Removal of Mn^{2+} ions from synthetic wastewater by electrocoagulation process. Desalination 260, 23–28.

Shaheen, S.M., Hooda, P.S. and Tsadilas, C.D. (2014). Opportunities and challenges in the use of coal fly ash for soil improvements – a review. Journal of Environmental Management 145, 249–267.

Sheedy, M. (1998). Case studies in applying recoflo ion-exchange technology. Journal of Minerals, Metals and Materials Society 50, 66–69.

Sheedy, M. and Parujen, P. (2012). Acid separation for impurity control and acid recycle using short bed ion exchange. In: Wang, S., Dulrizac, J.E., Free, M.L., Hwang, J.Y. and Kim, D. (Editors), T.T. Chen Honorary Symposium on Hydrometallurgy, Electrometallurgy and Materials Characterization, TMS 2012 Annual Meeting and Exhibition Orlando, FL.

Sephton, M.G. and Webb, J.A. (2017). Application of Portland cement to control acid mine drainage generation from waste rocks. Applied Geochemistry 81, 143–154.

Sephton, M.G., Webb, J.A. and McKnight, S. (2017). Applications of Portland cement blended with fly ash and acid mine drainage treatment sludge to control acid mine drainage generation from waste rocks. Applied Geochemistry 103, 1–14.

Sheoran, A.S. and Sheoran, V. (2006). Heavy metal removal mechanism of acid mine drainage in wetlands: A critical review. Minerals Engineering 19, 105–116.

Shepherd, J. (2017). Ochre and biochar: technologies for phosphorus capture and re-use. PhD thesis, University of Edinburgh.

Shin, C.H., Kim, J.Y., Lee, H.S., Kim, H.S., Mohapatra, D., Ahn, J.W. and Bae, W. (2009). Recovery of nitric acid from waste etching solution using solvent extraction. Journal of Hazardous Materials 163, 729–734.

Shirazi, M.M.A., Bastani, D., Kargari, A. and Tabatabaei, M. (2013). Characterization of polymeric membranes for membrane distillation using atomic force microscopy. Desalination and Water Treatment 2013, 6003–6008.

Shirazi, M. M. A., Kargari, A. and Tabatabaei, M. (2014a). Evaluation of commercial PTFE membranes in desalination by direct contact membrane distillation. Chemical Engineering and Processing 76, 16–25.

Shirazi, M.M.A., Kargari, A., Tabatabaei, M., Ismail, A.F. and Matsuura, T. (2014b). Assessment of atomic force microscopy for characterization of PTFE membranes for membrane distillation (MD) process. Desalination and Water Treatment 2014, 1–10.

Sibiliski, U. (2001). AngloGold desalination pilot plant project. In: Conference on Environmental Responsible Mining in South Africa. CSIR, Pretoria, South Africa.

Sibrell, P.L., Montgomery, G.A., Ritenour, K.L. and Tucker, T.W. (2009). Removal of phosphorus from agricultural wastewaters using adsorption media prepared from acid mine drainage sludge. Water Research 43(8), 2240–2250.

Sibrell, P.L. and Tucker, T.W. (2012). Fixed bed sorption of phosphorus from wastewater using iron oxide-based media derived from acid mine drainage. Water, Air and Soil Pollution 223, 5105–5117.

Sierra, C., Saiz, J.R.Á. and Gallego, J.L.R. (2013). Nanofiltration of acid mine drainage in an abandoned mercury mining area. Water, Air, and Soil Pollution 224(10), 1734.

Silva, R.A., Menezes, J.C.S.S., Lopes, F.A., Kirchheim, A.P. and Schneider, I.A.H. (2017). Synthesis of a goethite pigment by selective precipitation of iron from acidic coal mine drainage. Mine Water and the Environment 36, 386–392.

Silva, R.A., Secco, M.P., Lermen, R.T., Schneider, I.A.H., Hidalgo, G.E.N. and Sampaio, C.H. (2019). Optimizing the selective precipitation of iron to produce yellow pigment from acid mine drainage. Minerals Engineering 135, 111–117.

Simate, G.S. (2012). The treatment of brewery wastewater using carbon nanotubes synthesized from carbon dioxide carbon source. PhD Thesis, University of the Witwatersrand, Johannesburg.

Simate, G.S. (2015). The use of carbon nanotubes in the treatment of water and wastewater. In: Thakur, V.K. and Thakur, M.K. (Editors), Chemical Functionalization of Carbon Nanomaterials: Chemistry and Applications. CRC Press, Boca Raton, FL, pp. 705–717.

Simate, G.S., Cluett, J., Iyuke, S.E., Musapatika, E.T., Ndlovu, S., Walubita, L.F. and Alvarez, A.E. (2011). The treatment of brewery wastewater for reuse: state of art. Desalination 273, 235–247.

Simate, G.S., Iyuke, S.E., Ndlovu, S. and Heydenrych, M. (2012). The heterogeneous coagulation and flocculation of brewery wastewater using carbon nanotubes. Water Research 46(4), 1185–1197.

Simate, G.S., Maledi, N., Ochieng, A., Ndlovu, S., Zhang, J. and Walubita, L.F. (2016). Coal-based adsorbents for water and wastewater treatment. Journal of Environmental Chemical Engineering 4(2), 2291–2312.

Simate, G.S. and Ndlovu, S. (2008). Bacterial leaching of nickel laterites using chemolithotrophic microorganisms: identifying influential factors using statistical design of experiments. International Journal of Mineral Processing 88, 31–36.

Simate, G.S. and Ndlovu, S. (2014). Acid mine drainage: challenges and opportunities. Journal of Environmental Chemical Engineering 2(3), 1785–1803.

Singh, H. and Mishra, B.K. (2017). Assessment of kinetics behavior of electrocoagulation process for the removal of suspended solids and metals from synthetic water. Environmental Engineering Research 22(2), 141–148.

Skousen, J. (2014). Overview of acid mine drainage treatment with chemicals. In: Jacobs, J.A., Lehr, J.H. and Testa, S.M. (2014). Acid Mine Drainage, Rock Drainage, and Acid Sulfate Soils: Causes, Assessment, Prediction, Prevention, and Remediation, 1st Edition. John Wiley & Sons, London.

Skousen, J.G., Sextone, A. and Ziemkiewicz, P.F. (2000). Acid mine drainage control and treatment. Available at http://citeseerx.ist.psu.edu/viewdoc/download?doi=10.1.1.488.6818&rep=rep1&type=pdf [Accessed 13 July 2020].

Skousen, J., Yang, J.E., Lee, J. and Ziemkiewicz, P. (2013). Review of fly ash as a soil amendment. Geosystem Engineering 16, 249–256.

Smith, J.P. (1999). The purification of polluted mine water. In: Fernández-Rubio, R. (Editor), Proceedings of the International Symposium on Mine, Water and Environment for the 21st Century, 1999 IMWA Congress, Seville, Spain.

Smith, K.S., Figueroa, L.A. and Plumlee, G.S. (2013). Can treatment and disposal costs be reduced through metal recovery? In: Brown, A., Figueroa, L. and Wolkersdorfer, C. (Editors), Reliable Mine Water Technology, Proceedings of the International Mine Water Association 2013 Annual Conference, Denver, CO, Volume I, pp. 729–734.

Song, K., Meng, Q., Shu, F. and Ye, Z. (2013). Recovery of high purity sulphuric acid from waster acid in toluene nitration process by rectification. Chemosphere 90, 1558–1562.

Srinivasan, V. and Subbaiyan, M. (1989). Electroflotation studies on Cu, Ni, Zn, and Cd with ammonium dodecyl dithiocarbamate. Separation Science and Technology 24(1–2), 145–150.

Stanković, V., Božić, D., Manasijević, I. and Bogdanović, G. (2008). Direct electrowinning of copper from minewaters. In: First Regional Symposium on Electrochemistry of South-East Europe, Rovinj, Croatia, pp. 116–117.

Stanković, V., Božić, D., Gorgievski M. and Bogdanović, G. (2009). Heavy metal ions adsorption from mine waters by sawdust. Chemical Industry and Chemical Engineering Quarterly 15, 237–249.

Tavlarides, L.L. and Doerkar, N.V. (1997a). Chemically active ceramic compositions with a thio and amine moiety, US. Patent No. 5,616,533.

Tavlarides, L.L. and Doerkar, N.V. (1997b). Chemically active ceramic compositions with a phospho acid moiety, U.S. Patent No. 5,612,175.

Tavlarides, L.L. and Doerkar, N.V. (1997c). Chemically active ceramic compositions with a pyrogallol moiety, US. Patent No. 5,624,881.

Tavlarides, L.L. and Doerkar, N.V. (1997d). Chemically active ceramic compositions with a hydroxyquinoline moiety, U.S. Patent No. 5,668,079 (Sept. 16, 1997).

Tay, J.H. and Show, K.Y. (1994). Municipal wastewater sludge as cementitious and blended cement materials. Cement and Concrete Composites 16, 39–48.

Taylor, J., Pape, S. and Murphy, N. (2005). A summary of passive and active treatment technologies for acid and metalliferous drainage (AMD). Available at https://www.earthsystems.com.au/wp-content/uploads/2012/02/AMD_Treatment_Technologies_06.pdf [Accessed 13 July 2020].

Thermo Scientific (2007). Ion exchange chromatography. Available at http://tools.thermofisher.com/content/sfs/brochures/TR0062-Ion-exchange-chrom.pdf [Accessed 22 May 2020].

Tjus, K., Bergström, R., Fortkamp, U., Forsberg, K. and Rasmuson, A. (2006). Development of a recovery system for metals and acids from pickling baths using nanofiltration and crystallization. Available at https://www.ivl.se/download/18.343dc99d14e8bb0f58b74e0/1445515632292/B1692.pdf [Accessed 30 March 2020].

Tomaszewska, M. (2000). Membrane distillation – examples of applications in technology and environmental protection. Polish Journal of Environmental Studies 9(1), 27–36.

Tomaszewska, M., Grayta, M. and Morawski, A. W. (2001). Recovery of hydrochloric acid from metal pickling solutions by membrane distillation. Separation and Purification 23, 591–600.

Tran, T.K., Leu, H.J., Chiu, K.F. and Lin, C.Y. (2017a). Electrochemical treatment of heavy metal-containing wastewater with the removal of COD and heavy metal ions. Journal of the Chinese Chemical Society 64(5), 1–10.

Tran, T.K., Chiu, K.F., Lin, C.Y. and Leu, H.J. (2017b). Electrochemical treatment of wastewater: selectivity of the heavy metals removal process. International Journal of Hydrogen Energy 42, 27741–27748.

Trumm, D. (2010). Selection of active and passive treatment systems for AMD – flow charts for New Zealand conditions. New Zealand Journal of Geology and Geophysics 53(2–3), 195–210.

Tsang, D.C.W., Olds, W.E., Weber, P.A. and Yip, A.C.K. (2013). Soil stabilization using AMD sludge, compost and lignite: TCLP leachability and continuous acid leaching. Chemosphere 93, 2839–2847.

Turner, M.D., Laurence, R.L. and Conner, W.C. (2000). Microwave radiation's influence on sorption and competitive sorption of zeolites. AIChE Journal 46(4).

Turhanen, P. and Vepsäläinen, J. (2013). New method for efficient removal of uranium and other heavy metals from water. Available at https://phys.org/news/2013-12-method-efficient-uranium-heavy-metals.html [Accessed 19 May 2020].

van der Bruggen, B., Vandecasteele, C., van Gestel, T., Doyen, W. and Leysen, R. (2003). A review of pressure-driven membrane processes in wastewater treatment and drinking water production. Environmental Progress 22(1), 46–56.

Veglio, F., Quaresima, R., Fornari, P. and Ubaldini, S. (2003). Recovery of valuable metals from electronic and galvanic industrial wastes by leaching and electrowinning. Waste Management 23, 245–252.

Venkatasaravanan, R., Ramesh, S.T. and Gunasheela, M. (2016). Removal of heavy metals from acid mine drainage (AMD) contaminated with high concentrations of Fe, Zn, and Cu using electrocoagulation. Advanced Porous Materials 4(1), 1–8.

Vasudevan, S., Lakshmi, J. and Packiyam, M. (2010). Electrocoagulation studies on removal of cadmium using magnesium electrode. Journal of Applied Electrochemistry 40, 2023–2032.

Ubaldini, S., Luptakova, A., Fornari, P. and Yoplac, E. (2013). Application of innovative remediation processes to mining effluents contaminated by heavy metals. Available at https://www.e3s-conferences.org/articles/e3sconf/abs/2013/01/e3sconf_ichm13_25001/e3sconf_ichm13_25001.html [Accessed 16 May 2020].

Uça, D. (2017). Sequential precipitation of heavy metals using sulfide-laden bioreactor effluent in a pH controlled system. Mineral Processing and Extractive Metallurgy Review 38(3), 162–167.

Unger-Lindig, Y., Merkel, B. and Schipek, M. (2010). Carbon dioxide treatment of low density sludge: a new remediation strategy for acidic mining lakes? Environmental Earth Sciences 60, 1711–1722.

Vaclav, P. and Eva, G. (2005). Desalting of acid mine drainage by reverse osmosis method – field tests. Available at http://www.imwa.info/docs/imwa_2005/IMWA2005_052_Pisa.pdf [Accessed 25 July 2020].

Walker, D.J. and Hurl, S., (2002). The reduction of heavy metals in a storm water wetland. Ecological Engineering 18(4), 407–414.

Wang, C.B. and Zhang, W.X. (1997). Synthesizing nanoscale iron particles for rapid and complete dechlorination of TCE and PCBs. Environmental Science and Technology 31, 2154–2156.

Wang, P. and Chung, T.S. (2015). Recent advances in membrane distillation processes: Membrane development, configuration design and application exploring. Journal of Membrane Science 474, 39–56.

Wang, W., Xu, Z. and Finch, J. (1996). Fundamental study of ambient temperature process in the treatment of acid mine drainage. Environmental Science and Technology 30, 2604–2608.

Wang, X.S., Wang, J. and Sun, C. (2006). Removal of copper (II) ions from aqueous solutions using natural kaolinite. Adsorption Science and Technology 24(6), 517–530.

Wang, Y.R., Tsang, D.C.W., Olds, W.E. and Weber, P.A. (2014). Utilizing acid mine drainage sludge and coal fly ash for phosphate removal from dairy wastewater. Environmental Technology34, 3177–3182.

Waters, J.C., Santomartino, S., Cramer, M., Murphy, N. and Taylor, J.R. (2003). Acid rock drainage treatment technologies identifying appropriate solutions. Proceedings of the 6th International Conference on Acid Rock Drainage, 12–18 July, Cairns, Queensland, Australia, pp. 831–843.

Wei, X. and Viadero, R.C. (2007a). Adsorption and pre-coat filtration studies of synthetic dye removal by acid mine drainage sludge. Journal of Environmental Engineering 133, 633–640.

Wei, X. and Viadero, R.C. (2007b). Synthesis of magnetite nanoparticles with ferric iron recovered from acid mine drainage: implications for environmental engineering. Colloids and Surfaces A: Physicochemical and Engineering Aspects 294, 280–286.

Wei, X., Viadero, R.C. and Bhojappa, S. (2008). Phosphorus removal by acid mine drainage sludge from secondary effluents of municipal wastewater treatment plants. Water Research 42(13), 3275–3284.

Wei, X., Viadero, R.C. and Buzby, K.M. (2005). Recovery of iron and aluminum from acid mine drainage by selective precipitation. Environmental Engineering Science 22(6), 745–755.

Wen, Q., Wu, Y., Zhao, L., Sun, Q. and Kong, F. (2010). Electricity generation and brewery wastewater treatment from sequential anode-cathode microbial fuel cell. Journal of Zhejiang University: Science B 11(2), 87–93.

Weng, C.H., Lin, D.F. and Chiang, P.C. (2003). Utilization of sludge as brick materials. Advances in Environmental Research 7, 679–685.

West, A., Kratochvil, D. and Fatula, P. (2011). Sulfate removal from acid mine drainage for potential water re-use. In: Lucey, J.T. (Editor), 72nd Annual International Water Conference, 13–17 November 2011, Orlando, FL.

White, D.A. and Asfar-Siddique, A. (1997). Removal of manganese and iron from drinking water using hydrous manganese dioxide. Solvent Extraction and Ion Exchange15(6), 1133–1145.

Wisniewski, J. and Wisniewska, G. (1999). Water and acid recovery from the rinse after metal etching operations. Hydrometallurgy 53(2), 105–119.

Wu, C.H. (2007). Adsorption of reactive dye onto carbon nanotubes: equilibrium, kinetics and thermodynamics. Journal of Hazardous Materials 144, 93–100.

Wuana, R.A. and Okieimen, F.E. (2011). Heavy metals in contaminated soils: a review of sources, chemistry, risks and best available strategies for remediation. International Scholarly Research Network Ecology (2011), 1–20.

Yah, C.S., Simate, G.S. and Iyuke, S.E. (2012). Nanoparticles toxicity and their routes of exposures. Pakistan Journal of Pharmaceutical Sciences 25(2), 477–491

Ying, X. and Fang, Z. (2006). Experimental research on heavy metal wastewater treatment with dipropyl dithiophosphate. Journal of Hazardous Materials 137(3), 1636–1642.

Zeta-Meter Inc. (1993). Everything You Need to Know About Coagulation and Flocculation-Guide. Zeta-Meter Inc., Staunton, VA.

Zhang, W. (2003). Nanoscale iron particles for environmental remediation: an overview. Journal of Nanoparticle Research 5, 323–332.

Zhang, M.K., Liu, Z.Y. and Wang, H. (2010). Use of single extraction methods to predict bioavailability of heavy metals in polluted soils to rice. Communications in Soil Science and Plant Analysis 41(7), 820–831.

Zhang, G., Zhao, Q., Jiao, Y., Wang, K., Lee, D. J. and Ren, N. (2012). Efficient electricity generation from sewage sludge using biocathode microbial fuel cell. Water Research 46, 43–52.

Zheng, Y., Gao, X.L., Wang, X.Y., Li, Z.X., Wang, Y.H. and Gao, C.J. (2015). Application of electrodialysis to remove copper and cyanide from simulated and real gold mine effluents. RSC Advances 5(26), 19807–19817.

Zhong, C. M., Xu, Z.L., Fang, X. H. and Cheng, L. (2007). Treatment of acid mine drainage (AMD) by ultra-low-pressure reverse osmosis and nanofiltration. Environmental Engineering Science 24(9), 1297–1306.

Zinck, J. (2006). Disposal, reprocessing and reuse options for acidic drainage treatment sludge. In: Barnhisel, R.I. (Editor), 7th International Conference on Acid Rock Drainage (ICARD). American Society of Mining and Reclamation (ASMR), Lexington, KY.

Index

Note: Locators in *italics* represent figures and **bold** indicate tables in the text.